Lecture Notes in Computer Sc

Edited by G. Goos, J. Hartmanis, and J. v.

T0250660

Springer
Berlin
Heidelberg
New York
Barcelona
Hong Kong
London
Milan
Paris
Tokyo

Peter Druschel   Frans Kaashoek
Antony Rowstron (Eds.)

Peer-to-Peer Systems

First International Workshop, IPTPS 2002
Cambridge, MA, USA, March 7-8, 2002
Revised Papers

Springer

Series Editors

Gerhard Goos, Karlsruhe University, Germany
Juris Hartmanis, Cornell University, NY, USA
Jan van Leeuwen, Utrecht University, The Netherlands

Volume Editors

Peter Druschel
Rice University
MS 132, Houston, TX 88005, USA
E-mail: druschel@cs.rice.edu

Frans Kaashoek
MIT Laboratory of Computer Science
200 Technology Square, Cambridge MA, USA
E-mail: kaashoek@lcs.mit.edu

Antony Rowstron
Microsoft Research Ltd.
7 J J Thomson Avenue, Cambridge, CB3 0FB, UK
E-mail: antr@microsoft.com

Cataloging-in-Publication Data applied for

Bibliograhpic information published by Die Deutsche Bibliothek
Die Deutsche Bibliothek lists this publication in the Deutsche Nationalbibliografie;
detailed bibliographic data is available in the Internet at http://dnb.ddb.de

CR Subject Classification (1998): C.2, H.3, H.4, D.4, F.2.2, E.1

ISSN 0302-9743
ISBN 3-540-44179-4 Springer-Verlag Berlin Heidelberg New York

Springer-Verlag Berlin Heidelberg New York
a member of BertelsmannSpringer Science+Business Media GmbH

http://www.springer.de

© Springer-Verlag Berlin Heidelberg 2002
Printed in Germany

Typesetting: Camera-ready by author, data conversion by Boller Mediendesign
Printed on acid-free paper   SPIN: 10873895   06/3142   5 4 3 2 1 0

Preface

Peer-to-peer has emerged as a promising new paradigm for large-scale distributed computing. The International Workshop on Peer-to-Peer Systems (IPTPS) aimed to provide a forum for researchers active in peer-to-peer computing to discuss the state of the art and to identify key research challenges.

The goal of the workshop was to examine peer-to-peer technologies, applications, and systems, and also to identify key research issues and challenges that lie ahead. In the context of this workshop, peer-to-peer systems were characterized as being decentralized, self-organizing distributed systems, in which all or most communication is symmetric.

The program of the workshop was a combination of invited talks, presentations of position papers, and discussions covering novel peer-to-peer applications and systems, peer-to-peer infrastructure, security in peer-to-peer systems, anonymity and anti-censorship, performance of peer-to-peer systems, and workload characterization for peer-to-peer systems. To ensure a productive workshop environment, attendance was limited to 55 participants.

Each potential participant was asked to submit a position paper of 5 pages that exposed a new problem, advocated a specific solution, or reported on actual experience. We received 99 submissions and were able to accept 31. Participants were invited based on the originality, technical merit, and topical relevance of their submissions, as well as the likelihood that the ideas expressed in their submissions would lead to insightful technical discussions at the workshop.

A digest of the discussions that took place at the workshop is provided in the first chapter. Thanks to Richard Clayton for editing this digest based on notes taken at the workshop by himself and George Danezis. We would like to thank the steering committee for their role in initiating the workshop, MIT for hosting the workshop, Neena Lyall for local arrangements and for taking care of countless logistical details, Kevin Fu for setting up the secure Web registration site, and last but not least, the program committee for selecting a superb technical program.

Finally, we wish to thank all participants for making IPTPS 2002 a great success. Plans are currently underway for IPTPS 2003, which is sure to build on the success of this first workshop.

July 2002   Peter Druschel, Frans Kaashoek, and Antony Rowstron

Organization

Steering Committee

Peter Druschel	Rice University, USA
Frans Kaashoek	MIT, USA
Antony Rowstron	Microsoft Research, UK
Scott Shenker	ICIR, Berkeley, USA
Ion Stoica	UC Berkeley, USA

Program Committee

Ross Anderson	Cambridge University, UK
Roger Dingledine	Reputation Technologies, Inc., USA
Peter Druschel (co-chair)	Rice University, USA
Steve Gribble	University of Washington, USA
David Karger	MIT, USA
John Kubiatowicz	UC Berkeley, USA
Robert Morris	MIT, USA
Antony Rowstron (co-chair)	Microsoft Research, UK
Avi Rubin	AT&T Labs Research, USA
Scott Shenker	ICIR, Berkeley, USA
Ion Stoica	UC Berkeley, USA

Organizing Chairs

Frans Kaashoek	MIT, USA
Antony Rowstron	Microsoft Research, UK

Sponsors

Microsoft Research	http://www.research.microsoft.com/

Table of Contents

Anonymous Overlays

Applications

Evaluation

Searching and Indexing

Data Management

Author Index

Workshop Report for IPTPS'02
1st International Workshop on Peer-to-Peer Systems
7-8 March 2002 – MIT Faculty Club, Cambridge, MA, USA

Richard Clayton

University of Cambridge, Computer Laboratory, Gates Building,
JJ Thompson Avenue, Cambridge CB3 0FD, United Kingdom
`richard.clayton@cl.cam.ac.uk`

Attendees were welcomed to the Workshop by Frans Kaashoek and Peter Druschel who reported that although the original plan had been to invite 35 people to a $1\frac{1}{2}$ day event, 99 papers had been submitted. The workshop had therefore been scaled up to 50 authors with 32 position papers, 26 of which would be presented over a two day program.

Session 1: DHT Routing Protocols: State of the Art and Future Directions *Chair: Scott Shenker*

Scott Shenker gave a brief overview of the problem space, noting that a great deal of work was being done on Distributed Hash Tables (DHTs) that were location independent routing algorithms. Given a key, they would find the right node.

Ben Y.Zhao, Yitao Duan, Ling Huang, Anthony D. Joseph, John D. Kubiatowitz, **"Brocade: Landmark Routing on Overlay Networks", presented by Ben Y. Zhao.** Recent advances in high-dimension routing research guarantee a sub-linear number of hops to locate nodes. However, they assume a uniformity of mesh that does not map well to the Internet. Brocade extends Tapestry to address this by eliminating hops across the wide area and by preventing traffic going through congested stub nodes. "Super-nodes" are spotted within connected domains and a secondary overlay layer is then used to transfer traffic between these super-nodes. Brocade is not useful for intra-domain messages, so it is necessary to classify traffic by destination. Old results for locality are cached and the super-node keeps a "cover net", an authoritative list of which nodes are local. The local super-nodes are found by snooping peer-to-peer traffic or by consulting the DNS. The super-nodes find each other by using Tapestry, which has built-in proximity metrics. The super-nodes do not need to be incredibly powerful. Simulations show that with 220M hosts, 20K AS's and 10% of the super-nodes "coming and going" then about 9% of the super-nodes' CPU time is spent dealing with changes. Publishing data about the changes

P. Druschel, F. Kaashoek, and A. Rowstron (Eds.): IPTPS 2002, LNCS 2429, pp. 1–21, 2002.
© Springer-Verlag Berlin Heidelberg 2002

uses about 160K bits/sec in bandwidth, and super-nodes need only about 2MB of storage. Simulations also show that the scheme halves delay for short hops and improves overall bandwidth usage.

Discussion: Q: Doesn't RON show the benefit of routing around problems in the wide area? A: Brocade doesn't preclude this and, naturally, Tapestry allows for routing around problems in its discovery protocols. Q: What about the cost of the Tapestry protocol itself? A: We try to optimize "behind the curtain" in the background so as not to affect performance. Also a lot of the behavior is local and nodes are replaced by a proximity network. Bottom line is we're getting a performance improvement of 50%, which is worth having and in the limit, you can always fall back into normal routing.

David Liben-Nowell, Hari Balakrishnan, David Karger "Observations on the Dynamic Evolution of Peer-to-Peer Networks", presented by David Liben-Nowell. The existing approach to describing DHTs is to carefully define an ideal state and show that whilst in that state the network has good properties. However, since this ideal state never happens "who cares?" Networks are dynamic and nodes come and go very frequently. One has to prove that the system gets back to an ideal state "soon after all the changes". Perhaps it would be better to define an "almost ideal state" that would be achievable in practice. One would require this state to be maintainable and for it to have good properties like fast search, good balancing and lookups that succeed when they should. With this approach traditional performance measures become meaningless. Because the network never stabilizes, the time to recovery and number of maintenance messages is infinite. A better measure would be to use the doubling time (until the network is twice the size) or the halving time (until $n/2$ nodes fail). The "half life" is the smaller of these and measures how long it is until only half of the system is as it was before. It is possible to prove that all peer-to-peer systems must be notified of $\Omega(log\ n)$ node changes every half-life and that systems will stay connected if nodes discover $log\ n$ new nodes per half-life. In Chord it is possible to find a protocol that runs in $O(log^2 n)$ messages – suggesting that further work is needed. It would also be helpful to understand the notion of half-life more clearly and measure some real values in the field.

Discussion: Q: Isn't joining or leaving from real networks somewhat bursty? A: If the changes aren't massive the proposed Chord protocol will cope. Q: What about things like BGP level events changing connectivity? A: The model may be broad enough to cope with this. More work is called for.

Sylvia Ratnasamy, Scott Shenker, Ion Stoica, "Routing Algorithms for DHTs: Some Open Questions", presented by Scott Shenker. Routing algorithms for DHTs have more commonalities than differences. This is an illustrative list of what the open research questions seem to be, along with some initial answers:

Q1: What path lengths can you get with $O(1)$ neighbors?

Viceroy seems to manage $O(log\ n)$ and Small Worlds $O(log^2 n)$.

Q2: Does this cause other things to go wrong?

We don't yet know.

Q3: What are the costs and dynamics of full recovery? Viz: if some nodes fail can you still reach the ones that remain?

It seems that most systems are remarkably good, losing 2% of routes with a 20% node failure rate. However, maybe routing isn't the issue, but data replication is. Perhaps we will have lost 20% of the data?

Q4: Can we characterize the effectiveness of proximity (geographic) routing?

We've not yet got a good model to show that this works, although it is clear that it does.

Q5: Is proximity neighbor selection significantly better than proximity routing?

Yes, a little better. But there's no good model yet.

Q6: If we had the full n^2 latency matrix would one do optimal neighbor selection in algorithms not based on Plaxton trees?

Q7: Can we choose identifiers in a *1-D* keyspace that adequately captures the geographic nature of nodes?

Q8: Does geographic layout have an impact on resilience, hot-spots and other aspects of performance?

We expect load balancing to be hard!

Q9: Can the two local techniques of proximity routing and proximity neighbor selection achieve most of the benefit of global geographic layout?

We don't yet have a way of doing this comparison.

Q10: Nodes have varied performance, by several orders of magnitude. If powerful nodes pretend to be multiple less "able" nodes is this "cloning" effective?

Q11: How can we redesign routing algorithms to exploit heterogeneity?

and from this list, the burning questions are Q2, Q9 and Q11.

Discussion: Q: Doesn't the improvement from proximity routing depend on the number of nodes? A: We're looking for asymptotic behavior. Q: Doesn't load balancing interact poorly with security? A: It would be very useful for papers to clearly indicate how they felt security properties emerged from their routing properties. If the system is extended or optimized one would like to check that the security properties still hold. Q: Can you separate out two levels of behavior, one for correctness and one for speed? A: We tried to do this with CAN. Q: Aren't we being unfriendly by designing our own routing level which is moving away from the underlying TCP? A: This may not be unfriendly, but the way forward may be to re-examine the notion of proximity to relate it to a graph of the actual system, rather than to a mere count of hops.

Session 2: Deployed Peer-to-Peer Systems
Chair: Roger Dingledine/Steve Gribble

Matei Ripeanu, Ian Foster, **"Mapping the Gnutella Network: Macroscopic Properties of Large-Scale Peer-to-Peer Systems"**, **presented by Matei Ripeanu.** Gnutella is a large, fast growing peer-to-peer system. Further growth is being hindered by inefficient resource use: the overlay network does not match the underlying network infrastructure. This study examined the network during a period of rapid change from 1,000 nodes in November 2000 to 50,000 nodes in May 2001. By March 2002 the Gnutella network was about 500,000 nodes. The network grew partly because its users were prepared to tolerate high latency and low quality results for file searches. Also, DSL connected users were 25% of the network at the start of the study and about 40% at the end.

Gnutella was originally a power-law graph (with the number of nodes proportional to L^{-k} for some value of k), but by May 2001, at the end of the study, it had become bi-modal (there were too many nodes with low connectivity). Also, the average path length had grown only 25% rather than the expected 55%. This made Gnutella more resilient to random node failures. Each link was transferring 6-8 Kbytes/sec so the overall administrative traffic in May 2001 was about 1 Gbyte/sec, about 2% of the US "backbone" traffic levels. 90% of this volume is query and ping traffic. The overall topology does not match the Internet topology with 40% of the nodes in the 10 largest ASs and the "wiring" is essentially random.

Discussion: Q: Why were there some periods of very rapid change? A: Bearshare changed their protocol so that saturated nodes no longer answered pings. Over about a week Gnutella stopped being a power-law network as the Morpheus system started using it. Q: Are clients maturing and allowing more scaling? A: Gnutella is now a two-layer system, which was a change. Improvements have mainly come from the protocol changes (pings dropped from 50% of traffic to 5%) but also to some extent from better engineered clients.

Qin Lv, Sylvia Ratnasamy, Scott Shenker, **"Can Heterogeneity Make Gnutella Scalable?", presented by Sylvia Ratnasamy.** The most common application of fielded Peer-To-Peer systems is file sharing. The current solutions are unstructured, the overlay is ad hoc (you can connect as you wish) and files may be placed almost anywhere. The only approach possible is random probing and unfortunately, poorly scaling ways of doing this have been chosen. DHTs are very structured and very scalable so they are good for finding a "needle in a haystack". But perhaps the unstructured solutions are "good enough", especially when one is looking for "hay", i.e. material of which there are many copies in the network. DHTs are bad at keyword searches (because they do exact matches) and cope poorly with rapid changes of network membership. A system like Gnutella has no structure to lose if many nodes leave or crash.

Gnutella would perform better with a scalable search – multiple "flood to one" random walks give a big performance gain. Biasing this to deal with node heterogeneity (Gnutella nodes have differences of 4 orders of magnitude in available bandwidth) or transient load changes would also help with the desirable state of "big nodes do more work". The idea is for nodes to have a capacity measure and replicate files and dump traffic onto less full neighbors. Where traffic cannot be dumped, the sender is asked to slow down. Results from simulations are encouraging, suggesting that although having "super-nodes" is a good idea, it may not be necessary to be explicit about having exactly two levels of traffic.

Discussion: Q: Can you measure performance directly rather than relying on a neighbor telling you that you're slow? A: Yes, we're aiming to do that. Q: What are DHTs good for? A: "needles". Some problems are like that, though the type of improved Gnutella we want to build will be better at needles than Gnutella currently is. If DHTs were required to do keyword searches then they might be just as expensive.

Bryce Wilcox-O'Hearn, "Experiences Deploying a Large-Scale Emergent Network", presented by Bryce Wilcox-O'Hearn. Mojo Nation was originally inspired by Ross Anderson's Eternity paper in the mid 90s. In 1998 it was put onto a commercial footing as a distributed data haven, but ran out of money in 2001. Mnet is the open source descendant.

Mojo Nation was an ambitious and complex system that incorporated digital cash as payment for resources. The digital cash worked and most of the rest did not. It didn't scale (never reaching 10,000 simultaneous nodes). 20,000 new people tried it each month, but its half-life was less than an hour. There were significant problems with the first connection to the system, which was needed to locate any neighbors. When the system was publicized the ensuing wave of people overwhelmed the server ("Slashdot killed my network"). "Original introduction" is an important and non-trivial issue and other systems such as LimeWire can be seen to be having problems with it. Of those who did connect 80% came, looked and left forever. This appears to be because there was no content they cared about on the system. Data had to be explicitly published and nodes didn't do this. The content was distributed as a 16 of 32 "erasure code" scheme, but 80% of all nodes were offline most of the time, so files could not be recreated. There were also problems with ISPs ("my enemy") who throttled bandwidth, forbade servers or changed their IP addresses on a regular basis. The original design was tunable to suggest network neighbors based on reliability and closeness (Round Trip Time). This formula was progressively updated until eventually RTT was scored at an extremely low level of significance.

Discussion: Q: How important was the mojo? A: This Chaumian blinded digital cash determined who could store files. It was complicated to make it secure, and people were always trying to steal it by exploiting bugs. It was also hard to verify whether services were actually performed. You needed something to discriminate against newcomers Q: Do we need a lightweight rendezvous protocol to solve this original introduction problem? A: Yes, but it needs to have no single points of failure or control, because people will want to control it. Q: Why use shares rather than multiple copies of documents? A: Less disk usage overall. Q: Would low churn have helped? A: Yes, it did work well sometimes, when nodes weren't coming and going quite so much. Q: Can one actually run a commercial file storing service? A: I'm not the person to ask.

Session 3: Anonymous Overlays *Chair: Roger Dingledine*

Andrei Serjantov, **"Anonymizing Censorship Resistant Systems"**, **presented by Andrei Serjantov.** The idea of censorship resistance is to make it hard for someone more powerful to remove content from a distributed filestore. Anonymity is also important, for publishers, readers and whoever may be storing the file. Issues such as searching or efficiency are of less concern – it matters more that the material is available than that it took a long time to arrive.

The protocol assumes a DHT system that can store keys and inter-node communication via anonymous addressing "onions". The basic idea is to split the document into shares and ask for the item to be stored. Nodes called forwarders encrypt the shares, select the actual storage points and return identifiers that will later be used to identify the routes to these storage locations. The identifiers are combined by the publisher and the resulting address is then publicized out-of-band, perhaps by anonymously posting it to Usenet. Retrieval again uses the forwarders, but the stored shares are returned to the retriever via separate nodes that remove the encryption. The various roles in the storage/retrieval protocol have strong guarantees. The storers of the document are unaware what they are storing. The nodes that forward requests can deny doing this and the publisher and any retriever can deny almost anything. Future work will prove the properties of the system in a formal way and will demonstrate the resistance of the system to attacks.

Discussion: Q: Doesn't the forwarder know what's going on? A: They're just moving random numbers around. Q: Why not publish the storer onions directly? A: This would allow an attacker to take out the start of the MIX chain and deny access to the document. It's also possible to adjust the protocol to prevent the forwarders being vulnerable in this way.

Steven Hazel, Brandon Wiley, **"Achord: A Variant of the Chord Lookup Service for Use in Censorship Resistant Peer-to-Peer Publishing Systems"**, **presented by Brandon Wiley.** The idea was to create a Chord variant that would provide a DHT with anonymity properties. This ideas are #1 not to report intermediate lookup progress (so that the storer remains hidden), #2 make node discovery harder by restricting find_successor, #3 to be careful about how finger tables are updated by restricting the information flow. This third property proves to be hard because "stuff for my finger table that would be best for me" will vary and therefore, over time, nodes will learn more and more identifiers for participants in the Chord ring. Rate limiting merely delays this process. The other properties mean that nodes can hide, which makes them wonderfully anonymous, but may collapse the system to a small number of nodes. Finally, all these changes make Chord rather slower and this may mean that stability is rarely achieved.

Discussion: Q: Why do you want to restrict learning of other nodes? A: If you know the identity of a storer then you can censor them. Q: Can't you use TCP to ensure that nodes are who they say they are? A: Yes, that helps a lot with authentication. Q: Won't IPv6 mess this up by letting people choose their own IP addresses and enter the Chord ring close to what they want to learn about and censor? A: We need another identifier with the right properties. We aren't alone in wanting this!

David Mazières gave a talk on "Real World Attacks on Anonymizing Services". Anonymous speech can upset people, so they try to shut down the systems that propagate it. They try to exploit software vulnerabilities, they use the system to attack someone powerful enough to themselves attack the system, they try to marginalize the system, they attract spam to the system, and they try to make life intolerable for the operator. Some attacks can be defeated by short-term logging or by providing some logs to users. Overloading can be addressed by trying to force the attacker to put a human into the loop, so as to increase the cost. Content based attacks can be countered by ensuring that it is easy for people to ignore anonymous content and by never storing or serving objectionable content. These issues must all be factored into the design of any anonymous service, where the most precious resource will be human time. This can be used by the defenders to slow down attackers, but also by attackers to wear down the operator of the service. Future work is looking at Tangler, which will entangle multiple documents together to make them harder to censor.

Michael J. Freedman, Emil Sit, Josh Cates, Robert Morris, "Tarzan: A Peer-to-Peer Anonymizing Network Layer", presented by Michael Freedman. Tarzan provides a method for people to talk to servers without anyone knowing who they are. Millions of nodes will participate and bounce traffic off each other in an untraceable way. Peer-to-peer mechanisms are used to organize the nodes and the mechanism works at the IP layer so that existing applications will not need modification. Because so many nodes take part, it will not be practical to block them all. Because everyone is relaying for everyone else, there is no "network edge" at which to snoop traffic, and because relayed traffic cannot be distinguished from originated traffic there is plausible deniability.

When you join the system you make random requests from a Chord ring to determine a list of possible peers and select a small number of these to create a source routed UDP based tunnel with each hop secured by a symmetric key. The nodes on the path hold a flow identifier to determine what action to take with incoming traffic. NAT is used both within the tunnel and also at the far end where a connection is made to the true destination. Tarzan can also support anonymous servers. A C++ based prototype has been created that was able to saturate a 100Mbit/sec Ethernet. The overhead for setup is ∼20ms/hop and for packet forwarding ∼1ms/hop (each plus the data transmission time).

Discussion: Q: How well does TCP run over the tunnel? A: There are no detailed figures yet, and it is hard to say what effect rebuilding tunnels will have. Q: How capable does an observer have to be to break this system? A: Few are big enough. Q: How reliable does the PNAT at the end of the tunnel need to be? A: For things like SSH this is an issue. We may need to have a bit less heterogeneity in node selection. Q: What about cover traffic? A: We need traffic in the network to hide our traffic. So we create a sparse overlay network with this traffic and mix this with longer hops. Q: If you know two ends A and B can you say that you cannot link A with B? A: Yes, we hope to do this reasonably well, but it depends what "cannot" means.

Session 4: Applications I *Chair: Frans Kaashoek*

Steven Hand, Timothy Roscoe, "Mnemosyne: Peer-to-Peer Stegano-graphic Storage", presented by Steven Hand. Mnemosyne (pronounced *ne moz'nē*) concentrates on high value, small size, information. It is not an efficient global storage system, but is aimed instead at providing anonymity and anti-censorship properties. The basic idea is that blocks within the (highly distributed) system are filled with random data. The file is encrypted so that it too looks like noise and the data is then placed pseudo-randomly into the store. Collisions are clearly a problem, and although these could be dealt with by writing each block several times, it is preferable to use Rabin's Information Dispersal Algorithm instead. Traffic analysis is a problem, but it can be countered by routing data through many nodes, avoiding fetching all of the shares and by reading information that isn't required. Writes are more of a problem since specious writes would damage data unnecessarily. A working prototype has been created using Tapestry as an underlying peer-to-peer system; it runs at about 80Kbytes/sec reading and 160Kbytes/sec writing. The crypto aspects are currently being assessed. The longer-term aim is to use multiple DHTs and ensure they anonymize traffic. It is hoped to construct a commercial multi-homed data storage system.

Discussion: Q: Are locations chosen randomly? A: Not entirely, there is some directory and file structure present. Q: Do you need to check if data is still there? A: Yes, because after a while it disappears. For some levels of availability the files may need to be refreshed. However, disappearing may be an appropriate fit with an application such as a personal messaging system. Q: How does it compare with OceanStore? A: It's not solving the same problem.

Sameer Ajmani, Dwaine Clarke, Chuang-Hue Moh, Steven Richman, "ConChord: Cooperative SDSI Certificate Storage and Name Resolution", presented by Sameer Ajmani. SDSI is a proposed decentralized public key infrastructure that allows for mapping of principals (keys) to locally specified names and the use of chains of certificates to delegate trust to other organizations via the use of groups of principals. Name resolution becomes a key location problem that turns out to be hard because of the local namespaces, global loops and other complications. ConChord computes derived certificates to eliminate loops and make lookups fast, but this means significant extra work is needed for insertions and significant additional storage is required. This pays off if the number of resolutions is high as compared to insertions. A centralized resolver might be possible, but it turns out to be hard to locate servers

(SPKI/SDSI recommends embedding URLs into keys to fix this) and it is unclear where derived certificates are to be stored. A DHT looks like a good match to the problem both for lookup and storage, and trust issues don't arise because the certificates are all self-signed.

ConChord was developed using Chord as the DHT. Two main problems arose. Firstly, maintaining closure (derived certificates) while supporting concurrent updates proved difficult because of the lack of atomic updates or locking. The solution is to provide eventual consistency using periodic replays of the updates. Secondly, large datasets cause load imbalance, but traditional (CFS-style) data distribution does not support concurrent updates to those datasets. The solution here is to distribute the sets over multiple nodes but serialize updates through a single node.

A system has been built and evaluation is very promising with fast resolution, single lookups for membership checks and insertion that is quick enough. Resolvers can also use ConChord to share resolutions and thus save work. Future work will tackle replication of data, malicious clients and storage limiting.

Discussion: Q: What about revocation? A: Planned for future work. Revocation (and revalidation) lists can be stored in ConChord. SPKI specifies that proofs that contain revocable certificates must also contain the relevant CRL(s). Q: What about deletions (certificate expirations)? A: We serialize deletions through a single node, which simplifies this. Q: Is there a better solution than DHTs? A: We did need some extra code on the DHT nodes to do some application-specific stuff to provide invariants, handle expirations, etc. We'd like to soften the closure ideas as part of future work.

Russ Cox, Athicha Muthitacharoen, Robert Morris, **"Serving DNS Using Chord", presented by Russ Cox.** The idea was to rework the Domain Name Service so that it runs over a Chord, but unfortunately the result "sucked". This was unexpected because DNS was originally created to replace a distributed file called hosts.txt, but the new system meant that everyone had to become a DNS administrator, everyone needed a 24x7 running machine and data can now be locally correct, yet globally wrong. Peer-to-peer should address these issues by providing simple ways to distribute data to a permanently running system that has a single view of what is stored. The arrival of DNSSEC means that the data can be distributed to untrusted systems. Furthermore, DNS data is entirely "needles", so it ought to be a killer app for DHTs!

The idea was to lookup SHA-1{hostname, data type} on the Chord ring. But latency on cache misses is $O(log\ n)$ whereas DNS effectively has log base one million. Using Chord is five times slower on average. Robustness should be a plus, but DNS is already very robust and the new design introduces new vulnerabilities. Network outages also cause significant problems with no locality of storage for local names. Examining the O'Reilly "Bind and DNS" book shows thirteen common errors, but 9 are arguably bugs in BIND, 3 are at the protocol level (and reoccur here, albeit in different forms) and only 1 disappears entirely (there are no slave servers to make a mess of configuring). DNS has evolved over the last twenty years to include server-side computation; it's not just a distributed `hosts.txt` anymore. DHTs will be unable to replace distributed databases that include server-side computation, unless that can be expanded to include some form of mobile code. Questions also arise as to how much systems can be trusted to continue working without any incentives for doing this; there'd be no-one to blame if one's DNS disappeared.

Discussion: Q: So your conclusion is that DNS does not work well with DHTs? A:*(from the floor)* "I disagree that other DHTs would be as bad, Tapestry wouldn't have the same latency on lookup." "If you used Pastry then you would get less latency and a better fit." "Maybe DHTs are a good tool if they're being used for the wrong things?"

Session 5: Are We on the Right Track?
Chair: John Kubiatowicz

Stefan Saroiu, P. Krishna Gummadi, Steven D. Gribble, "Exploring the Design Space of Distributed and Peer-to-Peer Systems: Comparing the Web, TRIAD, and Chord/CFS", presented by Stefan Saroiu. Peer-to-peer has arrived, but will these systems stay? They do many things well but crucial "ilities" are missing: securability, composability and predictability. Because in DHTs a name is an address, this means that the name of content dictates which node it has to be placed upon, which might not be secure. Because routing is name based and servers are routers, you cannot trust routers more than servers. Because topology is dictated by keys you can surround and hijack content. Equally, it's hard to provide extra resources for "hot content" – you don't control your neighbors, but they are providing the Quality of Service. A Chord network with 20% modems has 80% slow paths. Moving forward it is necessary to enforce who publishes or participates, engineer according to specific load and value of content, and it must be possible to delegate, engineer responsibilities and isolate failures in a predictable way.

Discussion: Q: When the systems stop being overlays and start becoming infrastructure does this change things? A: Yes, that's exactly what I hope to influence. Q: For censorship resistance you may not want controllability? A: You need to think before losing that sort of property.

Pete Keleher, Bobby Bhattacharjee, Bujor Silaghi, "Are Virtualized Overlay Networks Too Much of a Good Thing?", presented by Bobby Bhattacharjee. Material is published by one node on a DHT, but stored by another. The virtualization provided by the overlay network gives relatively short paths to any node and load balancing is straightforward. These are clean elegant abstractions with provable properties. However, locality of access is lost. One cannot prefetch, search nearby or do other useful things with application specific information or with names that are naturally hierarchical. The choices are to add locality back at higher levels, use a higher granularity of exports, or to get rid of the virtualization.

TerraDir is a non-virtualized overlay directory service that assumes a static rooted hierarchical namespace. It caches paths and also returns a digest of everything else at an accessed node. This allows more/less specific queries to be done within the digest without further communication. Another important design feature is that the higher something is placed in the hierarchy, the more it is replicated. The system has been simulated. It was found that higher levels did more work, but as cache was added load balancing improved, as did the latency on requests. The system was resilient with >90% successful searches with 30% of systems failed. Current work is looking at load adaptive replication (where the main issue is consistency) and at improving searches where there is no good match between the search and the name hierarchy.

Discussion: Q: Why are there only 32K nodes in your simulation? A: We only had 24 SPARCs, though they did all have loads of RAM. We wanted the simulation to finish!

Session 6: Searching and Indexing *Chair: Robert Morris*

Adriana Iamnitchi, Matei Ripeanu, Ian Foster, "Locating Data in (Small-World?) Peer-to-Peer Scientific Collaborations", presented by Adriana Iamnitchi Scientific collaborations are characterized by groups of users sharing files and mainly reading them. Other group members will also wish to see any new files, so there is strong group locality and also time locality, in that the same file may be requested multiple times. It is an open problem as to whether scientific collaborations exhibit particular patterns and whether they can be exploited to create self-configuring networks that match the collaboration network characteristics. The Fermi high-energy physics collaboration was studied. It consists of 1000+ physicists at 70 institutions in 18 countries. Examining file usage it could be seen that path lengths were similar to a random network, but there was significant clustering of connections. This is the classic definition of a "small world".

A small world can be seen as a network of loosely connected clusters. The idea was to build a search system to take advantage of usage patterns. The search combines information dissemination within clusters and query routing/flooding among clusters. Gossip is used to maintain cluster membership and disseminate location info. Bloom filters are used to compress file location information. Requests are flooded to the same cluster and then to other clusters if requested information is not found locally. The system needs to adapt to the users' changing data interests, which is done by connecting nodes if they share many files in common or disconnecting them as the number of shared files drops. This ensures that the network mirrors the usage patterns. For future work it remains to be seen if there are general lessons or if this approach is domain specific.

Discussion: Q: What information is disseminated? The data itself by replicating files, or the file location? A: The data is large and dynamic as new files are inserted into the network, so copying and indexing is expensive. Q. Are there many new files? A. Yes, it is characteristic of scientific collaborations that new files are created. Q. Is there any software available or in development? A. This problem came from Grid computing where there is a lot of effort in sharing computational resources (not only files). The Globus toolkit `www.globus.org` is freely available software for creating and collaborating in Grids.

Matthew Harren, Joseph M. Hellerstein, Ryan Huebsch, Boon Thau Loo, Scott Shenker, Ion Stoica, "Complex Queries in DHT-based Peer-to-Peer Networks", presented by Ryan Huebsch. There is a poor match between databases and peer-to-peer systems. The former have strong semantics and powerful queries, the latter, flexibility, fault-tolerance and decentralization. However, query processing is better matched to peer-to-peer systems. A keyword search corresponds to a simple "canned" SQL query and the idea is to add more types of query. For example, a "join" can be done by hashing and hoping to get the data back from the node where a local join is performed. The motivation is to provide a decentralized system, not to provide improved performance or to replace relational databases.

A system called PIER has been developed on top of CAN. It has a simple API: "publish", "lookup", "multicast" (restricted to particular namespaces rather than the whole network), "lscan" (to retrieve data from a local namespace) and when new data is added a callback mechanism is used. Current work is on the effect of particular algorithms and looking for new types of operation. Several issues arise which are common across the peer-to-peer community: caching, using replication, and security. Specific database issues are when to pre-compute intermediate results, how to handle continuous queries and alerters, choice of performance metrics, and query optimization as it relates to routing.

Discussion: Q: What's the difference between this and distributed databases? A: They haven't taken off. What is actually built is parallel databases under one administrator, whereas this is decentralized. Q: Astrolabe focuses on continuous queries and aims to have constant loading. Won't these joins have a nasty complexity in terms of communications loading? A: This is the realm of performance metrics, and yes we're concerned about these. Q: Surely users will remove the application if its resource requirements are intrusive? A: It might well be important to see what users will tolerate, we haven't looked at this yet. Q: Is this a read-only system? A: One reads the raw data, but it is also necessary to keep track of indexes and optimize them.

David Karger gave a talk on "Text Retrieval in Peer-to-Peer Systems". The traditional approach to text retrieval is to observe that the majority of the information base is text. The user formulates a text query (often inaccurately), the system processes the corpus and extracts documents. The user then refines the query and the procedure iterates.

The procedure has two metrics, "recall" is the percentage of relevant documents in the corpus that are retrieved and "precision" is the percentage of retrieved documents that have relevance. There is always a trade-off between performance on these two metrics. The procedure is speeded up by pre-processing, since users notice delays of more than 0.5sec and give up after 10 seconds. Searching the web has non-traditional aspects. People care more about precision than about recall, and they are prepared to wait rather longer for results.

Boolean keyword searches are done by using inverted indexes and performing intersections and joins. They can also be done by direct list merging, which gives a complexity of $O(size\ of\ list\ of\ smallest\ term)$. Problems arise with "synonymy" (several words for the same thing – fixed by using a thesaurus to increase recall but lower precision), and "polysemy" (one word means several things, increasing precision but lowering recall – unfortunately asking for the user's assistance may just confuse).

Further issues are "harsh cutoffs" (relevant documents are missed because they don't contain all the required keywords – fixed by quorum systems such as Altavista's) and "uniform influence" (no allowance in the scoring for multiple usage of a word, and rare terms have no special influence). These latter problems can be addressed by vector space models where each co-ordinate has a value expressing occurrence and the "dot product" measures similarity. These systems inherently provide rankings of results and are easy to expand with synonyms. In some systems, if the user indicates that a particular document located by a first search is relevant then a refined search can be done with hundreds or thousands of terms taken from that document.

Google has a corpus of about three billion pages, average size 10K (i.e. 30TB of data). Their inverted index is of similar size. They use a boolean vector space model. Experience has been that most queries are only two terms and queries have a Zipf distribution so caching helps a bit. Their scoring scheme also looks at links, raising the relevance of highly linked pages and observing that the anchor text in the link may be a better description of the contents than the actual page. Google uses a server farm of several thousand machines. In principle a moderate sized peer-to-peer system of perhaps ~30,000 nodes should be able to perform the same function...

An obvious design would be to partition the documents between the nodes and each node then builds the inverted index for its documents. The results then need to be merged, which creates an n^2 connectivity problem. Alternatively, a DHT could be used to partition the terms and a query is done by talking to appropriate nodes. The drawback is that the lists could be very long which creates a bandwidth problem as the lists are merged. An improvement would be to create an inverted index on term pairs (i.e. pre-answer all two term queries), however this would generate $n^2/2$ pairs and for a Google sized database would create ~15,000TB of index. However, the idea does work well for some special cases where n is small (song titles) or where a windowing approach can be used (relevant documents will tend to have the terms near to each other).

However, the economics favor a centralized approach and Google already uses a type of peer-to-peer design in their data center where they can exploit high inter-machine bandwidths. Their main bottleneck is content providers restricting their rate of "crawl" across the web, which can make their index old and out-of-date. Distributing the crawler function will not assist, but making the content providers into peers might assist, assuming their input could be trusted.

Distributed systems do differ from "big iron" in that it is possible to gain a feeling of privacy and to provide some anonymity from the partition of control and knowledge. Expertise networks such as MIT's Haystack or HP's Shock try to route questions to experts and the challenge is to index the experts' abilities.

Session 7: Security in Peer-to-Peer Systems
Chair: Steve Gribble

Emil Sit, Robert Morris, **"Security Considerations for Peer-to-Peer Distributed Hash Tables", presented by Emil Sit.** In a distributed peer-to-peer system there is no "Trusted Computing Base". Peers can only trust themselves and the protocols need to be designed to allow peers to verify the correct operation of others. A reasonable adversary model is that any node can be malicious, discarding traffic or sending malicious packets. Malicious nodes can collude with each other, but they do not have the ability to intercept and read traffic for arbitrary users.

One idea is to develop system invariants and verify them. For example, in Chord you expect to halve the distance to your goal on each step, so this should be verified. If recursive queries were allowed then this might improve latency but a malicious node could cause incorrect results or cause traffic to loop forever. If progress is being monitored then bad behavior can be detected. If one wishes to know whether a node is the correct endpoint – and hence its answer cannot be bettered – then the rule assigning keys to nodes needs to be verifiable so that one can check that the value hashes correctly to the node and that the node is placed correctly according to some relatively hard to forge value such as an IP address. Malicious nodes may conspire to suck nodes or queries into an "evil ring" which does not contain all the participants in the system. New nodes being added into the DHT needs to cross-check the answers with other nodes to ensure they are consistent. Independent routes to resources (such as in CAN) can be utilized for checking. Another important problem is that malicious nodes may fail to create the replicas that they should and this may not be detected if a single node is responsible for this action.

General design principles for security are: #1 define verifiable system invariants, and verify them; #2 allow the querier to observe lookup progress; #3 assign keys to nodes in a verifiable way; #4 be wary of server selection in routing; #5 cross-check routing tables using random queries; and #6 avoid single points of responsibility.

Discussion: Q: What do you do when you detect an error? A: You may be able to route in another way. Q: Can you tell what the spacing in a Chord ring should be? A: Yes, you can look at the number of participants or just look at the spacing of your own successors.

John R. Douceur, **"The Sybil Attack", presented by John R. Douceur.**
In large distributed systems you can make assumptions about what percentage
of the participants are conspiring against you. However, it is hard to substanti-
ate these assumptions and in fact, all other participants may be under a single
opponent's control. The "Sybil Attack" (named after a 1973 book on multiple
personalities) is an attempt to break down the determination of identity distinctness, where an identity is an abstract notion connected in a provable way
to a persistent item such as a public key.

An obvious source of identity information is a trustworthy authority such
as Verisign or, more subtly when using IP addresses or DNS, ICANN. Another
source of information would be yourself, in that you can test if two remote
systems can do something that they couldn't manage if they were a single entity.
You might also be able to get others to assist in this testing and accept identities
that others have endorsed. The type of tests that could be attempted would be
a communications resource challenge (can they handle large volumes of traffic),
a storage resource challenge (can they store lots of data) or a computational
resource challenge (can they do sums very fast). There are problems with such
tests – the challenges must require simultaneous solving by all the identities
(because otherwise the opponent can solve one problem at a time) and of course
the opponent may command more resources than a standard node anyway. You
might hope to leverage something from other people's view of identities, but
of course the identities who are vouching for each other may all be under the
opponents control. The paper formalizes all of this, but the conclusion is that
distinctness can only be verified by means of a certificating authority or by
measurement of some, yet to be discovered, uniform constraint. At the moment
identity verification does not scale.

Discussion: Q: You can't really fix this at the system level. In World War I
the entire German spy network in the UK was run by the British. In World War
II the Germans faked the SOE network in The Netherlands. You won't be able
to tell if the rest of the network isn't the 50,000 people at the NSA. A: Worse
than that, it could be just one person. Q: Where did you get all the marvelous
animated graphics you've used? A: Standard with PowerPoint 2002.

Jared Saia, Amos Fiat, Steve Gribble, Anna Karlin, Stefan Saroiu,
**"Dynamically Fault-Tolerant Content Addressable Networks", pre-
sented by Jared Saia.** Napster was shut down by legal attacks on a central
server and research shows that Gnutella would shatter if a relative handful of
peers were removed. It is easy to shut down a single machine, which has limited bandwidth or a limited number of lawyers. However, it is harder to remove
large numbers of machines. The Deletion Resistant Network (DRN) is a scalable, distributed, peer-to-peer system that after the removal of $\frac{2}{3}$ of the peers by
an omniscient adversary who can choose which to destroy, 99% of the rest can
access 99% of the remaining data. However, DRN is only robust against a static
attack. If all the original peers are removed, then the system fails even if many
new peers have joined. The Dynamic DRN has stronger properties against the

same adversary, and for a fixed n and $\epsilon > 0$, if there are $O(n)$ data items in the network then in any period where $(1 - 2\epsilon)n$ peers are deleted and n join, then with high probability all but the fraction ϵ of the live peers can access a $1 - \epsilon$ fraction of the content.

These networks are based on butterfly networks (a constant degree version of a hypercube). The network properties can be proved probabilistically using expander graphs. Peer join time and search time require $O(log\ n)$ messages, but in the whole system $O(log^3 n)$ storage and $O(log^3 n)$ messages are needed, ie: the network has some very desirable robustness properties, but the time and space bounds remain comparable with other systems.

Discussion: Q: What happens when the system becomes so big that you need a new level of butterfly linkage? A: We don't yet have any good ideas how to deal with large changes in size. Best we can do is $O(n^2)$ messages and $O(n)$ broadcasts. Q: This structure protects the network, but what about the data items? A: We're not protecting any particular data item against attacks directed specifically at it, but they can be duplicated.

Session 8: Applications II *Chair: Ion Stoica*

Sridhar Srinivasan, Ellen Zegura, **"Network Measurement as a Cooperative Enterprise", presented by Sridhar Srinivasan.** Network measurement is undertaken to improve performance and to assess the utilization of resources. The challenge is to deploy an Internet-wide service, keep the overhead low (to avoid perturbing the network) and to be assured about the accuracy of reported values. M-Coop is a peer-to-peer measurement architecture. Each peer has an area of responsibility (AOR) which it reports upon, ideally at least as small as an AS. An overlay network is constructed, with peers selecting neighbors in adjacent ASs. A second overlay network is built within the AS, the nodes of which peer entirely within the AS and have AORs of parts of the AS. When measurement queries are made they are passed over the overlay network and it is the data collected from the experiences of these packets which is reported back as the metric. Data is also given a "trust" component, which is a measure of past reliability. This is done by comparing results with other measurements of the link to a neighbor as well as "pings" of nearby machines.

Future challenges are to improve the composition of measurements (the path through the overlay may not be the same as the general experience for other traffic); to deal with colluding malicious nodes; to verify all measurements; to assess what level of participation is needed to get a good answer to queries; and to determine how useful the information will be in practice.

Discussion: Q: Latency is easy to verify but bandwidth isn't? A: Exactly so.

Venkata N. Padmanabhan, Kunwadee Sripanidkulchai, **"The Case for Cooperative Networking", presented by Kunwadee Sripanidkulchai.** CoopNet is a peer-to-peer system that is intended to complement and work in conjunction with existing client-server architectures. Even when making minimal assumptions about peer participation it provides a way of dealing with "flash crowds" on web sites such as occurred on news sites on 9/11 or the notorious Slashdot effect. The bottleneck in such cases is not disk transfer, since everyone is fetching the same "hot" content. CPU cycles are an issue, though changing from dynamic to static content fixes this. The main problem is bandwidth, so the idea is to serve the content from co-operating peers, with clients being redirected to an appropriate peer – the redirection being a relatively small (1%) bandwidth imposition. The co-operating peers announce themselves as they fetch the content by means of an HTTP pragma, and the server can then record their identities in case their assistance is needed. Returning appropriate alternative sites is a complex problem. Using BGP prefix clusters is lightweight but crude and various better schemes are under investigation. Simulations using the traces of the MSNBC website on 9/11 show that good results can be obtained with just 200 co-operating peers. The peers were busy just 20% of the time and the bandwidth they had to donate was low. Unfortunately, there were long tails on some of the resource usage distributions and further work is needed to minimize these.

Discussion: Q: How does this compare with DHTs? A: This is push technology oriented. Q: When do you decide to start redirecting? A: Preferably just as you start to be overloaded Q: How does this compare with systems such as Scribe? A: Those systems require you to commit what you belong to ahead of time, rather than dynamically. Q: Aren't there concerns with privacy here? A: This is only for popular documents and the server is always in control of when it redirects.

Ion Stoica, Dan Adkins, Sylvia Ratnasamy, Scott Shenker, Sonesh Surana, Shelley Zhuang, **"Internet Indirection Infrastructure", presented by Ion Stoica.** Today's Internet has a point-to-point communications abstraction. It doesn't work well for multicast, anycast or mobility. Existing solutions to these problems change IP (into mobile IP or IP multicast). These solutions are hard to implement whilst maintaining scalability, they do not interoperate or compose, and people may not be incentivized to provide them. The result has been provision of facilities at the application layer (such as Narada, Overcast, Scattercast...) but efficiency is hard to achieve. It should be noted that all previous schemes have used indirection, so perhaps there should be an indirection layer as an overlay network placed over the IP layer.

The service model is "best efforts" and data is exchanged by name. To receive a packet a trigger is maintained in the overlay network by the end-point that owns it. This trigger will know how to reach the end point. This scheme is capable of supporting many different types of services. Mobility is simple, the end-point tells the trigger where it has moved to. In multicast many hosts all insert the same named trigger. For anycast there is an exact match on the first part of a name and a longest prefix match on the last part. Composable services can be done by having a stack of triggers and sending packets off to each service in turn (e.g. sending data via an HTML→WML converter before delivering it to a wireless device). These stacks can be built by both sender and receiver without the other needing to be aware of the processing. Load balancing or location proximity can be expressed by putting semantics into red tape bits in the trigger. An early prototype has been implemented based on Chord. Each trigger is stored on a server and the DHT is used to find the best matching trigger. The results can be cached and further packets sent to the server directly.

Discussion: Q: What about authorizations? A: You can have public and private identifiers, and this leads to the idea of changing from a public to a private trigger once authorization has been given. Q: Does this system defend against Denial of Service attacks? A: At the old layer, yes, but there are new possibilities at the overlay layer such as creating circular routes. Q: Perhaps you are putting too much into this new layer? A: Many problems can be solved in a new way by using this indirection layer. For example you can now seriously consider providing a reliable multicast.

Tyron Stading, Petros Maniatis, Mary Baker, "Peer-to-Peer Caching Schemes to Address Flash Crowds", presented by Tyron Stading. Noncommercial sites do not usually expect flash crowds and do not have the resources to pay for commercial systems to mitigate their effects. "Backslash" is a peer-to-peer collaborative web mirroring system suitable for use by collectives of websites to provide protection against unusually high traffic loads. The idea is to create a load balancing system that will usually direct requests to the main site for page content. If it perceives that an overload is about to occur, the pages are rewritten into the load-balancing collective and further requests are served from there until the overload condition finishes.

The collective is based on a DHT (currently CAN) and the data (currently assumed to be static) is distributed to members of the collective using cache diffusion techniques. The idea is that popular content will be pushed out to more nodes. This can produce a "bubble effect" where the inner nodes of the collective become idle and the outer nodes do all the serving. To prevent this, some probabilistic forwarding of requests is done to try and use all of the collective nodes. The system has been simulated handling two flash crowds at once and it worked well in balancing the load. However, with too much diffusion agility the second flash crowd caused competition for entries and performance suffered. The probabilistic forwarding also worked less well than expected. Future work will simulate the system at a higher fidelity and will look at the effect of cache invalidation for changing content.

Discussion: Q: Who is going to use this service? What are the incentives? A: We concentrated on non-profits because they are more likely to work to help each other. Q: Do you need to copy ahead of time? A: Yes. If you don't copy the files before the flash crowd gets huge then there's no guarantee the files will be available.

Session 9: Data Management *Chair: David Karger*

Robbert Van Renesse, Kenneth Birman, "Scalable Management and Data Mining Using Astrolabe", presented by Robbert Van Renesse. Astrolabe takes snapshots of the global state of a system and distributes summaries to its clients. Its hierarchy gives it scalability, the use of mobile SQL gives it flexibility, robustness is achieved by using epidemic (Gossip) protocols and security comes from its use of certificates.

Astrolabe has a DNS-like domain hierarchy with domain names identified by path names within the hierarchy. Each domain has an attribute list called a MIB which identifies a domain, lists its Gossip contacts, lists server addresses for data access and indicates how many local hosts there are. These MIBs can be aggregated into parent domains. There is a simple API of Get_MIB(*domain_name*), Get_Children(*domain_name*) and Set_Attr(*domain_name, attribute, value*). All hosts hold their own MIB and also the MIBs of sibling domains, giving a storage requirement of $O(log\ n)$ per host. The MIBs are extensible with further information as required. Standard SQL queries are gossiped to summarize data into parents. This allows aggregation queries to be made such as "where is the highest loaded host" or "which domains have subscribers who are interested in this topic" or "have all hosts received the latest software update".

The system works by passing messages using a simple epidemic protocol that uses randomized communications between nearby hosts. This is fast (latency grows $O(log\ n)$ with probabilistic guarantees on maxima, assuming trees with constant branching factors) and is robust even in the face of denial-of-service attacks. Failure is detected when timestamps in MIB copies do not update and new domains are found by gossiping, occasional broadcasts and by configuration files.

Discussion: Q: Would an overlay network with a general tree structure be better suited? A: We're doing aggregation, and DHTs don't help with that. We started by trying to build a scalable multicast system and we needed ways to get past firewalls etc. The routing is all proximity based and every aspect of the system reflects locality. Q: What is being gained by the massive replication of data? A: Results are available in a single step. Q: Aren't some queries expensive? A: We restrict join queries to make the load manageable.

Nancy Lynch, Dahlia Malkhi, David Ratajczak, "Atomic Data Access in Content Addressable Networks", presented by David Ratajczak. This work shows that atomicity is useful and achievable in a peer-to-peer DHTs with practical fault-tolerance. Atomicity can be seen as consistency with the effect of having a single copy of a piece of data being accessed serially. It is a crucial property where there are multiple writers and can be used as a building block for many distributed primitive operations. The other important property that is demonstrated is "liveness", viz that any submission to an active node is guaranteed a response. However, one cannot achieve atomicity or liveness in an asynchronous failure-prone system and of course peer-to-peer systems are dynamic and an unfortunate sequence of events can disconnect the network. The asynchronous property means that you need to use timeouts to detect failures and this can cause problems if the original request is still active. Therefore, it is assumed that communication links are reliable FIFO channels and that node failures do not occur. This can be done by local fault-tolerant, redundant, hardware (a "replica group") which turns failures into graceful disengagement from the network.

The API for the system is Join(), Update() and Leave(), with the Leave function now being a non-trivial operation. The guarantees provided include atomicity and liveness. The algorithms are in the paper and repay careful study. Future work will determine if consecutive nodes in the DHT ring can form a replica group, with a multi-level structure around the ring. It looks as if setting thresholds for splitting large replica groups and merging small ones can be set in the $O(log\ n)$ region – and then some good properties will result. At present the algorithm presented does not fail especially gracefully and more details of the fault-tolerance properties need to be worked out. The system will need to be built before it can be fully understood.

Discussion: Q: Isn't it hard to leave whilst you're receiving constant updates? A: In practice, all the operations in one node are in one thread. You can prove that you will eventually make progress and leave (or indeed join) the network.

Yan Chen, Randy Katz, John Kubiatowicz, "Dynamic Replica Placement for Scalable Content Delivery", presented by Yan Chen. Content Distribution Networks (CDNs) attempt to position data so as to have useful things in local "replicas", so as to improve the user experience of the web and of streaming media, whilst minimizing resource consumption. The problem is how to choose replica locations and then keep them up-to-date, viz: an adaptive cache coherence system is required.

Previous work concentrated on static replica placement, assuming that the clients' locations and access patterns were known in advance. Data transfer cannot be done by IP multicast because this is impractical from one domain to another and application layer multicast (ALM) fails to scale. The usual solution is to replicate the root of the ALM system, but this suffers from consistency problems and communication overhead. This work uses a peer-to-peer overlay location service for scalability and locality of search. Tapestry is used because it already has some locality within it.

The system simultaneously creates a tree for disseminating data and decides upon replica placement. The idea is to search for qualified local replicas first and then place new replicas on the Tapestry overlay path. Two algorithms were investigated. A naïve scheme allows a node that is holding the data to decide whether it is a suitable parent to hold the data given the identity of the requesting client and if so it then puts the replica as close to the client as possible. The smart scheme is prepared to consider its parent, siblings and server children as possible parents as well and then chooses the node with the lightest load. It then places the replica a long way from the client. This latter scheme has a higher overhead in messages, but results in better placement, with fewer replicas required. It performs almost as well as in the ideal case where all requests are known in advance. Future work will evaluate the system with diverse topologies and real workloads. Dynamic deletion and insertion will be considered so that the system can adapt as user interests change, and the system will be integrated into the OceanStore project.

Discussion: Q: How important is Tapestry to this system? A: Tapestry is providing proximity properties, identifying potential replica placement points, and improving the scalability of the CDN for update dissemination so that each node only has to maintain states for its parent and direct children.

Thanks

The workshop ended with thanks being expressed to the organizers and in particular to Frans Kaashoek, who in turn expressed his gratitude to Neena Lyall for all her work on the logistics of running the workshop and Microsoft Research for their generous financial support.

Acknowledgements

My attendance at the workshop was due to the financial assistance of the Cambridge MIT Institute.

Thanks to George Danezis for comparing this account to his own detailed notes of the event and pointing out everything I'd forgotten to include. Thanks also to the other workshop participants for their corrections to my original inaccurate reporting of their contributions.

Observations on the Dynamic Evolution of
Peer-to-Peer Networks

David Liben-Nowell, Hari Balakrishnan, and David Karger*

Laboratory for Computer Science
Massachusetts Institute of Technology
{dln,hari,karger}@lcs.mit.edu
http://pdos.lcs.mit.edu/chord

Abstract. A fundamental theoretical challenge in peer-to-peer systems is proving statements about the evolution of the system while nodes are continuously joining and leaving. Because the system will operate for an infinite time, performance measures based on runtime are uninformative; instead, we must study the *rate* at which nodes consume resources in order to maintain the system state.

This "maintenance bandwidth" depends on the rate at which nodes tend to enter and leave the system. In this paper, we formalize this dependence. Having done so, we analyze the Chord peer-to-peer protocol. We show that Chord's maintenance bandwidth to handle concurrent node arrivals and departures is near optimal, exceeding the lower bound by only a logarithmic factor. We also outline and analyze an algorithm that converges to a correct routing state from an arbitrary initial condition.

1 Introduction

Peer-to-peer (P2P) routing protocols like CAN [4], Chord [7], Pastry [5], and Tapestry [8] induce a connected overlay network across the Internet, with a rich structure that enables efficient key lookups. The typical approach to the design of such overlays goes roughly as follows. First, an "ideal" overlay structure is specified, under which key lookups are efficient. Then, a protocol is specified that allows nodes to join or leave the network, properly rearranging the ideal overlay to account for their presence or absence. Finally, fault tolerance may be discussed: one can show that the ideal overlay can still route efficiently even after the failure of some fraction of the nodes.

Such an approach ignores the fact that a P2P network is a continuously evolving system. The join protocol may work well if joins happen sequentially, but what if many happen concurrently? The ideal overlay may tolerate faults, but once those faults occur, the overlay is no longer ideal. So what happens as the faults continue to accumulate over time?

---

* This research was sponsored by the Defense Advanced Research Projects Agency (DARPA) and the Space and Naval Warfare Systems Center, San Diego, under contract N66001-00-1-8933, by NSF contract CCR-9624239, and by a Packard Foundation fellowship.

P. Druschel, F. Kaashoek, and A. Rowstron (Eds.): IPTPS 2002, LNCS 2429, pp. 22–33, 2002.

To cope with these problems, any realistic P2P system must implement some kind of *maintenance protocol* that continuously repairs the overlay as nodes come and go, ensuring that the overlay remains globally connected and supports efficient lookups. In analyzing this maintenance protocol, we must recognize that the system is unlikely ever to be in its ideal state. Thus, we must show that lookups and joins (and the maintenance protocol itself) occur correctly even in the imperfect overlay.

Because a P2P system is intended to be running continuously and system membership is dynamic, the time taken to maintain the system's state is not a proper measure of resource usage; rather, what matters is how much resource bandwidth is consumed by nodes in maintaining control information in the form of routing tables and other such data structures.

This paper investigates the per-node network bandwidth consumed by maintenance protocols in P2P networks. We are motivated by the observation that this property—which addresses how much work each node must do in the interests of providing connectivity and a good topological structure—may be an important factor in determining the long-term viability of large-scale, dynamic P2P systems. For instance, if the per-node bandwidth consumed by these maintenance protocols were to grow fairly rapidly (e.g., linearly) as the network size increases, then a system would quickly overwhelm the access bandwidths of its participants and become impractical.

Any node joining the network must send at least some number of housekeeping messages to let other nodes know of its presence, to provide basic connectivity. Additional messages are usually required to update routing table information on nodes, so that efficient lookups can then occur. Similarly, because nodes may fail without any notification, each node must periodically monitor the state of some or all of its neighbors, consuming network bandwidth.[1]

We can ask a number of questions in this framework. At what rate must each node in the system do work in order to keep the system in a "good" state? How much work is required simply to provide a connected structure where lookups are correct? How much work is required to provide a richer structure where lookups are correct and also fast?

To answer these questions, we make two kinds of observations about P2P maintenance protocols. First, we give lower bounds on the maintenance protocol bandwidth for connectivity in any P2P network as nodes join and leave. We characterize this lower bound using the notion of *half-life*, which essentially measures the time for replacement of half the nodes in the network by new arrivals. We show that per-node maintenance protocol bandwidth is lower-bounded by $\Omega(\log N)$ per half-life for any P2P system that wishes to remain connected

---

[1] Alternatively, a node may detect failures only when it actually needs to contact a neighbor; however, this merely defers the network traffic for finding a new neighbor until the old one fails. It also raises the risk that all of a node's neighbors fail without the failures being noticed, permanently disconnecting that node from the network.

with high probability.[2] Second, we analyze the maintenance protocol used by Chord [7], a P2P routing protocol. We show that Chord consumes bandwidth only logarithmically larger than our lower bound. It is noteworthy our system provides fast lookup at a resource cost not much greater than the minimum necessary merely to maintain connectivity. Critical to this analysis is a demonstration that Chord's join, lookup, and maintenance protocols work correctly even when the system is not in its idealized stable state.

This style of evolutionary analyses of P2P networks has not been well-developed. Many P2P systems focus on models in which nodes join and depart only in a well-behaved fashion, allowing maintenance to happen only at the time of arrival and departure. We believe this kind of well-behaved model is unrealistic. Other protocols allow for the possibility of unexpected failures, and show that the system is still well-structured after such failures occur. These analyses, however, assume that the system begins in an ideal starting state, and do not show how the system returns to this ideal state after the failures; thus, accumulation of failures over time eventually disrupts the system. (See, e.g., [2, 4, 5, 7, 8].)

Recently, Saia et al. [6] have explored the use of a butterfly network in a P2P setting. Their system retains good routing structure even after the *adversarial* removal of a constant fraction of the nodes, and they show how to maintain their network as nodes fail, as long as the number of nodes joining the network is always sufficiently larger than the number of failures. In their system, keeping a well-structured network as nodes join is more difficult, and their results to do not apply to a steady state in which the number of nodes in the system remains constant.

Perhaps the closest to our evolutionary analysis is the recent work of Pandurangan et al. [3], who study a centralized, flooding-based P2P protocol. Using a Poisson arrival/departure model, they show that their protocol results in an overlay network whose diameter remains logarithmic, with high probability. However, their scheme does not solve the problem of *routing* within the P2P network: to find the node responsible for a given data item, they propose flooding the network, requiring $\Omega(N)$ messages. Also, their system requires a central server to guarantee connectivity.

We believe that our evolutionary analysis, with its recognition that the ideal state will rarely occur, is crucial for proper understanding of P2P protocols in practice.

This note summarizes work which is reported in full in [1].

---

[2] Throughout this paper, *with high probability* (abbreviated *whp*) means with probability at least $1-1/N$. In a system with N nodes, events that happen with probability exceeding $1/N$ become "expected" (i.e., the expected number of nodes at which the event occurs exceeds one). Thus, it is standard in theoretical analysis to attempt to ensure that bad events occur with probability at most $1/N$. As is usual, the proof shows how parameters can be varied to achieve any desired bound on the failure probability, e.g., $1/N^2$ as opposed to $1/N$.

2   A Half-Life Lower Bound

In this section, we give a general lower bound for the bandwidth of maintenance messages in P2P systems, based on the rate of node joins and departures.

Definition 1. *Consider a P2P system with N nodes alive at time t. The doubling time at time t is time that it takes for N additional nodes to arrive. The halving time at time t is the time that elapses before half of the nodes alive at time t depart. The half-life at time t is the smaller of the doubling and halving times at time t. Finally, the half life of the entire system is the minimum half-life over all times t.*

Intuitively, a half-life of τ means that after time $t + \tau$, only half the state of the system can be extrapolated from its state at time t.

Half-life is a coarse measure of the rate of change of a system, and does not impose any specific conditions on the particular fine-grained pattern of arrivals and departures; a half-life of τ can result from a steady stream of node joins and failures or from the simultaneous joins or failures of a massive number of nodes. Although there are some pathological situations in which the half-life is not a meaningful measure (e.g., the simultaneous failure of almost all nodes in the system), we believe that the concept of half-life is a useful and general characterization of the rate of change of P2P systems in a wide variety of circumstances.

As a specific example, consider a Poisson model of arrivals/departures [3]: nodes arrive according to a Poisson process with rate λ, while each node in the system departs independently according to an exponential distribution with rate parameter μ (i.e., expected node lifetime is $1/\mu$). If there are N nodes in the system at time t, then the expected doubling time is N/λ and the expected halving time is $(1/\mu)\ln 2$. (The probability p that a node fails in time τ is $1 - e^{-\mu\tau}$; setting $\tau = (1/\mu)\ln 2$ makes $p = 1/2$.) The half life is then $\min((\ln 2)/\mu, N/\lambda)$.

If λ and μ are fixed and the system is in a steady state, then the arrival rate of λ must be balanced by the departure rate of $N\mu$ (each of N nodes is leaving at rate μ), implying $N = \lambda/\mu$. Then the doubling time is $1/\mu$ and halving time and half-life are both $(1/\mu)\ln 2$. This reflects a general property: in any system where the number of nodes is stable, the doubling time, halving time, and half-life are all equal to within constant factors.

Using this Poisson model, we derive a lower bound on the rate at which bandwidth must be consumed to maintain connectivity of the P2P network.

Theorem 2. *Consider any P2P system with any initial configuration. Suppose that there is some node n that, on average, receives notification about fewer than k new nodes per τ time.*

Then there is a sequence of joins and leaves with half-life τ and a time t so that node n is disconnected from the network by time t with probability $(1 - \frac{1}{e-1})^k$.

Corollary 3. *Consider any N-node P2P network that remains connected with high probability for every sequence of joins and leaves with half-life τ.*

Then every node must be notified with an average of $\Omega(\log N)$ new nodes per τ time.

The corollary follows from the theorem by setting $(1 - 1/(e - 1))^k = 1/N$ in the theorem. The intuition behind the theorem is as follows. In a half-life, the probability that any particular node in the network fails is $1/2$. Thus, if any node has fewer than $\log N$ neighbors, then the probability that they all fail during this half life is larger than $1/N$. So each node must maintain a set of $\log N$ neighbors. In each half-life, then, each node loses about $(\log N)/2$ neighbors; it must replace its failed neighbors to remain connected in the next half-life.[3]

3   A Dynamic Model for Chord

This section outlines and analyzes two maintenance protocols in Chord. The first is *weak stabilization* from [7], which maintains a small amount of correct routing information in the face of concurrent arrivals and departures. The second is *strong stabilization*, which ensures a correct routing overlay from an arbitrary initial condition.

3.1   Background on Chord

Chord nodes[4] and keys are hashed into a random location on the unit circle; a key is assigned to the first node encountered moving clockwise from it. Each node knows its *successor node*—the node immediately following it on the circle—which allows correct lookup of any key k by walking around the circle until reaching k's successor. We speed this search using *fingers*: $n.finger[i]$ is the first node following $n + 2^i$ on the identifier circle. Intuitively, any node always has a finger pointing halfway to any destination, so that a sequence of $\log N$ "halvings" of the distance take us to the key. Each node u also maintains its *predecessor*, the node closest to u that has u as its successor.

Each node n periodically executes a *weak stabilization* procedure to maintain the desired routing invariants: it contacts its successor s, and if $s.predecessor = p$ falls between nodes n and s, then node n sets $n.successor := p$. To maintain finger pointers, each node n periodically searches for improved fingers by running $find\_successor(n + 2^{i-1})$ for each finger i.

---

[3] Note that this does not require that each node u learn about $\Omega(\log N)$ nodes in every half-life, since u may receive a message containing information about many new nodes; instead, it requires that u receive information about new nodes at an *average rate* of $\Omega(\log N)$ per half-life.

[4] For load balancing, each "real" Chord node maintains $\log N$ *virtual nodes* with different identifiers; For simplicity, we analyze work per *virtual* node. A system that does not need perfect load balancing can run one virtual node per real node; one that does require load balancing will need to do $\log N$ times as much work.

```
// ask node n to find the successor of id
n.find_successor(id)
    if (id ∈ (n, n.successor])
        return n.successor;
    else
        n′ := closest_preceding_node(id);
        return n′.find_successor(id);
```

```
// join the system using information from node n′.
n.join(n′)
    predecessor := nil;
    s := n′.find_successor(n);
    build_fingers(s);
    successor := s;
```

```
// periodically refresh finger table entries.
n.fix_fingers()
    build_fingers(n);
```

```
// update finger table via searches by node n′.
n.build_fingers(n′)
    // get first non-trivial finger entry.
    i₀ := ⌊log(successor − n)⌋ + 1;
    for each i ≥ i₀ index into finger[];
    finger[i] := n′.find_successor(n + 2^{i−1});
```

where $i_0 := \lfloor \log(successor - n) \rfloor + 1$ and $finger[i] := n'.find\_successor(n + 2^{i-1})$.

```
// search the local table for the highest predecessor of id
n.closest_preceding_node(id)
    for i := m downto 1
        if (finger[i] ∈ (n, id))
            return finger[i];
    return n;
```

```
// periodically verify n's immediate successor,
// and tell the successor about n.
n.stabilize()
    x := successor.predecessor;
    if (x ∈ (n, successor))
        successor := x;
    successor.notify(n);
```

```
// n′ thinks it might be our predecessor.
n.notify(n′)
    if (predecessor = nil or n′ ∈ (predecessor, n))
        predecessor := n′;
```

```
// update successor list using successor's successor list.
n.fix_successor_list()
    ⟨s₁, . . . , sₖ⟩ := successor.successor_list;
    successor_list := ⟨successor, s₁, s₂, . . . , sₖ₋₁⟩;
```

Fig. 1. Pseudocode for the Chord P2P system.

A node departing the Chord ring can cause disconnection of the ring because another node may no longer be able to contact its successor. To prevent this disconnection, each node keeps a *successor list* of the first $\Theta(\log N)$ nodes following it on the ring. A node n maintains its successor list by repeatedly fetching the successor list of $s = n.successor$, removing its last entry, and prepending s to it. If node s fails, then n sets $n.successor$ to the next node on its successor list. Node n also periodically confirms that its predecessor has not failed; if so, it sets $n.predecessor = $ **nil**.

See Figure 1 for pseudocode.

A note on our model. For simplicity, we limit ourselves to a synchronous model of stabilization. We can thus refer to a *round* of stabilization. With mild complications, we can handle (without an increase in running time) a network with a reasonable degree of asynchrony, where machines operate at roughly the same rate, and messages take roughly consistent times to reach their destinations.

In the remainder of this work, we consider independent, random joins and failures. Because Chord identifiers are generated randomly, any correlations among failures or joins in the physical world disappear in the logical Chord world. Thus adversarial real node failures translate to random Chord node failures. This contrasts with the stronger notion of adversarial failures of Saia et al. [6]—our model handles an adversary oblivious to the structure of the Chord overlay, while their network is robust against an omniscient adversary.

3.2 The Ring-like State in Chord

The state of a correct Chord ring can be characterized as follows. Each node has exactly one successor, so the graph defined by successor pointers is a *pseudoforest*, a graph in which all components are directed trees pointing towards a root cycle (instead of a root node). We will limit our consideration to connected networks, where the graph is a *pseudotree*. The network is (weakly) stable when all nodes are in the cycle. For each cycle node u, there is a tree rooted at u which we call u's *appendage*, denoted $\mathcal{A}_u$. We insist that a node u joining the system invoke $u.join(n)$ for an existing node n that is already on the cycle.

Definition 4. *A Chord network with successor lists of length $\Theta(\log N)$ is ring-like if, for some c,*

1. *Each cycle node's successor is the cycle node with the next-highest identifier. The nodes in each appendage $\mathcal{A}_u$ fall between u and u's cycle predecessor. Every node's path of successor pointers to the cycle has increasing identifiers.*
2. *Every node u that joined the network at least $c \log^2 N$ rounds ago is "good": u is on the cycle and u never lies between $v + 2^i$ and $v.finger[i]$, for any v and i.*
3. *At least a third of the nodes are good.*
4. *Any $\log N$ consecutive appendages $\mathcal{A}_u$ contain only $O(\log N)$ nodes in total.*

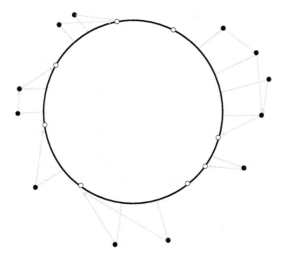

Fig. 2. An example of the ring-like state—unfilled nodes are on the cycle, filled nodes are in appendages.

5. *Nodes that failed at least $c \log^2 N$ rounds ago are not contained in any succes-*
   sor lists, and no more than a quarter of the nodes in any successor list have
   failed at all. Successor lists are consistent—no u.successor_list skips over a
   live node that is contained in (u.predecessor).successor_list—and include all
   nodes that joined the cycle at least $c \log^2 N$ rounds ago.

An example is given in Figure 2.

The ring-like state is the "normal" operating condition of a Chord network. Our main result is that a Chord network in the ring-like state remains in the ring-like state, as long as nodes send $\Omega(\log^2 N)$ messages before N new nodes join or $N/2$ nodes fail.

Theorem 5. *Start with a network of N nodes in the ring-like state with succes-sor lists of length $\Theta(\log N)$, and allow N random joins and $N/2$ random failures at arbitrary times over at least $c \log^2 N$ rounds. Then, with high probability, we end up in the ring-like state.*

Intuitively, the theorem follows because appendages are not too big, and not too many nodes join them. Thus over $c \log^2 N$ rounds, the appendage nodes have time to join the cycle.

Theorem 6. *In the ring-like state, lookups require $O(\log N)$ time.*

This theorem follows from Properties 2 and 3 of Definition 4. For every node u and i, the pointer $u.finger[i]$ is accurate with respect to good nodes. Thus our analysis showing logarithmic time search when all fingers are correct can be easily adapted to show that, in logarithmically many steps, a *find_successor*(k) search ends up at the last good node n preceeding key k. Since at least a third of the nodes in the network are good, there are, with high probability, only $O(\log N)$ non-good nodes between n and the successor of k. Even passing over these one-by-one using successor pointers requires only logarithmically many additional steps.

The correctness of lookups is somewhat subtle in this dynamic setting since, e.g., searches by nodes on the cycle will only return other nodes on the cycle (even if the "correct" answer is on an appendage). However, lookups arrive at a "correct" node, in the following sense: each *find_successor*(k) is correct at the instant that it terminates, i.e., yields a node v that is responsible for a key range including k. If v does not hold the key k, one of the following cases holds: (1) k is not yet available because it is being held at a node in an appendage (but, by Property 2, it will join the cycle within a half-life); (2) v is on the ring and responsible for the key k, but is in the process of transferring keys from its successor (but this transfer will complete quickly, and then v will have key k); or (3) v was previously responsible for the key k, but has since transferred k to another node. We can handle (3) by modifying the algorithm to have each node maintain a copy of all transferred data for one half-life after the transfer.

3.3   Strong Stabilization

The previous section proved, given our model, that Chord's stabilization protocol maintains a state in which routing is done correctly and quickly. But, fearful of bugs in an implementation, or a breakdown in our model,[5] we now wish to take a more cautious view. In this section, we extend the Chord protocol to one that will stabilize the network from an *arbitrary* state, even one not reachable by correct operation of the protocol. This protocol does not reconnect a disconnected network; we rely on some external means to do so.

This approach is in keeping with our focus on the behavior of our system *over time*. Over a sufficiently long period of time, extremely unlikely events (such as the simultaneous failure of all nodes in a successor list) can happen. We need to cope with them.

A Chord network is *weakly stable* if, for all nodes u in the network, we have ($u.successor$).$predecessor = u$ and *strongly stable* if, in addition, for each node u, there is no node v so that $u < v < u.successor$. A *loopy* network is one which is weakly but not strongly stable; see Figure 3. Previous Chord protocols guaranteed weak stability only; however, such networks can be globally inconsistent—e.g., no node u in Figure 3 has the correct *successor*(u). The result of this scenario is that $u.find\_successor(q) \neq v.find\_successor(q)$ for some nodes u and v and some query q, and thus data available in the network will appear unavailable to some nodes.

---

[5] For example, a node might be out of contact for so long that some nodes believe it to have failed, while it remains convinced that it is alive. Such inconsistent opinions could lead the system to a strange state.

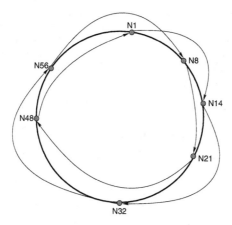

Fig. 3. An example of a network that is weakly stable but not strongly stable.

The previous Chord stabilization protocol guarantees that all nodes have indegree and outdegree one, so a weakly stable network consists of a topological cycle, but one in which successors might be incorrect. For a node u, call u's *loop* the set of nodes found by following successor pointers starting from u and continuing until we reach a node w so that $successor(w) \geq u$. In a loopy network, there is a node u so that u's loop is a strict subset of u's component; here, lookups may not be correct.

The fundamental stabilization operation by which we unfurl a loopy cycle is based upon *self-search*, wherein a node u searches for itself in the network. If the network is loopy, then a self-search from u traverses the circle once and then finds the first node on the loop succeeding u—i.e., the first node w found by following successor pointers so that $predecessor(w) < u < w$. We extend our previous stabilization protocol by allowing each node u to maintain a second successor pointer. This second successor is generated by self-search, and improved in exactly the same way as in the previous protocol. See Figure 4.

Theorem 7. *A connected Chord network strongly stabilizes within $O(N^2)$ rounds if no nodes join it, and in $O(N^3)$ rounds if there are no joins and at most $O(N)$ failures occur over $\Omega(\log N)$ rounds.*

Corollary 8. *A connected loopy Chord network strongly stabilizes within $O(N^2)$ rounds with no failures, and $O(N^3)$ rounds if there are at most $O(N)$ failures occur over $\Omega(\log N)$ rounds.*

The requirement on the failure rate exists solely to allow us to maintain a successor list with sufficiently many live nodes, and thus maintain connectivity. The intuition for the theorem is that cycles are the only configurations which are

```
n.join(n′)                               n.stabilize()
  on_cycle := false;                       u := successor[0].find_successor(n);
  predecessor := nil;                       on_cycle := (u = n);
  s := n′.find_successor(n);                 if (successor[0] = successor[1]
  while (¬ s.on_cycle) do                          and u ∈ (n, successor[1]))
    s := s.find_successor(n′);                   successor[1] := u;
  successor[0] := s;                         for (i := 0, 1)
  successor[1] := s;                            update_and_notify(i);
```

```
n.update_and_notify(i)
  s := successor[i]
  x := s.predecessor;
  if (x ∈ (n, s))
    successor[i] := x;
  s.notify(n);
```

Fig. 4. Pseudocode for strong stabilization.

not improved by weak stabilization, and self-search turns any loopy cycle into a
"non-cycle" by adding a second successor pointer. Therefore, the only configuration not improved by these two operations taken together is a non-loopy (i.e.,
strongly stable) cycle.

The corollary follows because a loopy Chord network will never permit any
new nodes to join until its loops merge—in a loopy network, for all u, we have
$u.on\_cycle =$ **false**, since u's self-search never returns u in a loopy network.
Thus, no node attempting to join can ever find a node s on the cycle to choose
as its successor.

While the runtime of our strong stabilization protocol is large, recall that
strong stabilization needs to be invoked *only* when the system gets into a pathological state. Such pathologies ought to be extremely rare, which means that
the lengthy recovery is a small fraction of the overall lifetime of the system. For
example, if pathological states occur only once every N^4 rounds, then the system
will only be spending a $1/N$ fraction of its time on strong stabilization. Nonetheless, it would clearly be preferable to develop a strong stabilization protocol that,
like weak stabilization, simply executes at a low rate in the background, rather
than bringing everything else to a halt for lengthy periods.

4   Conclusion

We have described the operation of Chord in a general model of evolution involving joins and departures. We have shown that a limited amount of housekeeping
work per node allows the system to resolve queries efficiently. There remains the
possibility of reducing this housekeeping work by logarithmic factors. Our current scheme postulates that the half life of the system is known; an interesting

question is whether the correct maintenance rate can be learned from observation of the behavior of neighbors. Another area to address is recovery from pathological situations. Our protocol exhibits slow recovery from certain pathological "disorderings" of the Chord ring. Although it is of course impossible to recover from total disconnection, an ideal protocol would recover quickly from any state in which the system remained connected.

References

[1] BALAKRISHNAN, H., KARGER, D. R., AND LIBEN-NOWELL, D. Analysis of the evolution of peer-to-peer systems. In *Proc. PODC 2002*. To appear.

[2] FIAT, A., AND SAIA, J. Censorship resistant peer-to-peer content addressable networks. In *Proc. SODA 2001*.

[3] PANDURANGAN, G., RAGHAVAN, P., AND UPFAL, E. Building low-diameter peer-to-peer networks. In *Proc. FOCS 2001*.

[4] RATNASAMY, S., FRANCIS, P., HANDLEY, M., KARP, R., AND SHENKER, S. A scalable content-addressable network. In *Proc. SIGCOMM 2001*.

[5] ROWSTRON, A., AND DRUSCHEL, P. Pastry: Scalable, distributed object location and routing for large-s cale peer-to-peer systems. In *Proc. Middleware 2001*.

[6] SAIA, J., FIAT, A., GRIBBLE, S., KARLIN, A. R., AND SAROIU, S. Dynamically fault-tolerant content addressable networks. This volume.

[7] STOICA, I., MORRIS, R., KARGER, D., KAASHOEK, M. F., AND BALAKRISHNAN, H. Chord: A scalable peer-to-peer lookup service for internet applications. In *Proc. SIGCOMM 2001*.

[8] ZHAO, B., KUBIATOWICZ, J., AND JOSEPH, A. Tapestry: An infrastructure for fault-tolerant wide-area location and routing. Tech. Rep. UCB/CSD-01-1141, Computer Science Division, U. C. Berkeley, Apr. 2001.

Brocade: Landmark Routing
on Overlay Networks

Ben Y. Zhao, Yitao Duan, Ling Huang, Anthony D. Joseph, and
John D. Kubiatowicz

Computer Science Division, U. C. Berkeley
{ravenben, duan, hlion, adj, kubitron}@cs.berkeley.edu

Abstract. Recent work such as Tapestry, Pastry, Chord and CAN provide efficient location utilities in the form of overlay infrastructures. These systems treat nodes as if they possessed uniform resources, such as network bandwidth and connectivity. In this paper, we propose a systemic design for a secondaryoverlay of super-nodes which can be used to deliver messages directly to the destination's local network, thus improving route efficiency. We demonstrate the potential performance benefits by proposing a name mapping scheme for a Tapestry-Tapestry secondary overlay, and show preliminary simulation results demonstrating significant routing performance improvement.

1   Introduction

Existing peer-to-peer overlay infrastructures such as Tapestry [11], Chord [8], Pastry [6] and CAN [4] demonstrated the benefits of scalable, wide-area lookup services for Internet applications. These architectures make use of name-based routing to route requests for objects or files to a nearby replica. Applications built on such systems ([2], [3], [7]), depend on reliable and fast message routing to a destination node, given some unique identifier.

Due to the theoretical approach taken in these systems, however, they assume that most nodes in the system are uniform in resources such as network bandwidth and storage. This results in messages being routed on the overlay with minimum consideration to actual network topology and differences between node resources.

In *Brocade*, we propose a secondary overlay to be layered on top of these systems, that exploits knowledge of underlying network characteristics. The secondary overlay builds a location layer between "supernodes," nodes that are situated near network access points, such gateways to administrative domains. By associating local nodes with their nearby "supernode," messages across the wide-area can take advantage of the highly connected network infrastructure between these supernodes to shortcut across distant network domains, greatly improving point-to-point routing distance and reducing network bandwidth usage.

In this paper, we present the initial architecture of a brocade secondary overlay on top of a Tapestry network, and demonstrate its potential performance

P. Druschel, F. Kaashoek, and A. Rowstron (Eds.): IPTPS 2002, LNCS 2429, pp. 34–44, 2002.

benefits by simulation. Section 2 briefly describes Tapestry routing and location, Section 3 describes the design of a Tapestry brocade, and Section 4 present preliminary simulation results. Finally, we discuss related work and conclude in Section 5.

2 Tapestry Routing and Location

Our architecture leverages Tapestry, an overlay location and routing layer presented by Zhao, Kubiatowicz and Joseph in [11]. Tapestry is one of several recent projects exploring the value of wide-area decentralized location services ([4], [6], [8]). It allows messages to locate objects and route to them across an arbitrarily-sized network, while using a routing map with size logarithmic to the network namespace at each hop. We present here a brief overview of the relevant characteristics of Tapestry. A detailed discussion of its algorithms, fault-tolerant mechanisms and simulation results can be found in [11].

Each Tapestry node can take on the roles of *server* (where objects are stored), *router* (which forward messages), and *client* (origins of requests). Objects and nodes have names independent of their location and semantic properties, in the form of random fixed-length bit-sequences with a common base (e.g., 40 Hex digits representing 160 bits). The system assumes entries are roughly evenly distributed in both node and object namespaces, which can be achieved by using the output of secure one-way hashing algorithms, such as SHA-1.

2.1 Routing Layer

Tapestry uses local routing maps at each node, called *neighbor maps*, to incrementally route overlay messages to the destination ID digit by digit (e.g., ***8 $\Longrightarrow$ **98 $\Longrightarrow$ *598 $\Longrightarrow$ 4598 where *'s represent wildcards). This approach is similar to longest prefix routing in the CIDR IP address allocation architecture [5]. A node N has a neighbor map with multiple levels, where each level represents a matching suffix up to a digit position in the ID. A given level of the neighbor map contains a number of entries equal to the base of the ID, where the ith entry in the jth level is the ID and location of the closest node which ends in "i"+suffix($N, j-1$). For example, the 9th entry of the 4th level for node 325AE is the node closest to 325AE in network distance which ends in 95AE.

When routing, the nth hop shares a suffix of at least length n with the destination ID. To find the next router, we look at its $(n+1)$th level map, and look up the entry matching the value of the next digit in the destination ID. Assuming consistent neighbor maps, this routing method guarantees that any existing unique node in the system will be found within at most $Log_b N$ logical hops, in a system with an N size namespace using IDs of base b. Since every neighbor map level assumes that the preceding digits all match the current node's suffix, it only needs to keep a small constant size (b) entries at each route level, yielding a neighbor map of fixed size $b \cdot Log_b N$. Figure 1 shows an example of hashed-suffix routing.

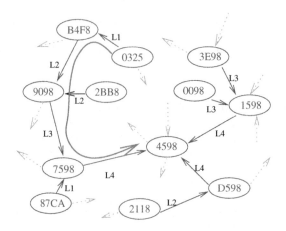

Fig. 1. *Tapestry routing example.* Path taken by a message from node 0325 for node 4598 in Tapestry using hexadecimal digits of length 4 (65536 nodes in namespace).

2.2   Data Location

Tapestry employs this infrastructure for data location. Each object is associated with one or more *Tapestry location roots* through a distributed deterministic mapping function. To advertise or publish an object O, the server S storing the object sends a publish message toward the Tapestry location root for that object. At each hop along the way, the publish message stores location information in the form of a mapping <Object-ID(O), Server-ID(S)>. Note that these mappings are simply pointers to the server S where O is being stored, and not a copy of the object itself. Where multiple objects exist, each server maintaining a replica publishes its copy. A node N that keeps location mappings for multiple replicas keeps them sorted in order of distance from N.

During a location query, clients send messages directly to objects via Tapestry. A message destined for O is initially routed towards O's root from the client. At each hop, if the message encounters a node that contains the location mapping for O, it is redirected to the server containing the object. Otherwise, the message is forward one step closer to the root. If the message reaches the root, it is guaranteed to find a mapping for the location of O. Note that the hierarchical nature of Tapestry routing means at each hop towards the root, the number of nodes satisfying the next hop constraint decreases by a factor equal to the identifier base (e.g. octal or hexadecimal) used in Tapestry. For nearby objects, client search messages quickly intersect the path taken by publish messages, resulting in quick search results that exploit locality. These and other properties are analyzed and discussed in more detail in [11].

3   Brocade Base Architecture

Here we present the overall design for the brocade overlay proposal, and define the design space for a single instance of the brocade overlay. We further clarify the design issues by presenting algorithms for an instance of a Tapestry on Tapestry brocade.

To improve point to point routing performance on an overlay, a brocade system defines a secondary overlay on top of the existing infrastructure, and provides a shortcut routing algorithm to quickly route to the local network of the destination node. This is achieved by finding nodes which have high bandwidth and fast access to the wide-area network, and tunnelling messages through an overlay composed of these "supernodes."

In overlay routing structures such as Tapestry [11], Pastry [6], Chord [8] and Content-Addressable Networks [4], messages are often routed across multiple autonomous systems (AS) and administrative domains before reaching their destinations. Each overlay hop often incurs long latencies within and across multiples AS's, consuming bandwidth along the way. To minimize both latency and network hops and reduce network traffic for a given message, brocade attempts to determine the network domain of the destination, and route directly to that domain. A "supernode" acts as a landmark for each network domain. Messages use them as endpoints of a tunnel through the secondary overlay, where messages would emerge near the local network of the destination node.

Before we examine the performance benefits, we address several issues necessary in constructing and utilizing a brocade overlay. We first discuss the construction of a brocade: how are supernodes chosen and how is the association between a node and its nearby supernode maintained? We then address issues in brocade routing: when and how messages find supernodes, and how they are routed on the secondary overlay.

3.1   Brocade Construction

The key to brocade routing is the tunnelling of messages through the wide area between landmark nodes (supernodes). The selection criteria are that supernodes have significant processing power (in order to route large amounts of overlay traffic), minimal number of IP hops to the wide-area network, and high bandwidth outgoing links. Given these requirements, gateway routers or machines close to them are attractive candidates. The final choice of a supernode can be resolved by an election algorithm between Tapestry nodes with sufficient resources, or as a performance optimizing choice by the responsible ISP.

Given a selection of supernodes, we face the issue of determining one-way mappings between supernodes and normal tapestry nodes for which they act as landmarks in Brocade routing. One possibility is to exploit the natural hierarchical nature of network domains. Each network gateway in a domain hierarchy can act as a brocade routing landmark for all nodes in its subdomain not covered by a more local subdomain gateway. We refer to the collection of these overlay nodes as the supernode's *cover set*. An example of this mapping is shown in

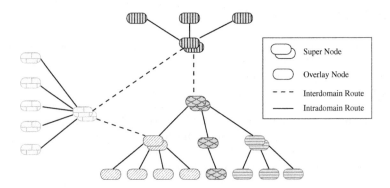

Fig. 2. Example of Supernode Organization

Figure 2. Supernodes keep up-to-date member lists of their cover sets, which are used in the routing process, as described below.

A secondary overlay can then be constructed on supernodes. Supernodes can have independent names in the brocade overlay, with consideration to the overlay design, e.g. Tapestry location requires names to be evenly distributed in the namespace.

3.2   Brocade Routing

Here we describe mechanisms required for a Tapestry-based brocade, and how they work together to improve long range routing performance. Given the complexity and latency involved in routing through an additional overlay, three key issues are: how are messages filtered so that only long distance messages are directed through the brocade overlay, how messages find a local supernode as entry to the brocader, and how a message finds the landmark supernode closest to the message destination in the secondary overlay.

Selective Utilization The use of a secondary overlay incurs a non-negligible amount of latency overhead in the routing. Once a message reaches a supernode, it must search for the supernode nearest to the destination node before routing to that domain and resuming Tapestry routing to the destination. Consequently, only messages that route outside the reach of the local supernode benefit from brocade routing.

We propose a naive solution by having each supernode maintain a listing of all Tapestry nodes in its cover set. We expect the node list at supernodes to be small, with a maximum size on the order of tens of thousands of entries. When a message reaches a supernode, the supernode can do an efficient lookup (via hashtable) to determine whether the message is destined for a local node, or whether brocade routing would be useful.

Finding Supernodes For a message to take advantage of brocade routing, it must be routed to a supernode on its way to its destination. How this occurs plays a large part in how efficient the resulting brocade route is. There are several possible approaches. We discuss three possible options here, and evaluate their relative performance in Section 4.

Naive A naive approach is to make brocade tunnelling an optional part of routing, and consider it only when a message reaches a supernode as part of normal routing. The advantage is simplicity. Normal nodes need to do nothing to take advantage of brocade overlay nodes. The disadvantage is that it severely limits the set of supernodes a message can reach. Messages can traverse several overlay hops before encountering a supernode, reducing the effectiveness of the brocade overlay.

IP-snooping In an alternate approach, supernodes can "snoop" on IP packets to determine if they are Tapestry messages. If so, supernodes can parse the message header, and use the destination ID to determine if brocade routing should be used. The intuition is that because supernodes are situated near the edge of local networks, any Tapestry message destined for an external node will likely cross its path. This also has the advantage that the source node sending the message need not know about the brocade supernodes in the infrastructure. The disadvantage is difficulty in implementation, and possible limitations imposed on regular traffic routing by header processing.

Directed The most promising solution is for overlay nodes to find the location of their local supernode, by using DNS resolution of a well-known name, e.g. `supernode.cs.berkeley.edu`, or by an expanding ring search. Once a new node joins a supernode's cover set, state can be maintained by periodic beacons. To reduce message traffic at supernodes, nodes keep a local *proximity cache* to "remember" local nodes they have communicated with. For each new message, if the destination is found in the proximity cache, it is routed normally. Otherwise, the node sends it directly to the supernode for routing. This is a proactive approach that takes advantage of any potential performance benefit brocade can offer. It does, however, require state maintenance, and the use of explicit fault-tolerant mechanisms should a supernode fail.

Landmark Routing on Brocade Once an inter-domain message arrives at the sender's supernode, brocade needs to determine the supernode closest to the message destination. This can be done by organizing the brocade overlay as a Tapestry network. As described in Section 2.2 and [11], Tapestry location allows nodes to efficiently locate objects given their IDs. Recall that each supernode keeps a list of all nodes inside its cover set. In the brocade overlay, each supernode advertises the IDs on this list as IDs of objects it "stores." When a supernode tries to route an outgoing inter-domain message, it uses Tapestry to search for an object with an ID identical to the message destination ID. By finding the object on the brocade layer, the source supernode has found the message destination's supernode, and forwards the message directly to it. The destination supernode then resumes normal overlay routing to the destination.

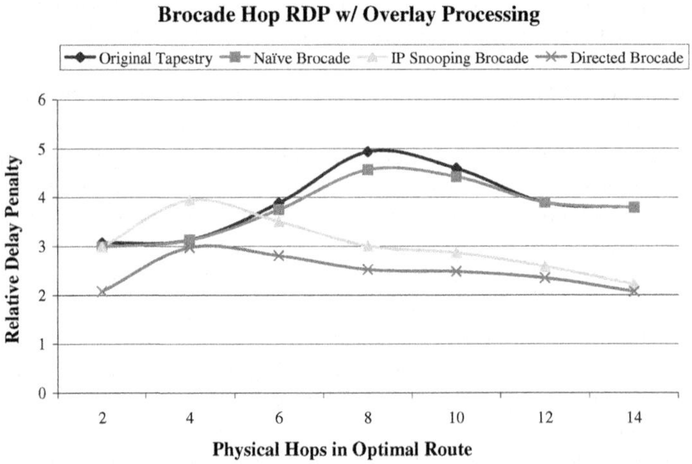

Brocade Hop RDP w/ Overlay Processing

Fig. 3. Hop-based RDP

Note these discussions make the implicit assumption that on average, inter-domain routing incurs much higher latencies compared to intra-domain routing. This, in combination with the distance constraints in Tapestry, allows us to assert that intra-domain messages will never route outside the domain. This is because the destination node will almost always offer the closest node with its own ID. This also means that once a message arrives at the destination's supernode, it will quickly route to the destination node.

4   Evaluation of Base Design

In this section, we present some analysis and initial simulation results showing the performance improvement possible with the use of brocade. In particular, we simulate the effect brocade routing has on point to point routing latency and bandwidth usage. For our experiments, we implemented a two layer brocade system inside a packet-level simulator that used Tapestry as both the primary and secondary overlay structures. The packet level simulator measured the progression of single events across a large network without regard to network effects such as congestion or retransmission.

Before presenting our simulation results, we first offer some back-of-the-envelope numerical support for why brocade supernodes should scale with the size of AS's and the rate of nodes entering and leaving the Tapestry. Given the size of the current Internet around 204 million nodes[1], and 20000 AS's, we estimate the size of an average AS to be around 10,000 nodes. Also, our current

---
[1] Source: http://www.netsizer.com/

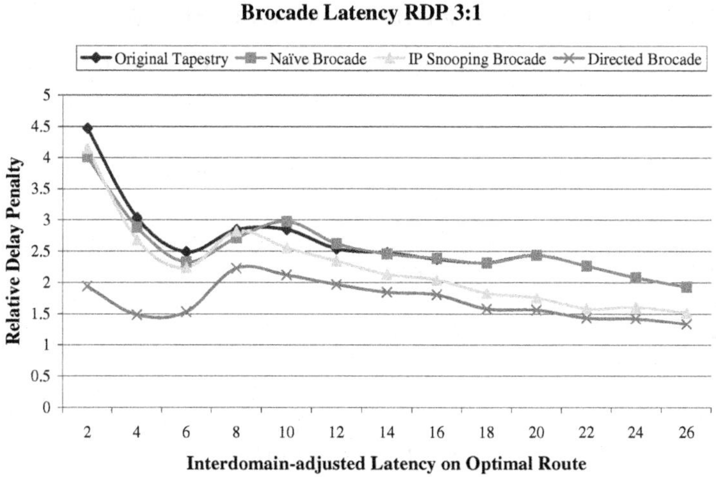

Brocade Latency RDP 3:1

Fig. 4. Weighted latency RDP, ratio 3:1

implementation of Tapestry on a PIII 800Mhz node achieves throughput of 1000 messages/second. In a highly volatile AS of 10000 nodes, where 10% of nodes enter or leave every minute, roughly 1.7% of the supernode processing power is used for handling the "registration" of new nodes.

We used in our experiments GT-ITM [10] transit stub topologies of 5000 nodes. We constructed Tapestry networks of size 4096, and marked 16 transit stubs as brocade supernodes. We then measured the performance of pair-wise communication paths using original Tapestry and all three brocade algorithms for finding supernodes (Section 3.2). We include four total algorithms: 1. original Tapestry, 2. naive brocade, 3. IP-snooping brocade, 4. directed brocade. For brocade algorithms, we assume the sender knows whether the destination node is local, and only uses brocade for inter-domain routing.

We use as our key metric a modified version of Relative Delay Penalty (RDP) [1]. Our modified RDP attempts to account for the processing of an overlay message up and down the protocol stack by adding 1 hop unit to each overlay node traversed. Each data point is generated by averaging the routing performance on 100 randomly chosen paths of a certain distance. In the RDP measurements, the sender's knowledge of whether the destination is local explains the low RDP values for short distances, and the spike in RDP around the average size of transit stub domains.

We measured the hop RDP of the four routing algorithms. For each pair of communication endpoints A and B, hop RDP is a ratio of # of hops traversed using brocade to the ideal hop distance between A and B. As seen in Figure 3, all brocade algorithms improve upon original Tapestry point to point routing. As expected, naive brocade offers minimal improvement. IP snooping improves

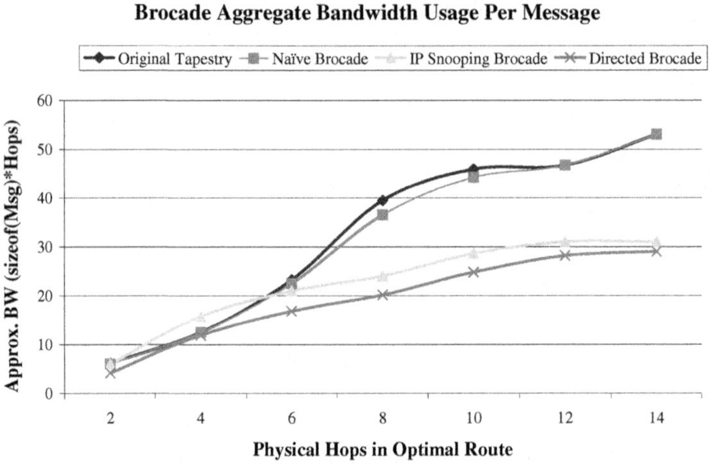

Brocade Aggregate Bandwidth Usage Per Message

Fig. 5. Aggregate bandwidth used per message

the hop RDP substantially, while directed brocade provides the most significant improvement in routing performance. For paths of moderate to long lengths, directed brocade reduces the routing overhead by more than 50% to near optimal levels (counting processing time). The small spike in RDP for IP snooping and directed brocade is due to the Tapestry location overhead in finding landmarks for destinations in nearby domains.

Figure 3 makes a simple assumption that all physical links have the same latency. To account for the fact that interdomain routes have higher latency, Figure 4 shows an RDP where each interdomain hop counts as 3 hop units of latency. We see that IP snooping and directed brocade still show the drastic improvement in RDP found in the simplistic topology results. We note that the spike in RDP experienced by IP snooping and directed brocade is exacerbated by the effect of higher routing time in interdomain traffic making Tapestry location more expensive. We also ran this test on several transit stub topologies with randomized latencies direct from GT-ITM, with similar results.

Finally, we examine the effect of brocade on reducing overall network traffic, by measuring the aggregate bandwidth taken per message delivery, using units of (sizeof(Msg) * hops). The result in Figure 5 shows that IP snooping brocade and directed brocade dramatically reduce bandwidth usage per message delivery. This is expected, since brocade forwards messages directly to the destination domain, and reduces message forwarding on the wide-area.

While certain decisions in our design are Tapestry specific, we believe similar design decisions can be made for other overlay networks ([4], [6], [8]), and these results should apply to brocade routing on those networks as well.

5   Related Work and Status

In related work, the Cooperative File System [2] leverages nodes with more resources by allowing them to host additional virtual nodes in the system, each representing one quantum of resource. This quantification is directed mostly at storage requirements, and CFS does not propose a mechanism for exploiting network topology knowledge. Our work is also partially inspired by the work on landmark routing [9], where packets are directed to a node in the landmark hierarchy closest to the destination before local routing.

While we present an architecture here using Tapestry at the lower level, the brocade overlay architecture can be generalized on top of any peer-to-peer network infrastructure. The presented architecture works as is on top of the Pastry [6] network. We are currently exploring brocades on top of CAN [4] and Chord [8]. We are implementing brocade in the Tapestry/OceanStore code base, and are experimenting with alternative efficient mechanisms for locating landmark nodes.

In conclusion, we have proposed the use of a secondary overlay network on a collection of well-connected "supernodes," in order to improve point to point routing performance on peer-to-peer overlay networks. The brocade layer uses Tapestry location to direct messages to the supernode nearest to their destination. Simulations show that brocade significantly improves routing performance and reduces bandwidth consumption for point to point paths in a wide-area overlay. We believe brocade is an interesting enhancement that leverages network knowledge for enhanced routing performance.

References

[1] CHU, Y., RAO, S. G., AND ZHANG, H. A case for end system multicast. In *Proceedings of ACM SIGMETRICS* (June 2000), pp. 1–12.

[2] DABEK, F., KAASHOEK, M. F., KARGER, D., MORRIS, R., AND STOICA, I. Wide-area cooperative storage with CFS. In *Proceedings of SOSP* (October 2001), ACM.

[3] KUBIATOWICZ, J., ET AL. OceanStore: An architecture for global-scale persistent storage. In *Proceedings of ACM ASPLOS* (November 2000), ACM.

[4] RATNASAMY, S., FRANCIS, P., HANDLEY, M., KARP, R., AND SCHENKER, S. A scalable content-addressable network. In *Proceedings of SIGCOMM* (August 2001), ACM.

[5] REKHTER, Y., AND LI, T. An architecture for IP address allocation with CIDR. RFC 1518, http://www.isi.edu/in-notes/rfc1518.txt, 1993.

[6] ROWSTRON, A., AND DRUSCHEL, P. Pastry: Scalable, distributed object location and routing for large-scale peer-to-peer systems. In *Proceedings of IFIP/ACM Middleware 2001* (November 2001).

[7] ROWSTRON, A., AND DRUSCHEL, P. Storage management and caching in PAST, a large-scale, persistent peer-to-peer storage utility. In *Proceedings of SOSP* (October 2001), ACM.

[8] STOICA, I., MORRIS, R., KARGER, D., KAASHOEK, M. F., AND BALAKRISHNAN, H. Chord: A scalable peer-to-peer lookup service for internet applications. In *Proceedings of SIGCOMM* (August 2001), ACM.

[9] TSUCHIYA, P. F. The landmark hierarchy: A new hierarchy for routing in very large networks. *Computer Communication Review 18*, 4 (August 1988), 35–42.

[10] ZEGURA, E. W., CALVERT, K., AND BHATTACHARJEE, S. How to model an internetwork. In *Proceedings of IEEE INFOCOM* (1996).

[11] ZHAO, B. Y., KUBIATOWICZ, J. D., AND JOSEPH, A. D. Tapestry: An infrastructure for fault-tolerant wide-area location and routing. Tech. Rep. UCB/CSD-01-1141, UC Berkeley, EECS, 2001.

Routing Algorithms for DHTs: Some Open Questions

Sylvia Ratnasamy[1,2], Ion Stoica[1], and Scott Shenker[2]

[1] University of California at Berkeley, CA, USA
[2] International Computer Science Institute, Berkeley, CA, USA

1   Introduction

Even though they were introduced only a few years ago, peer-to-peer (P2P) filesharing systems are now one of the most popular Internet applications and have become a major source of Internet traffic. Thus, it is extremely important that these systems be scalable. Unfortunately, the initial designs for P2P systems have significant scaling problems; for example, Napster has a centralized directory service, and Gnutella employs a flooding-based search mechanism that is not suitable for large systems.

In response to these scaling problems, several research groups have (independently) proposed a new generation of scalable P2P systems that support a *distributed hash table* (DHT) functionality; among them are Tapestry [15], Pastry [6], Chord [14], and Content-Addressable Networks (CAN) [10]. In these systems, which we will call DHTs, files are associated with a key (produced, for instance, by hashing the file name) and each node in the system is responsible for storing a certain range of keys. There is one basic operation in these DHT systems, lookup(key), which returns the identity (*e.g.*, the IP address) of the node storing the object with that key. This operation allows nodes to put and get files based on their key, thereby supporting the hash-table-like interface.[1]

This DHT functionality has proved to be a useful substrate for large distributed systems; a number of projects are proposing to build Internet-scale facilities layered above DHTs, including distributed file systems [5, 7, 4], application-layer multicast [11, 16], event notification services [3, 1], and chat services [2]. With so many applications being developed in so short a time, we expect the DHT functionality to become an integral part of the future P2P landscape.

The core of these DHT systems is the routing algorithm. The DHT nodes form an overlay network with each node having several other nodes as neighbors. When a lookup(key) is issued, the lookup is routed through the overlay network to the node responsible for that key. The scalability of these DHT algorithms is tied directly to the efficiency of their routing algorithms.

Each of the proposed DHT systems listed above – Tapestry, Pastry, Chord, and CAN – employ a different routing algorithm. Usually discussion of DHT routing issues is in the context of one particular algorithm. And, when more than one is mentioned, they are often compared in competitive terms in an effort to determine which is "best". We think

---

[1] The interfaces of these systems are not all identical; some reveal only the put and get interface while others reveal the lookup(key) function directly. However, the above discussion refers to the underlying functionality and not the details of the API.

P. Druschel, F. Kaashoek, and A. Rowstron (Eds.): IPTPS 2002, LNCS 2429, pp. 45–52, 2002.

both of these trends are wrong. The algorithms have more commonality than differences, and each algorithm embodies some insights about routing in overlay networks. Rather than always working in the context of a single algorithm, or comparing the algorithms competitively, a more appropriate goal would be to combine these insights, and seek new insights, to produce even better algorithms. In that spirit we describe some issues relevant to routing algorithms and identify some open research questions. Of course, our list of questions is not intended to be exhaustive, merely illustrative.

As should be clear by our description, this paper is not about finished work, but instead is about a research agenda for future work (by us and others). We hope that presenting such a discussion to this audience will promote synergy between research groups in this area and help clarify some of the underlying issues. We should note that there are many other interesting issues that remain to be resolved in these DHT systems, such as security and robustness to attacks, system monitoring and maintenance, and indexing and keyword searching. These issues will doubtless be discussed elsewhere in this workshop. Our focus on routing algorithms is not intended to imply that these other issues are of secondary importance.

We first (very) briefly review the routing algorithms used in the various DHT systems in Section 2. We then, in the following sections, discuss various issues relevant to routing: state-efficiency tradeoff, resilience to failures, routing hotspots, geography, and heterogeneity.

2   Review of Existing Algorithms

In this section we review some of the existing routing algorithms. All of them take, as input, a key and, in response, route a message to the node responsible for that key. The keys are strings of digits of some length. Nodes have identifiers, taken from the same space as the keys (*i.e.*, same number of digits). Each node maintains a routing table consisting of a small subset of nodes in the system. When a node receives a query for a key for which it is not responsible, the node routes the query to the neighbor node that makes the most "progress" towards resolving the query. The notion of progress differs from algorithm to algorithm, but in general is defined in terms of some distance between the identifier of the current node and the identifier of the queried key.

Plaxton et al.:   Plaxton *et al.* [9] developed perhaps the first routing algorithm that could be scalably used by DHTs. While not intended for use in P2P systems, because it assumes a relatively static node population, it does provide very efficient routing of lookups. The routing algorithm works by "correcting" a single digit at a time: if node number 36278 received a lookup query with key 36912, which matches the first two digits, then the routing algorithm forwards the query to a node which matches the first three digits (*e.g.*, node 36955). To do this, a node needs to have, as neighbors, nodes that match each prefix of its own identifier but differ in the next digit. For a system of n nodes, each node has on the order of $O(\log n)$ neighbors. Since one digit is corrected each time the query is forwarded, the routing path is at most $O(\log n)$ overlay (or application-level) hops.

This algorithm has the additional property that if the n^2 node-node latencies (or "distances" according to some metric) are known, the routing tables can be chosen to minimize the expected path latency and, moreover, the latency of the overlay path between two nodes is within a constant factor of the latency of the direct underlying network path between them.

Tapestry: Tapestry [15] uses a variant of the Plaxton *et al.* algorithm. The modifications are to ensure that the design, originally intended for static environments, can adapt to a dynamic node population. The modifications are too involved to describe in this short review. However, the algorithm maintains the properties of having $O(\log n)$ neighbors and routing with path lengths of $O(\log n)$ hops.

Pastry: In Pastry [6], nodes are responsible for keys that are the closest numerically (with the keyspace considered as a circle). The neighbors consist of a *Leaf Set* L which is the set of $|L|$ closest nodes (half larger, half smaller). Correct, not necessarily efficient, routing can be achieved with this leaf set. To achieve more efficient routing, Pastry has another set of neighbors spread out in the key space (in a manner we don't describe here). Routing consists of forwarding the query to the neighboring node that has the longest shared prefix with the key (and, in the case of ties, to the node with identifier closest numerically to the key). Pastry has $O(\log n)$ neighbors and routes within $O(\log n)$ hops.

Chord: Chord [14] also uses a one-dimensional circular key space. The node responsible for the key is the node whose identifier most closely follows the key (numerically); that node is called the key's *successor*. Chord maintains two sets of neighbors. Each node has a *successor list* of k nodes that immediately follow it in the key space. Routing correctness is achieved with these lists. Routing efficiency is achieved with the *finger list* of $O(\log n)$ nodes spaced exponentially around the key space. Routing consists of forwarding to the node closest, but not past, the key; pathlengths are $O(\log n)$ hops.

CAN: CAN chooses its keys from a d-dimensional toroidal space. Each node is associated with a hypercubal region of this key space, and its neighbors are the nodes that "own" the contiguous hypercubes. Routing consists of forwarding to a neighbor that is closer to the key. CAN has a different performance profile than the other algorithms; nodes have $O(d)$ neighbors and pathlengths are $O(dn^{\frac{1}{d}})$ hops. Note, however, that when $d = \log n$, CAN has $O(\log n)$ neighbors and $O(\log n)$ pathlengths like the other algorithms.

3   State-Efficiency Tradeoff

The most obvious measure of the efficiency of these routing algorithms is the resulting pathlength. Most of the algorithms have pathlengths of $O(\log n)$ hops, while CAN has longer paths of $O(dn^{\frac{1}{d}})$. The most obvious measure of the overhead associated with keeping routing tables is the number of neighbors. This isn't just a measure of the state required to do routing but it is also a measure of how much state needs to be adjusted when nodes join or leave. Given the prevalence of inexpensive memory and the highly transient user populations in P2P systems, this second issue is likely to be much more important than the first. Most of the algorithms require $O(\log n)$ neighbors, while CAN requires only $O(d)$ neighbors.

Ideally, one would like to combine the best of these two classes of algorithms in hybrid algorithms that achieve short pathlengths with a fixed number of neighbors.

Question 1 *Can one achieve $O(\log n)$ pathlengths (or better) with $O(1)$ neighbors?*

One would expect that, if this were possible, that some other aspects of routing would get worse.

Question 2 *If so, are there other properties (such as those described in the following sections) that are made worse in these hybrid routing algorithms?*

4   Resilience to Failures

The above routing results refer to a perfectly functioning system with all nodes operational. However, P2P nodes are notoriously transient and the resilience of routing to failures is a very important consideration. There are (at least) three different aspects to resilience.

First, one needs to evaluate whether routing can continue to function (and with what efficiency) as nodes fail without any time for other nodes to establish other neighbors to compensate; that is the neighboring nodes know that a node has failed, but they don't establish any new neighbor relations with other nodes. We will call this *static resilience* and measure it in terms of the percentage of reachable key locations and of the resulting average path length.

Question 3 *Can one characterize the static resilience of the various algorithms? What aspects of these algorithms lead to good resilience?*

Second, one can investigate the resilience when nodes have a chance to establish some neighbors, but not all. That is, when nodes have certain "special" neighbors, such as the *successor list* or the *Leaf Set*, and these are re-established after a failure, but no other neighbors are re-established (such as the *finger set*). The presence of these special neighbors allow one to prove the correctness of routing, but the following question remains:

Question 4 *To what extent are the observed path lengths better than the rather pessimistic bounds provided by the presence of these special neighbors?*

Finally, one can ask how long it takes various algorithms to fully recover their routing state, and at what cost (measured, for example, by the number of nodes participating in the recovery or the number of control messages generated for recovery).

Question 5 *How long does it take, on average, to recover complete routing state? And what is the cost of doing so?*

A related question is:

Question 6 *Can one identify design rules that lead to shorter and/or cheaper recoveries?*

For instance, is symmetry (where the node neighbor relation is symmetric) important in restoring state easily? One could also argue that in the face of node failure, having the routing automatically send messages to the correct alternate node (*i.e.* the node that takes over the range of the identifier space that was previously held by the failed node) leads to quicker recovery.

5   Routing Hot Spots

When there is a hotspot in the query pattern, with a certain key being requested extremely often, then the node holding that key may become overloaded. Various caching and replication schemes have been proposed to overcome this *query hotspot* problem; the effectiveness of these schemes may vary between algorithms based on the fan-in at the node and other factors, but this seems to be a manageable problem. More problematic, however, is if a node is overloaded with too much routing traffic. These *routing hotspots* are harder to deal with since there is no local action the node can take to redirect the routing load. Some of the *proximity* techniques we describe below might be used to help here, but otherwise this remains an open problem.

Question 7 *Do routing hotspots exist and, if so, how can one deal with them?*

6   Incorporating Geography

The efficiency measure used above was the number of application-level hops taken on the path. However, the true efficiency measure is the end-to-end latency of the path. Because the nodes could be geographically dispersed, some of these application-level hops could involve transcontinental links, and others merely trips across a LAN; routing algorithms that ignore the latencies of individual hops are likely to result in high-latency paths. While the original "vanilla" versions of some of these routing algorithms did not take these hop latencies into account, almost all of the "full" versions of the algorithms make some attempt to deal with the geographic proximity of nodes. There are (at least) three ways of coping with geography.

Proximity Routing: Proximity routing is when the routing choice is based not just which neighboring node makes the "most" progress towards the key, but is also based on which neighboring node is "closest" in the sense of latency. Various algorithms implement proximity routing differently, but they all adopt the same basic approach of weighing progress in identifier space against cost in latency (or geography). Simulations have shown this to be a very effective tool in reducing the average path latency.

Question 8 *Can one formally characterize the effectiveness of these proximity routing approaches?*

Proximity Neighbor Selection: This is a variant of the idea above, but now the proximity criterion is applied when choosing neighbors, not just when choosing the next hop.

Question 9 *Can one show that proximity neighbor selection is always better than proximity routing? Is this difference significant?*

As mentioned earlier, if the n^2 node-pair distances (as measured by latency) are known, the Plaxton/Tapestry algorithm can choose the neighbors so as to minimize the expected overlay path latency. This is an extremely important property, that is (so far) the exclusive domain of the Plaxton/Tapestry algorithms. We don't whether other algorithms can adopt similar approaches.

Question 10 *If one had the full n^2 distance matrix, could one do optimal neighbor selection in algorithms other than Plaxton/Tapestry?*

Geographic Layout: In most of the algorithms, the node identifiers are chosen randomly (*e.g.* hash functions of the IP address, etc.) and the neighbor relations are established based solely on these node identifiers. One could instead attempt to choose node identifiers in a geographically informed manner.[2] An initial attempt to do so in the context of CAN was reported on in [12]; this approach was quite successful in reducing the latency of paths. There was little in the layout method specific to CAN, but the high-dimensionality of the key space may have played an important role; recent work [8] suggests that latencies in the Internet can be reasonably modeled by a d-dimension geometric space with $d \geq 2$. This raises the question of whether systems that use a one-dimensional key set can adequately mimic the geographic layout of the nodes.

Question 11 *Can one choose identifiers in a one-dimensional key space that will adequately capture the geographic layout of nodes?*

However, this may not matter because the geographic layout may not offer significant advantages over the two proximity methods.

Question 12 *Can the two* local *techniques of proximity routing and proximity neighbor selection achieve most of the benefit of* global *geographic layout?*

Moreover, these geographically-informed layout methods may interfere with the robustness, hotspot, and other properties mentioned in previous sections.

Question 13 *Does geographic layout have an impact on resilience, hotspots, and other aspects of performance?*

7   Extreme Heterogeneity

All of the algorithms start by assuming that all nodes have the same capacity to process messages and then, only later, add on techniques for coping with heterogeneity.[3] However, the heterogeneity observed in current P2P populations [13] is quite extreme, with differences of several orders of magnitude in bandwidth. One can ask whether the routing algorithms, rather than merely *coping* with heterogeneity, should instead use it to their *advantage*. At the extreme, a star topology with all queries passing through a single hub node and then routed to their destination would be extremely efficient, but would require a very highly capable nub node (and would have a single point of failure). But perhaps one could use the very highly capable nodes as mini-hubs to improve routing. In another position paper here, some of us argue that heterogeneity can be used to make Gnutella-like systems more scalable. The question is whether one could similarly modify the current DHT routing algorithms to exploit heterogeneity:

Question 14 *Can one redesign these routing algorithms to exploit heterogeneity?*

It may be that no sophisticated modifications are needed to leverage heterogeneity. Perhaps the simplest technique to cope with heterogeneity, one that has already been mentioned in the literature, is to *clone* highly capable nodes so that they could serve

---

[2] Note that geographic layout differs from the two above *proximity* methods in that here there is an attempt to affect the global layout of the node identifiers, whereas the proximity methods merely affect the local choices of neighbors and forwarding nodes.

[3] The authors of [13] deserve credit for bringing the issue of heterogeneity to our attention.

as multiple nodes; *i.e.*, a node that was 10 times more powerful than other nodes could function as 10 virtual nodes.[4] When combined with proximity routing and neighbor selection, cloning would allow nodes to route to themselves and thereby "jump" in key space without any forwarding hops.

Question 15 *Does cloning plus proximity routing and neighbor selection lead to significantly improved performance when the node capabilities are extremely heterogeneous?*

References

[1] A. ROWSTRON, A-M. KERMARREC, M. C., AND DRUSCHEL, P. Scribe: The design of a large-scale event notification infrastructure. In *Proceedings of NGC 2001* (Nov. 2001).

[2] BASED CHAT, C. http://jxme.jxta.org/demo.html, 2001.

[3] CABRERA, L. F., JONES, M. B., AND THEIMER, M. Herald: Achieving a global event notification service. In *Proceedings of the 8th IEEE Workshop on Hot Topics in Operating Systems (HotOS-VIII)* (Elmau/Oberbayern, Germany, May 2001).

[4] DABEK, F., KAASHOEK, M. F., KARGER, D., MORRIS, R., AND STOICA, I. Wide-area cooperative storage with CFS. In *Proceedings of the 18th ACM Symposium on Operating Systems Principles (SOSP '01)* (To appear; Banff, Canada, Oct. 2001).

[5] DRUSCHEL, P., AND ROWSTRON, A. Past: Persistent and anonymous storage in a peer-to-peer networking environment. In *Proceedings of the 8th IEEE Workshop on Hot Topics in Operating Systems (HotOS 2001)* (Elmau/Oberbayern, Germany, May 2001), pp. 65–70.

[6] DRUSCHEL, P., AND ROWSTRON, A. Pastry: Scalable, distributed object location and routing for large-scale peer-to-peer systems. In *Proceedings of the 18th IFIP/ACM International Conference on Distributed Systems Platforms (Middleware 2001)W* (Nov 2001).

[7] KUBIATOWICZ, J., BINDEL, D., CHEN, Y., CZERWINSKI, S., EATON, P., GEELS, D., GUMMADI, R., RHEA, S., WEATHERSPOON, H., WEIMER, W., WELLS, C., AND ZHAO, B. OceanStore: An architecture for global-scale persistent storage. In *Proceeedings of the Ninth international Conference on Architectural Support for Programming Languages and Operating Systems (ASPLOS 2000)* (Boston, MA, November 2000), pp. 190–201.

[8] NG, E., AND ZHANG, H. Towards global network positioning. In *Proceedings of ACM SIGCOMM Internet Measurement Workshop 2001* (Nov. 2001).

[9] PLAXTON, C., RAJARAMAN, R., AND RICHA, A. Accessing nearby copies of replicated objects in a distributed environment. In *Proceedings of the ACM SPAA* (Newport, Rhode Island, June 1997), pp. 311–320.

[10] RATNASAMY, S., FRANCIS, P., HANDLEY, M., KARP, R., AND SHENKER, S. A scalable content-addressable network. In *Proc. ACM SIGCOMM* (San Diego, CA, August 2001), pp. 161–172.

[11] RATNASAMY, S., HANDLEY, M., KARP, R., AND SHENKER, S. Application-level Multicast using Content-Addressable Networks. In *Proceedings of NGC 2001* (Nov. 2001).

[12] RATNASAMY, S., HANDLEY, M., RICHARDKARP, AND SHENKER, S. Topologically-aware overlay construction and server selection. In *Proceedings of Infocom '2002* (Mar. 2002).

[13] SAROIU, S., GUMMADI, K., AND GRIBBLE, S. A measurement study of peer-to-peer file sharing systems. In *Proceedings of Multimedia Conferencing and Networking* (San Jose, Jan. 2002).

---

[4] This technique has already been suggested for some of the algorithms, and could easily be applied to the others. However, in some algorithms it would require alteration in the way the node identifiers were chosen so that they weren't tied to the IP address of the node.

[14] STOICA, I., MORRIS, R., KARGER, D., KAASHOEK, M. F., AND BALAKRISHNAN, H. Chord: A scalable peer-to-peer lookup service for internet applications. In *Proceedings of the ACM SIGCOMM '01 Conference* (San Diego, California, August 2001).

[15] ZHAO, B. Y., KUBIATOWICZ, J., AND JOSEPH, A. Tapestry: An infrastructure for fault-tolerant wide-area location and routing. Tech. Rep. UCB/CSD-01-1141, University of California at Berkeley, Computer Science Department, 2001.

[16] ZHUANG, S., ZHAO, B., JOSEPH, A. D., KATZ, R. H., AND KUBIATOWICZ, J. Bayeux: An architecture for wide-area, fault-tolerant data dissemination. In *Proceedings of NOSS-DAV'01* (Port Jefferson, NY, June 2001).

Kademlia: A Peer-to-Peer Information System Based on the XOR Metric*

Petar Maymounkov and David Mazières

New York University
{petar,dm}@cs.nyu.edu
http://kademlia.scs.cs.nyu.edu

Abstract. We describe a peer-to-peer distributed hash table with provable consistency and performance in a fault-prone environment. Our system routes queries and locates nodes using a novel XOR-based metric topology that simplifies the algorithm and facilitates our proof. The topology has the property that every message exchanged conveys or reinforces useful contact information. The system exploits this information to send parallel, asynchronous query messages that tolerate node failures without imposing timeout delays on users.

1 Introduction

This paper describes Kademlia, a peer-to-peer distributed hash table (DHT). Kademlia has a number of desirable features not simultaneously offered by any previous DHT. It minimizes the number of configuration messages nodes must send to learn about each other. Configuration information spreads automatically as a side-effect of key lookups. Nodes have enough knowledge and flexibility to route queries through low-latency paths. Kademlia uses parallel, asynchronous queries to avoid timeout delays from failed nodes. The algorithm with which nodes record each other's existence resists certain basic denial of service attacks. Finally, several important properties of Kademlia can be formally proven using only weak assumptions on uptime distributions (assumptions we validate with measurements of existing peer-to-peer systems).

Kademlia takes the basic approach of many DHTs. Keys are opaque, 160-bit quantities (e.g., the SHA-1 hash of some larger data). Participating computers each have a node ID in the 160-bit key space. ⟨key,value⟩ pairs are stored on nodes with IDs "close" to the key for some notion of closeness. Finally, a node-ID-based routing algorithm lets anyone efficiently locate servers near any given target key.

Many of Kademlia's benefits result from its use of a novel XOR metric for distance between points in the key space. XOR is symmetric, allowing Kademlia participants to receive lookup queries from precisely the same distribution of nodes contained in their routing tables. Without this property, systems such

---

* This research was partially supported by National Science Foundation grants CCR 0093361 and CCR 9800085.

P. Druschel, F. Kaashoek, and A. Rowstron (Eds.): IPTPS 2002, LNCS 2429, pp. 53–65, 2002.

as Chord [5] do not learn useful routing information from queries they receive. Worse yet, asymmetry leads to rigid routing tables. Each entry in a Chord node's finger table must store the precise node preceding some interval in the ID space. Any node actually in the interval would be too far from nodes preceding it in the same interval. Kademlia, in contrast, can send a query to any node within an interval, allowing it to select routes based on latency or even send parallel, asynchronous queries to several equally appropriate nodes.

To locate nodes near a particular ID, Kademlia uses a single routing algorithm from start to finish. In contrast, other systems use one algorithm to get near the target ID and another for the last few hops. Of existing systems, Kademlia most resembles Pastry's [1] first phase, which (though not described this way by the authors) successively finds nodes roughly half as far from the target ID by Kademlia's XOR metric. In a second phase, however, Pastry switches distance metrics to the numeric difference between IDs. It also uses the second, numeric difference metric in replication. Unfortunately, nodes close by the second metric can be quite far by the first, creating discontinuities at particular node ID values, reducing performance, and complicating attempts at formal analysis of worst-case behavior.

2   System Description

Our system takes the same general approach as other DHTs. We assign 160-bit opaque IDs to nodes and provide a lookup algorithm that locates successively "closer" nodes to any desired ID, converging to the lookup target in logarithmically many steps.

Kademlia effectively treats nodes as leaves in a binary tree, with each node's position determined by the shortest unique prefix of its ID. Figure 1 shows the position of a node with unique prefix 0011 in an example tree. For any given node, we divide the binary tree into a series of successively lower subtrees that don't contain the node. The highest subtree consists of the half of the binary tree not containing the node. The next subtree consists of the half of the remaining tree not containing the node, and so forth. In the example of node 0011, the subtrees are circled and consist of all nodes with prefixes 1, 01, 000, and 0010 respectively.

The Kademlia protocol ensures that every node knows of at least one node in each of its subtrees, if that subtree contains a node. With this guarantee, any node can locate any other node by its ID. Figure 2 shows an example of node 0011 locating node 1110 by successively querying the best node it knows of to find contacts in lower and lower subtrees; finally the lookup converges to the target node.

The remainder of this section fills in the details and makes the lookup algorithm more concrete. We first define a precise notion of ID closeness, allowing us to speak of storing and looking up ⟨key, value⟩ pairs on the k closest nodes to the key. We then give a lookup protocol that works even in cases where no

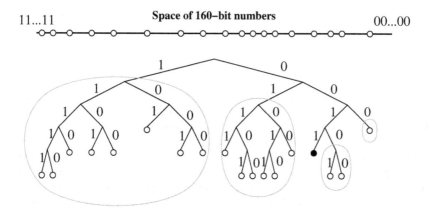

Fig. 1: Kademlia binary tree. The black dot shows the location of node $0011\cdots$ in the tree. Gray ovals show subtrees in which node $0011\cdots$ must have a contact.

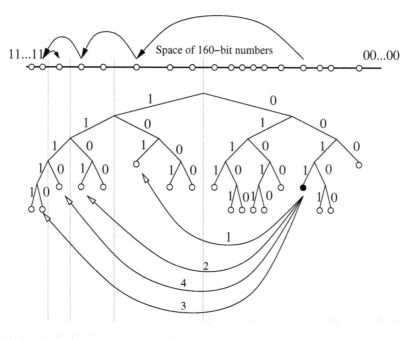

Fig. 2: Locating a node by its ID. Here the node with prefix 0011 finds the node with prefix 1110 by successively learning of and querying closer and closer nodes. The line segment on top represents the space of 160-bit IDs, and shows how the lookups converge to the target node. Below we illustrate RPC messages made by 1110. The first RPC is to node 101, already known to 1110. Subsequent RPCs are to nodes returned by the previous RPC.

node shares a unique prefix with a key or some of the subtrees associated with a given node are empty.

2.1   XOR Metric

Each Kademlia node has a 160-bit node ID. Node IDs are currently just random 160-bit identifiers, though they could equally well be constructed as in Chord. Every message a node transmits includes its node ID, permitting the recipient to record the sender's existence if necessary.

Keys, too, are 160-bit identifiers. To assign ⟨key,value⟩ pairs to particular nodes, Kademlia relies on a notion of distance between two identifiers. Given two 160-bit identifiers, x and y, Kademlia defines the distance between them as their bitwise exclusive or (XOR) interpreted as an integer, $d(x, y) = x \oplus y$.

We first note that XOR is a valid, albeit non-Euclidean metric. It is obvious that that $d(x, x) = 0$, $d(x, y) > 0$ if $x \neq y$, and $\forall x, y : d(x, y) = d(y, x)$. XOR also offers the triangle property: $d(x, y) + d(y, z) \geq d(x, z)$. The triangle property follows from the fact that $d(x, y) \oplus d(y, z) = d(x, z)$ and $\forall a \geq 0, b \geq 0 : a + b \geq a \oplus b$.

We next note that XOR captures the notion of distance implicit in our binary-tree-based sketch of the system. In a fully-populated binary tree of 160-bit IDs, the magnitude of the distance between two IDs is the height of the smallest subtree containing them both. When a tree is not fully populated, the closest leaf to an ID x is the leaf whose ID shares the longest common prefix of x. If there are empty branches in the tree, there might be more than one leaf with the longest common prefix. In that case, the closest leaf to x will be the closest leaf to ID $\tilde{x}$ produced by flipping the bits in x corresponding to the empty branches of the tree.

Like Chord's clockwise circle metric, XOR is *unidirectional*. For any given point x and distance $\Delta > 0$, there is exactly one point y such that $d(x, y) = \Delta$. Unidirectionality ensures that all lookups for the same key converge along the same path, regardless of the originating node. Thus, caching ⟨key,value⟩ pairs along the lookup path alleviates hot spots. Like Pastry and unlike Chord, the XOR topology is also symmetric ($d(x, y) = d(y, x)$ for all x and y).

2.2   Node State

Kademlia nodes store contact information about each other to route query messages. For each $0 \leq i < 160$, every node keeps a list of ⟨IP address, UDP port, Node ID⟩ triples for nodes of distance between 2^i and 2^{i+1} from itself. We call these lists k-buckets. Each k-bucket is kept sorted by time last seen—least-recently seen node at the head, most-recently seen node at the tail. For small values of i, the k-buckets will generally be empty (as no appropriate nodes will exist). For large values of i, the lists can grow up to size k, where k is a system-wide replication parameter. k is chosen such that any given k nodes are very unlikely to fail within an hour of each other (for example $k = 20$).

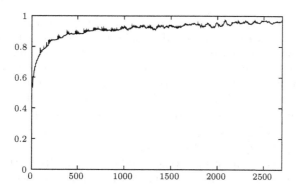

Fig. 3: Probability of remaining online another hour as a function of uptime. The x axis represents minutes. The y axis shows the the fraction of nodes that stayed online at least x minutes that also stayed online at least $x + 60$ minutes.

When a Kademlia node receives any message (request or reply) from another node, it updates the appropriate k-bucket for the sender's node ID. If the sending node already exists in the recipient's k-bucket, the recipient moves it to the tail of the list. If the node is not already in the appropriate k-bucket and the bucket has fewer than k entries, then the recipient just inserts the new sender at the tail of the list. If the appropriate k-bucket is full, however, then the recipient pings the k-bucket's least-recently seen node to decide what to do. If the least-recently seen node fails to respond, it is evicted from the k-bucket and the new sender inserted at the tail. Otherwise, if the least-recently seen node responds, it is moved to the tail of the list, and the new sender's contact is discarded.

k-buckets effectively implement a least-recently seen eviction policy, except that live nodes are never removed from the list. This preference for old contacts is driven by our analysis of Gnutella trace data collected by Saroiu et. al. [4]. Figure 3 shows the percentage of Gnutella nodes that stay online another hour as a function of current uptime. The longer a node has been up, the more likely it is to remain up another hour. By keeping the oldest live contacts around, k-buckets maximize the probability that the nodes they contain will remain online.

A second benefit of k-buckets is that they provide resistance to certain DoS attacks. One cannot flush nodes' routing state by flooding the system with new nodes. Kademlia nodes will only insert the new nodes in the k-buckets when old nodes leave the system.

2.3   Kademlia Protocol

The Kademlia protocol consists of four RPCs: PING, STORE, FIND_NODE, and FIND_VALUE. The PING RPC probes a node to see if it is online. STORE instructs a node to store a ⟨key, value⟩ pair for later retrieval.

FIND_NODE takes a 160-bit ID as an argument. The recipient of a the RPC returns ⟨IP address, UDP port, Node ID⟩ triples for the k nodes it knows about closest to the target ID. These triples can come from a single k-bucket, or they may come from multiple k-buckets if the closest k-bucket is not full. In any case, the RPC recipient must return k items (unless there are fewer than k nodes in all its k-buckets combined, in which case it returns every node it knows about).

FIND_VALUE behaves like FIND_NODE—returning ⟨IP address, UDP port, Node ID⟩ triples—with one exception. If the RPC recipient has received a STORE RPC for the key, it just returns the stored value.

In all RPCs, the recipient must echo a 160-bit random RPC ID, which provides some resistance to address forgery. PINGs can also be piggy-backed on RPC replies for the RPC recipient to obtain additional assurance of the sender's network address.

The most important procedure a Kademlia participant must perform is to locate the k closest nodes to some given node ID. We call this procedure a *node lookup*. Kademlia employs a recursive algorithm for node lookups. The lookup initiator starts by picking α nodes from its closest non-empty k-bucket (or, if that bucket has fewer than α entries, it just takes the α closest nodes it knows of). The initiator then sends parallel, asynchronous FIND_NODE RPCs to the α nodes it has chosen. α is a system-wide concurrency parameter, such as 3.

In the recursive step, the initiator resends the FIND_NODE to nodes it has learned about from previous RPCs. (This recursion can begin before all α of the previous RPCs have returned). Of the k nodes the initiator has heard of closest to the target, it picks α that it has not yet queried and resends the FIND_NODE RPC to them.[1] Nodes that fail to respond quickly are removed from consideration until and unless they do respond. If a round of FIND_NODEs fails to return a node any closer than the closest already seen, the initiator resends the FIND_NODE to all of the k closest nodes it has not already queried. The lookup terminates when the initiator has queried and gotten responses from the k closest nodes it has seen. When $\alpha = 1$, the lookup algorithm resembles Chord's in terms of message cost and the latency of detecting failed nodes. However, Kademlia can route for lower latency because it has the flexibility of choosing any one of k nodes to forward a request to.

Most operations are implemented in terms of the above lookup procedure. To store a ⟨key,value⟩ pair, a participant locates the k closest nodes to the key and sends them STORE RPCs. Additionally, each node re-publishes ⟨key,value⟩ pairs as necessary to keep them alive, as described later in Section 2.5. This ensures persistence (as we show in our proof sketch) of the ⟨key,value⟩ pair with very high probability. For Kademlia's current application (file sharing), we also require the original publisher of a ⟨key,value⟩ pair to republish it every 24 hours. Otherwise, ⟨key,value⟩ pairs expire 24 hours after publication, so as to limit stale index information in the system. For other applications, such as digital

---

[1] Bucket entries and FIND replies could be augmented with round trip time estimates for use in selecting the α nodes.

certificates or cryptographic hash to value mappings, longer expiration times may be appropriate.

To find a ⟨key,value⟩ pair, a node starts by performing a lookup to find the k nodes with IDs closest to the key. However, value lookups use FIND_VALUE rather than FIND_NODE RPCs. Moreover, the procedure halts immediately when any node returns the value. For caching purposes, once a lookup succeeds, the requesting node stores the ⟨key,value⟩ pair at the closest node it observed to the key that did not return the value.

Because of the unidirectionality of the topology, future searches for the same key are likely to hit cached entries before querying the closest node. During times of high popularity for a certain key, the system might end up caching it at many nodes. To avoid "over-caching," we make the expiration time of a ⟨key,value⟩ pair in any node's database exponentially inversely proportional to the number of nodes between the current node and the node whose ID is closest to the key ID.[2] While simple LRU eviction would result in a similar lifetime distribution, there is no natural way of choosing the cache size, since nodes have no *a priori* knowledge of how many values the system will store.

Buckets are generally kept fresh by the traffic of requests traveling through nodes. To handle pathological cases in which there are no lookups for a particular ID range, each node refreshes any bucket to which it has not performed a node lookup in the past hour. Refreshing means picking a random ID in the bucket's range and performing a node search for that ID.

To join the network, a node u must have a contact to an already participating node w. u inserts w into the appropriate k-bucket. u then performs a node lookup for its own node ID. Finally, u refreshes all k-buckets further away than its closest neighbor. During the refreshes, u both populates its own k-buckets and inserts itself into other nodes' k-buckets as necessary.

2.4   Routing Table

Kademlia's basic routing table structure is fairly straight-forward given the protocol, though a slight subtlety is needed to handle highly unbalanced trees. The routing table is a binary tree whose leaves are k-buckets. Each k-bucket contains nodes with some common prefix of their IDs. The prefix is the k-bucket's position in the binary tree. Thus, each k-bucket covers some range of the ID space, and together the k-buckets cover the entire 160-bit ID space with no overlap.

Nodes in the routing tree are allocated dynamically, as needed. Figure 4 illustrates the process. Initially, a node u's routing tree has a single node— one k-bucket covering the entire ID space. When u learns of a new contact, it attempts to insert the contact in the appropriate k-bucket. If that bucket is not full, the new contact is simply inserted. Otherwise, if the k-bucket's range includes u's own node ID, then the bucket is split into two new buckets, the old contents divided between the two, and the insertion attempt repeated. If a k-bucket with a different range is full, the new contact is simply dropped.

---

[2] This number can be inferred from the bucket structure of the current node.

Routing table for a node whose ID is 00..00

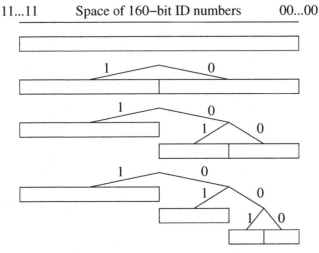

Fig. 4: Evolution of a routing table over time. Initially, a node has a single k-bucket, as shown in the top routing table. As the k-buckets fill, the bucket whose range covers the node's ID repeatedly splits into two k buckets.

One complication arises in highly unbalanced trees. Suppose node u joins the system and is the only node whose ID begins 000. Suppose further that the system already has more than k nodes with prefix 001. Every node with prefix 001 would have an empty k-bucket into which u should be inserted, yet u's bucket refresh would only notify k of the nodes. To avoid this problem, Kademlia nodes keep all valid contacts in a subtree of size at least k nodes, even if this requires splitting buckets in which the node's own ID does not reside. Figure 5 illustrates these additional splits. When u refreshes the split buckets, all nodes with prefix 001 will learn about it.

2.5 Efficient Key Re-publishing

To ensure the persistence of key-value pairs, nodes must periodically republish keys. Otherwise, two phenomena may cause lookups for valid keys to fail. First, some of the k nodes that initially get a key-value pair when it is published may leave the network. Second, new nodes may join the network with IDs closer to some published key than the nodes on which the key-value pair was originally published. In both cases, the nodes with a key-value pair must republish it so as once again to ensure it is available on the k nodes closest to the key.

To compensate for nodes leaving the network, Kademlia republishes each key-value pair once an hour. A naïve implementation of this strategy would require many messages—each of up to k nodes storing a key-value pair would

Relaxed routing table for a node whose ID is 00..00

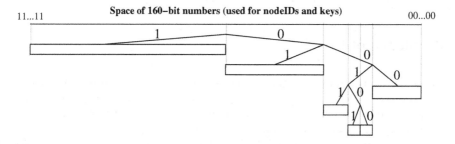

Fig. 5: **This figure examplifies the relaxed routing table of a node whose ID is 00 ... 00. The relaxed table may have small (expected constant size) irregularities in its branching in order to make sure it knows all contacts in the smallest subtree around the node that has at least k contacts.**

perform a node lookup followed by $k - 1$ STORE RPCs every hour. Fortunately, the republishing process can be heavily optimized. First, when a node receives a STORE RPC for a given key-value pair, it assumes the RPC was also issued to the other $k - 1$ closest nodes, and thus the recipient will not republish the key-value pair in the next hour. This ensures that as long as republication intervals are not exactly synchronized, only one node will republish a given key-value pair every hour.

A second optimization avoids performing node lookups before republishing keys. As described in Section 2.4, to handle unbalanced trees, nodes split k-buckets as required to ensure they have complete knowledge of a surrounding subtree with at least k nodes. If, before republishing key-value pairs, a node u refreshes all k-buckets in this subtree of k nodes, it will automatically be able to figure out the k closest nodes to a given key. These bucket refreshes can be amortized over the republication of many keys.

To see why a node lookup is unnecessary after u refreshes buckets in the subtree of size $\geq k$, it is necessary to consider two cases. If the key being republished falls in the ID range of the subtree, then since the subtree is of size at least k and u has complete knowledge of the subtree, clearly u must know the k closest nodes to the key. If, on the other hand, the key lies outside the subtree, yet u was one of the k closest nodes to the key, it must follow that u's k-buckets for intervals closer to the key than the subtree all have fewer than k entries. Hence, u will know all nodes in these k-buckets, which together with knowledge of the subtree will include the k closest nodes to the key.

When a new node joins the system, it must store any key-value pair to which it is one of the k closest. Existing nodes, by similarly exploiting complete knowledge of their surrounding subtrees, will know which key-value pairs the new node should store. Any node learning of a new node therefore issues STORE RPCs to transfer relevant key-value pairs to the new node. To avoid redundant

STORE RPCs, however, a node only transfers a key-value pair if it's own ID is closer to the key than are the IDs of other nodes.

3 Sketch of Proof

To demonstrate proper function of our system, we need to prove that most operations take $\lceil \log n \rceil + c$ time for some small constant c, and that a $\langle$key,value$\rangle$ lookup returns a key stored in the system with overwhelming probability.

We start with some definitions. For a k-bucket covering the distance range $[2^i, 2^{i+1})$, define the *index* of the bucket to be i. Define the *depth*, h, of a node to be $160 - i$, where i is the smallest index of a non-empty bucket. Define node y's *bucket height* in node x to be the index of the bucket into which x would insert y minus the index of x's least significant empty bucket. Because node IDs are randomly chosen, it follows that highly non-uniform distributions are unlikely. Thus with overwhelming probability the height of a any given node will be within a constant of $\log n$ for a system with n nodes. Moreover, the bucket height of the closest node to an ID in the kth-closest node will likely be within a constant of $\log k$.

Our next step will be to assume the invariant that every k-bucket of every node contains at least one contact if a node exists in the appropriate range. Given this assumption, we show that the node lookup procedure is correct and takes logarithmic time. Suppose the closest node to the target ID has depth h. If none of this node's h most significant k-buckets is empty, the lookup procedure will find a node half as close (or rather whose distance is one bit shorter) in each step, and thus turn up the node in $h - \log k$ steps. If one of the node's k-buckets is empty, it could be the case that the target node resides in the range of the empty bucket. In this case, the final steps will not decrease the distance by half. However, the search will proceed exactly as though the bit in the key corresponding to the empty bucket had been flipped. Thus, the lookup algorithm will always return the closest node in $h - \log k$ steps. Moreover, once the closest node is found, the concurrency switches from α to k. The number of steps to find the remaining $k - 1$ closest nodes can be no more than the bucket height of the closest node in the kth-closest node, which is unlikely to be more than a constant plus $\log k$.

To prove the correctness of the invariant, first consider the effects of bucket refreshing if the invariant holds. After being refreshed, a bucket will either contain k valid nodes or else contain every node in its range if fewer than k exist. (This follows from the correctness of the node lookup procedure.) New nodes that join will also be inserted into any buckets that are not full. Thus, the only way to violate the invariant is for there to exist $k + 1$ or more nodes in the range of a particular bucket, and for the k actually contained in the bucket all to fail with no intervening lookups or refreshes. However, k was precisely chosen for the probability of simultaneous failure within an hour (the maximum refresh time) to be small.

In practice, the probability of failure is much smaller than the probability of k nodes leaving within an hour, as every incoming or outgoing request updates nodes' buckets. This results from the symmetry of the XOR metric, because the IDs of the nodes with which a given node communicates during an incoming or outgoing request are distributed exactly compatibly with the node's bucket ranges.

Moreover, even if the invariant does fail for a single bucket in a single node, this will only affect running time (by adding a hop to some lookups), not correctness of node lookups. For a lookup to fail, k nodes on a lookup path must each lose k nodes in the same bucket with no intervening lookups or refreshes. If the different nodes' buckets have no overlap, this happens with probability 2^{-k^2}. Otherwise, nodes appearing in multiple other nodes' buckets will likely have longer uptimes and thus lower probability of failure.

Now we consider a ⟨key,value⟩ pair's recovery. When a ⟨key,value⟩ pair is published, it is populated at the k nodes, closest to the key. It is also re-published every hour. Since even new nodes (the least reliable) have probability $1/2$ of lasting one hour, after one hour the ⟨key,value⟩ pair will still be present on one of the k nodes closest to the key with probability $1 - 2^{-k}$. This property is not violated by the insertion of new nodes that are close to the key, because as soon as such nodes are inserted, they contact their closest nodes in order to fill their buckets and thereby receive any nearby ⟨key,value⟩ pairs they should store. Of course, if the k closest nodes to a key fail and the ⟨key,value⟩ pair has not been cached elsewhere, Kademlia will fail to store the pair and therefore lose the key.

4   Implementation Notes

In this section, we describe two important techniques we used to improve the performance of the Kademlia implementation.

4.1   Optimized Contact Accounting

The basic desired property of k-buckets is to provide LRU checking and eviction of invalid contacts without dropping any valid contacts. As described in Section 2.2, if a k-bucket is full, it requires sending a PING RPC every time a message is received from an unknown node in the bucket's range. The PING checks to see if the least-recently used contact in the k-bucket is still valid. If it isn't, the new contact replaces the old one. Unfortunately, the algorithm as described would require a large number of network messages for these PINGs.

To reduce traffic, Kademlia delays probing contacts until it has useful messages to send them. When a Kademlia node receives an RPC from an unknown contact and the k-bucket for that contact is already full with k entries, the node places the new contact in a *replacement cache* of nodes eligible to replace stale k-bucket entries. The next time the node queries contacts in the k-bucket, any unresponsive ones can be evicted and replaced with entries in the replacement cache. The replacement cache is kept sorted by time last seen, with the most recently seen entry having the highest priority as a replacement candidate.

A related problem is that because Kademlia uses UDP, valid contacts will sometimes fail to respond when network packets are dropped. Since packet loss often indicates network congestion, Kademlia locks unresponsive contacts and avoids sending them any further RPCs for an exponentially increasing backoff interval. Because at most stages Kademlia's lookup only needs to hear from one of k nodes, the system typically does not retransmit dropped RPCs to the same node.

When a contact fails to respond to 5 RPCs in a row, it is considered stale. If a k-bucket is not full or its replacement cache is empty, Kademlia merely flags stale contacts rather than remove them. This ensures, among other things, that if a node's own network connection goes down teporarily, the node won't completely void all of its k-buckets.

4.2   Accelerated Lookups

Another optimization in the implementation is to achieve fewer hops per lookup by increasing the routing table size. Conceptually, this is done by considering IDs b bits at a time instead of just one bit at a time. As previously described, the expected number of hops per lookup is $\log_2 n$. By increasing the routing table's size to an expected $2^b \log_{2^b} n$ k-buckets, we can reduce the number of expected hops to $\log_{2^b} n$.

Section 2.4 describes how a Kademlia node splits a k-bucket when the bucket is full *and* its range includes the node's own ID. The implementation, however, also splits ranges not containing the node's ID, up to $b - 1$ levels. If $b = 2$, for instance, the half of the ID space not containing the node's ID gets split once (into two ranges); if $b = 3$, it gets split at two levels into a maximum of four ranges, etc. The general splitting rule is that a node splits a full k-bucket if the bucket's range contains the node's own ID *or* the depth d of the k-bucket in the routing tree satisfies $d \not\equiv 0 \pmod{b}$. (The depth is just the length of the prefix shared by all nodes in the k-bucket's range.) The current implementation uses $b = 5$.

Though XOR-based routing resembles the first stage routing algorithms of Pastry [1], Tapestry [2], and Plaxton's distributed search algorithm [3], all three become more complicated when generalized to $b > 1$. Without the XOR topology, there is a need for an additional algorithmic structure for discovering the target within the nodes that share the same prefix but differ in the next b-bit digit. All three algorithms resolve this problem in different ways, each with its own drawbacks; they all require secondary routing tables of size $O(2^b)$ in addition to the main tables of size $O(2^b \log_{2^b} n)$. This increases the cost of bootstrapping and maintenance, complicates the protocols, and for Pastry and Tapestry complicates or prevents a formal analysis of correctness and consistency. Plaxton has a proof, but the system is less geared for highly fault-prone environments like peer-to-peer networks.

5 Summary

With its novel XOR-based metric topology, Kademlia is the first peer-to-peer system to combine provable consistency and performance, latency-minimizing routing, and a symmetric, unidirectional topology. Kademlia furthermore introduces a concurrency parameter, α, that lets people trade a constant factor in bandwidth for asynchronous lowest-latency hop selection and delay-free fault recovery. Finally, Kademlia is the first peer-to-peer system to exploit the fact that node failures are inversely related to uptime.

References

[1] A. Rowstron and P. Druschel. Pastry: Scalable, distributed object location and routing for large-scale peer-to-peer systems. *Accepted for Middleware, 2001*, 2001. `http://research.microsoft.com/~antr/pastry/`.

[2] Ben Y. Zhao, John Kubiatowicz, and Anthony Joseph. Tapestry: an infrastructure for fault-tolerant wide-area location and routing. Technical Report UCB/CSD-01-1141, U.C. Berkeley, April 2001.

[3] Andréa W. Richa C. Greg Plaxton, Rajmohan Rajaraman. Accessing nearby copies of replicated objects in a distributed environment. In *Proceedings of the ACM SPAA*, pages 311–320, June 1997.

[4] Stefan Saroiu, P. Krishna Gummadi and Steven D. Gribble. A Measurement Study of Peer-to-Peer File Sharing Systems. Technical Report UW-CSE-01-06-02, University of Washington, Department of Computer Science and Engineering, July 2001.

[5] Ion Stoica, Robert Morris, David Karger, M. Frans Kaashoek, and Hari Balakrishnan. Chord: A scalable peer-to-peer lookup service for internet applications. In *Proceedings of the ACM SIGCOMM '01 Conference*, San Diego, California, August 2001.

Efficient Peer-to-Peer Lookup Based on a Distributed Trie

Michael J. Freedman[1] and Radek Vingralek[2]*

[1] MIT Lab for Computer Science, 200 Technology Sq., Cambridge, MA 02139 USA
mfreed@lcs.mit.edu
[2] Oracle Corp., 500 Oracle Parkway, Redwood Shores, CA 94065 USA
radek.vingralek@oracle.com

Abstract. Two main approaches have been taken for distributed key-value lookup operations in peer-to-peer systems: broadcast searches [1, 2] and location-deterministic algorithms [5, 6, 7, 9]. We describe a third alternative based on a distributed trie. This algorithm functions well in a very dynamic, hostile environment, offering security benefits over prior proposals. Our approach takes advantage of working-set temporal locality and global key/value distribution skews due to content popularity. Peers gradually learn system state during lookups, receiving the sought values and/or internal information used by the trie. The distributed trie converges to an accurate network map over time. We describe several modes of information piggybacking, and conservative and liberal variants of the basic algorithm for adversarial settings. Simulations show efficient lookups and low failure rates.

1 Introduction

We describe a set of algorithms for key-based lookup in a distributed system consisting of a number of uniform *peers*. A lookup service is a necessary component for peer-to-peer file-sharing systems, which need to map filenames to the location of peers that store them. Such an algorithm may return the files themselves or the addresses of servers storing of files.

Most lookup algorithms deploy a *lookup structure* to efficiently locate values associated with keys. The lookup structure can be organized as an (extendible) hash table, a trie, a binary tree, a B-tree, etc. A distributed lookup algorithm trades off the efficiency of lookups for the maintenance of the lookup structure. Maintenance is initiated by either a key/value pair insert or by a membership change.

Lookups can be made very efficient by replicating the lookup structure on every peer. However, maintenance is slow as all peers must keep their lookup structure replicas consistent. To date, peer-to-peer systems have taken two approaches to reducing the maintenance costs.

---

* This work was partially done while both authors were employed by InterTrust Technologies, STAR Lab, 4750 Patrick Henry Drive, Santa Clara, CA 95054 USA.

P. Druschel, F. Kaashoek, and A. Rowstron (Eds.): IPTPS 2002, LNCS 2429, pp. 66–75, 2002.

The first approach, adopted by Gnutella [2] and Freenet [1], eliminates the lookup structure altogether and thus requires no maintenance. However, lookups are implemented by a broadcast-like search on all known peers, costing efficiency and scalability. (Freenet aims to reduce overhead by broadcasting first to neighbors that previously returned similar keys and replicating key/value pairs along the entire path between sender and receiver.)

The second approach partitions the lookup structure and distributes a subset of partitions on each peer. Since maintenance updates are typically localized, peers update only a smaller number of partition replicas. The system assigns partitions to peers either statically or dynamically. In the static-assignment approach, peer addresses are mapped to the key space and each node replicates only those partitions that are "close" to its address. Systems such as Chord [7] (mapping via consistent hashing), CAN [5] (routing on a d-torus), and Tapestry [9] and Pastry [6] (both related to Plaxton trees [4]) adopt this model. The static assignment enables efficient message routing, with the worst case message overhead limited by a logarithm of the network size (assuming no updates to the lookup structure). However, if peers join and leave the network rapidly, the lookup structure may never stabilize to a state that enjoys this provably logarithmic bound. Furthermore, the routing overhead does not adapt to the workload. Consequently, peers will always incur a higher message overhead to access "far away" parts of the key space. Moreover, the peers must dedicate resources for upkeep of the routing information that they do not use. Finally, it is easy for an attacker to target a specific part of the key space because she can determine (using the static mapping) the peers that hold the corresponding partitions of the lookup structure.

When the partitions are assigned dynamically, the distribution of the partitions can adapt to the workload by having peers replicate only the partitions that they frequently access. It is also harder to attack a particular part of the key space because it may be impossible to determine locations of all replicas of a particular partition without inspecting routing tables of all peers. However, the same property makes efficient routing more difficult.

Relaxing the consistency criteria for partition replicas can further reduce maintenance costs. However, it can happen that peers hold stale replicas if they update local lookup structures lazily [3, 8]. They can commit addressing errors when requesting values from peers that are unavailable or that no longer hold the values. To ensure that the extra cost of addressing errors does not approach the cost of a broadcast, peers must limit addressing errors by piggybacking the updates on other traffic. Then, peers reconcile conflicting updates to achieve replica convergence.

We present a set of algorithms that exploit both *dynamic partitioning* based on peers' access locality and *lazy updates* of the lookup structure to reduce maintenance cost. All algorithms piggyback trie state on lookup responses. Peers use timestamping to reconcile conflicting updates. The algorithms differ in the volume of the trie structure piggybacked and how aggressively the requester uses these partitions.

These algorithms do not preclude *strong anonymity* for either peers initiating a lookup or peers storing the corresponding value. The trie structure can index endpoints of fixed-length mix-net circuits, as opposed to the desired peers. The mix-net circuit will relay messages from an endpoint to an anonymized recipient. Therefore, our lookup algorithms (as well as Chord, CAN, Tapestry and Pastry) can provide anonymity at a constant cost of extra messages, by treating anonymity as a goal for the underlying communication channel.

2   System Model

The system consists of n peers that all implement the following interface:

- lookup(key). The callee sends to the caller the value associated with key if successful or a failure message. In both cases, the callee may piggyback additional state on the response (algorithm dependent).
- insert(key, value). The callee inserts a < key,value > pair into its lookup structure.
- join(). The callee sends to the caller initial state needed to bootstrap lookup operations.

Peers cannot update or delete values that were previously inserted, although they can re-insert new versions under a different (or even the same) key. The caller is responsible for ensuring key uniqueness, if required. Peers join the system only intermittently, where "join" is defined as an *initial* entry into the network, not a reconnection after failure. Peers can leave at any time or fail silently; no maintenance operations are necessary in either case.

Each peer stores a number of key/value pairs locally, that were either inserted or looked up by the peer. The peer also stores partitions of a lookup structure organized as a trie, as shown in Figure 1. A trie representation is insensitive to the insertion ordering. Consequently, it is easier to merge two incomparable versions of the lookup structure.

Internal trie nodes consist of 2^m *routing tables*. Each routing table consists of l *entries*. An entry consists of a peer address a and a timestamp t. Each level of trie node "consumes" m bits of the k-bit key. If the node is a leaf (defined as having depth $\lceil k/m \rceil$), then the entry in its matching routing table indicates that peer a was known at time t to hold the specified value. Otherwise, the entry at its ith routing table indicates that peer a was known at time t to hold a replica of the ith child of the node. The timestamps are generated locally by each peer; we assume loose clock synchronization (say, to within a few hours).

Each peer stores only the subset of trie nodes corresponding to its access pattern. However, all peers maintain the following invariant:

> *Ancestor invariant*: If a peer holds a trie node, it must hold all ancestors of the node.

We represent trie paths more compactly by explicitly relying on this invariant: If a peer appears in some routing table, it is known to hold not only the

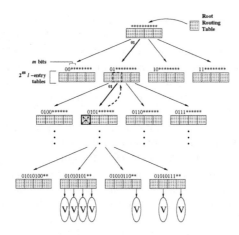

Fig. 1. Trie lookup structure ($k = 10, m = 2$).

node's child, but also an entire path down to that child. Peers naturally maintain this invariant by replacing trie nodes and routing table entries based on access frequency. Logically, nodes closer to the trie root are more widely replicated by peers, removing any single point of failure.

3   Algorithms

The algorithms we present for a distributed trie lookup share most of the basic steps. They differ only in what state is piggybacked on lookup responses and how this state is used. We first describe the basic framework shared by all algorithms and subsequently the differences.

In order to join the distributed system, a peer needs to know the address of at least one participating peer, called its *introducer*. The introducer responds to a join request with its own root routing table, as labeled in Figure 1. The new peer uses this root routing table as a bootstrap mechanism when it exhausts the entries within its own trie during a lookup.

Insertion is performed locally by inserting the key/value pair (and the corresponding path from the root of the trie) into local storage. Alternatively, the peer could send an insert request to other peers, similar to [1]. The peers may decline to store the inserted pair. This paper does not consider such an insert operation, in order to focus solely on the effect of piggybacking state in lookups.

The lookup caller first checks local storage for the value corresponding to the lookup key. If present, the lookup process terminates. Otherwise, the caller initiates a distributed lookup process. We present an example of `lookup(0101000000)` for comprehension.

Caller A searches its local trie for a routing table that most specifically matches lookup key. Such a routing table `010100*` is shown with a solid box in Figure 1. Subsequently, caller A then sends a lookup query to peer B, who has the latest timestamp in the routing table. Peer B is most recently known (to Peer A) to hold the child trie node 010100* (that A does not currently have). B returns the actual value if B holds it. B's response may either contain additional trie state or be a failure with no state. If B returns a deeper routing table, A drops down to that level and repeats this process. If B returns failure, caller A tries other peers in its routing table in decreasing timestamp order. If all such peers fail, A backtracks in its local trie and repeats the same process on the parent routing table. The figure shows a dashed line: routing table 010100* failed and is crossed out, and A backtracked to routing table 0101*.

Recall that each trie node's routing tables are of maximum size l. Once the caller starts backtracking, we enumerate larger "virtual" routing tables: an entry list containing all peers thought to hold the desired child trie node. Peer A's virtual view of 0101* is that level's actual routing table merged with all tables in its subtrie (minus the entries in table 010100* already contacted). By the ancestor invariant, the peers holding trie nodes in the subtrie must also have a copy of the higher-level routing table. Therefore, we effectively increase the size of the higher-level routing tables without additional storage. If an entry's routing table is full when new entries are installed, the least recent entry is evicted from this table and propagates up the trie according to timestamp.

Peers may backtrack their local tries during lookup up to the root routing table. If the value is not found at this time, the lookup process terminates with a failure.

3.1   Bounded, Unbounded, and Full Path Modes

The *bounded*, *unbounded*, and *full path* modes explore the tradeoff between the size of piggybacked trie state with the speed of convergence of peers' tries to an accurate network map.

In *bounded mode*, the callee responds to a lookup with its most specific routing table matching the key (or the value itself), provided its routing table is more specific (deeper) than the caller's current table. Otherwise, the callee responds with a failure. The caller integrates the more specific routing table into its local trie and proceeds with the lookup on this table.

In *unbounded mode*, the callee responds with its most specific routing table for the key, regardless of how this compares to the caller's current routing table. This additional state is useful to pre-fetch information about new peers or more recent locations of higher-level trie nodes to be used when backtracking. The caller integrates the returned routing table into its local table at that same depth, by selecting the l most recent distinct entries from the two tables.

In *full path mode*, the callee responds with the entire path (consisting of routing tables) from the root table to its most specific routing table for the key. The caller integrates the path into its local trie using the same mechanism as in the unbounded mode.

3.2 Conservative and Liberal Modes

Most peer-to-peer lookup algorithms are susceptible to malicious behavior. We describe one particular attack conceptually similar to DNS cache poisoning and propose a conservative mode to resist this attack.

A malicious peer can effectively suppress access to a value by falsely advertising the availability of some key (with a recent timestamp) and then dropping lookup requests. A set of l malicious peers can cause innocent peers to completely replace the entries in their routing tables with malicious peer addresses. While backtracking can help route around this problem by finding less-specific routing tables, these malicious peers have caused the system to lose efficiency.

We propose a verified-only update heuristic for a *conservative mode*. Namely, callers update their local trie with only the new entries that transitively led them to the desired value. This assumes that peers can verify the validity of returned data. For example, if peer B returns {C,D,E} to peer A at depth 1, peer C returns {F,G,H} to A at depth 2, and peer F returns the actual value, peer A updates his trie only with peers B, C, and F in the corresponding-depth routing tables. Conservative mode ensures that entries in our routing tables have performed useful work in the past. Therefore, we hope they will continue to be useful.

In *liberal mode*, callers immediately update their local tries with any piggy-backed state.

4 Preliminary Experimental Evaluation

We implemented a simulator to compare the performance of algorithms modes described in Section 3. We present *preliminary* simulation results in this section.

We simulated a system consisting of 200 peers. The tries maintained by the peers were characterized by parameters $k = 10$, $m = 2$ and $l = 10$. Each experiment started with an initial loading phase, when the peers inserted a total of 2,000 randomly-generated key/value pairs in the system. These pairs were distributed randomly among all peers. The keys were uniformly distributed in $[0, 2^k - 1]$. Subsequently, at each simulated time-step, we issued a lookup to a randomly-selected peer and dynamically changed the membership by removing peers or adding new ones with a probability 0.005.

For each lookup phase we collected the following statistics:

- *Message overhead.* We classify lookups as either local (*i.e.*, those that could be satisfied by a peer locally) or remote (*i.e.*, those that required sending lookup operations to other peers). For each remote lookup, we measured the number of lookup operations that were sent to other peers (transitively) in order to satisfy the request.
- *Failure probability.* For each lookup, we measured the probability of its failure. A lookup fails when the requesting peer's trie did not contain sufficient information to locate an existing key/value pair (even after contacting other peers).

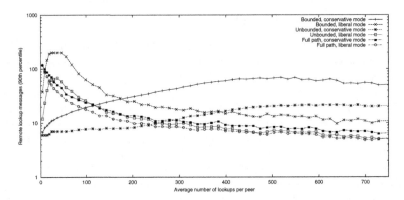

Fig. 2. The 90th percentile of the number of messages generated by a remote lookup.

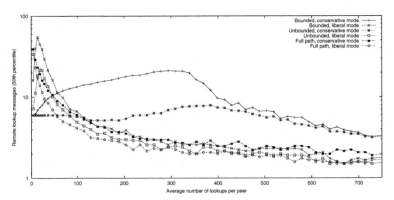

Fig. 3. The 50th percentile of the number of messages generated by a remote lookup.

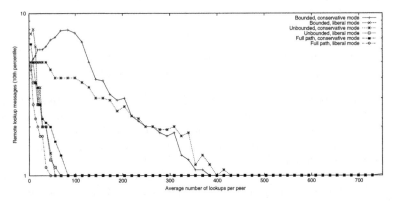

Fig. 4. The 10th percentile of the number of messages generated by a remote lookup.

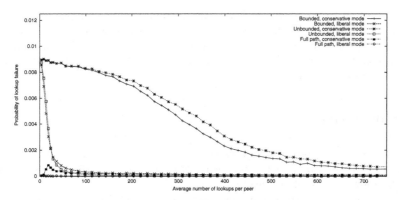

Fig. 5. The probability of lookup failure.

The message overheads of all algorithm modes are found in Figures 2, 3, and 4. The three figures respectively show the 90th, 50th, and 10th percentiles of the number of query/response message pairs (*i.e.*, lookup operations) that were generated by a remote lookup. The reported values are conservative, as the message overhead explicitly excludes lookups that were satisfied locally. We also did not allow peers to benefit from access locality by generating keys according to a skewed distribution. The failure probabilities of all algorithm modes are found in Figure 5.

Based on the limited test data, we make the following conjectures:

- The message overheads of all modes converge toward less than $\log n$ lookup operations per remote lookup. The failure probability of all modes converge toward zero. Thus peers seem capable of correctly learning a recent view of the distributed trie.
- There is an implicit trade-off between message size and convergence time to low message overhead. Full path mode converges the fastest, yet sends the largest messages. Bounded mode converges the slowest, but with the benefit of sending smaller messages.
- The bounded and unbounded conservative modes converge slower than their liberal counterparts. (The increases in message overhead for the conservative modes, as shown in Figure 2, is a consequence of the learning process: While a lookup might fail after a node contacts its few known peers at the beginning, it will exhaustively contact its increasing number of known peers as the experiment progresses.) Consequently, there is a performance cost for reducing risk of a system infiltration by malicious peers.
- The full path conservative and liberal modes converge with almost the same rates. Therefore, if peers exchange enough state, they can reduce the risk of malicious system infiltration with only small performance loss.

In addition to these experiments with dynamic membership, we ran a similar series of experiments with static membership. The results were practically the same as those reported in the included figures (we exclude the corresponding graphs due to space constraints). Therefore, we conjecture that all of the algorithm modes are relatively resilient to system membership changes.

5   Conclusions

We propose a new approach to key/value lookup in peer-to-peer systems based on a distributed trie. Compared to broadcast-based routing, the algorithms clearly lead to a lower message overhead.

Compared to the static-partitioning-based (or location-deterministic) approaches, these algorithms can deliver a lower message overhead, given sufficient time for peers to learn the distribution of frequently-accessed keys. Message overhead does not depend on the proximity between the peer address and the looked-up key, nor are peers required to maintain state about regions of the keyspace that they do not access. Our algorithms do not assume a globally known mapping of peers address space into the key space. Consequently, it is more difficult for an adversary to target all replicas of a particular key/value pair because it is impossible to determine all locations of a particular key/value pair without inspecting routing tables of all peers.

On the other hand, our algorithms can degenerate to a broadcast for peers with very stale views, while the static-partitioning-based approaches have logarithmic upper bounds on message overhead per lookup. (Although the logarithmic upper bounds assume absence of updates to the lookup structure.)

We believe that both static and dynamic partitioning algorithms have their merits. The static algorithms provide a better worst case behavior and typically require peers to store a smaller state. The dynamic algorithms can adapt to the workload and are more resilient to certain types of attacks. We conceive that both kinds of algorithms may be used in conjunction: A peer may send a limited number of lookup operations based on its local trie and, only if all failed, revert to a static-partitioning-based lookup. Symmetrically, if the static-partitioning algorithms seek to reduce their (fixed) message overhead by caching additional state, they can use our algorithms to maintain consistency of the cached state.

References

[1] Ian Clarke, Oscar Sandberg, Brandon Wiley, and Theodore Hong. Freenet: A distributed anonymous information storage and retrieval system. In *Proceedings of the Workshop on Design Issues in Anonymity and Unobservability*, pages 46–66, July 2000.

[2] Gnutella website. http://gnutella.wego.com.

[3] W. Litwin, M. Neimat, and D. Schneider. lh^* - linear hashing for distributed files. In *In Proceedings of the ACM SIGMOD Conference*, May 1993. Washington, DC.

[4] C. Plaxton, R. Rajaraman, and A. Richa. Accessing nearby copies of replicated objects in a distributed environment. In *Proceedings of the ACM SPAA*, pages 311–320, June 1997.

[5] S. Ratnasamy, P. Francis, M. Handley, R. Karp, and S. Shenker. A scalable content-addressable network. In *Proc. ACM SIGCOMM*, San Diego, 2001.

[6] Antony Rowstron and Peter Druschel. Pastry: Scalable, distributed object location and routing for large-scale peer-to-peer systems. In *Proceedings of the 18th IFIP/ACM International Conference on Distributed Systems Platforms (Middleware 2001)*, November 2001.

[7] Ion Stoica, Robert Morris, David Karger, M. Frans Kaashoek, and Hari Balakrishnan. Chord: A scalable peer-to-peer lookup service for internet applications. In *Proc. ACM SIGCOMM*, San Diego, 2001.

[8] R. Vingralek, Y. Breitbart, and G. Weikum. Distributed file organization with scalable cost/performance. In *In Proceedings of the ACM-SIGMOD International Conference on Management of Data*, May 1994. Minneapolis, MN.

[9] Ben Zhao, John Kubiatowicz, and Anthony Joseph. Tapestry: An infrastructure for fault-tolerant wide-area location and routing. Technical Report UCB/CSD-01-1141, Computer Science Division, U. C. Berkeley, April 2001.

Self-Organizing Subsets: From Each According to His Abilities, to Each According to His Needs

Amin Vahdat, Jeff Chase, Rebecca Braynard, Dejan Kostić, Patrick Reynolds,
and Adolfo Rodriguez*

Department of Computer Science
Duke University

Abstract. The key principles behind current peer-to-peer research include fully distributing service functionality among all nodes participating in the system and routing individual requests based on a small amount of locally maintained state. The goals extend much further than just improving raw system performance: such systems must survive massive concurrent failures, denial of service attacks, etc. These efforts are uncovering fundamental issues in the design and deployment of distributed services. However, the work ignores a number of practical issues with the deployment of general peer-to-peer systems, including i) the overhead of maintaining consistency among peers replicating mutable data and ii) the resource waste incurred by the replication necessary to counteract the loss in locality that results from random content distribution.

We argue that the key challenge in peer-to-peer research is not to distribute service functions among all participants, but rather to distribute functions to meet target levels of availability, survivability, and performance. In many cases, only a subset of participating hosts should take on server roles. The benefit of peer-to-peer architectures then comes from massive diversity rather than massive decentralization: with high probability, there is always some node available to provide the required functionality should the need arise.

1 Introduction

Peer-to-peer principles are fundamental to the concept of survivable, massive-scale Internet services incorporating large numbers—potentially billions—of heterogeneous hosts. Most recent peer-to-peer research systems distribute service functions (such as storage or indexing) evenly across all participating nodes [4, 6, 7, 8, 9, 12, 13]. At a high level, many of these efforts use a distributed hash table, with regions of the table mapped to each participant. The challenge then is to

* This research is supported in part by the National Science Foundation (EIA-9972879, ITR-0082912), Hewlett-Packard, IBM, Intel, and Microsoft. Braynard and Reynolds are supported by an NSF graduate fellowships and Vahdat is also supported by an NSF CAREER award (CCR-9984328).

P. Druschel, F. Kaashoek, and A. Rowstron (Eds.): IPTPS 2002, LNCS 2429, pp. 76–84, 2002.

locate the remote host responsible for a target region of the hash space in a scalable manner, while: i) adapting to changes in group membership, ii) achieving locality with the underlying IP network, and iii) caching content and/or request routing state so as to minimize the average number of hops to satisfy a request.

These recent efforts constitute important basic research in massively decentralized systems, and they have produced elegant solutions to challenging and interesting problems. However, these approaches seek massive decentralization as an end in itself, rather than as a means to the end of devising practical service architectures that are scalable, available, and survivable. From a practical standpoint, they address the wrong set of issues in peer-to-peer computing.

We suggest that distributing service functions across a carefully selected subset of nodes will yield better performance, availability, and scalability than massively decentralized approaches. The true opportunity afforded by peer-to-peer systems is not the ability to put everything everywhere. Rather, it is the opportunity to put anything anywhere. Why distribute an index across one million nodes when a well-chosen subset of one thousand can provide the resources to meet target levels of service performance and availability?

Given n participants in a peer-to-peer system, we argue that the best approach is not to evenly spread functionality across all n nodes, but rather to select a minimal subset of m nodes to host the service functions. This choice should reflect service load, node resources, predicted stability, and network characteristics, as well as overall system performance and availability targets. While it may turn out that $m = n$ in some cases, we believe that $m \ll n$ in most cases. Membership in the service subset and the mapping of service functions must adapt automatically to changes in load, participant set, node status, and network conditions, all of which may be highly dynamic. Thus we refer to this approach for peer-to-peer systems as *self-organizing subsets*.

One goal of our work is to determine the appropriate subset, m, of replicas required to deliver target levels of application performance and availability. The ratio of subset size m to the total number of nodes n can approximately be characterized by:

$$\frac{m}{n} = \frac{u}{dE}$$

where u is the sum of all service resources consumed by the n total hosts, d is the sum of all service resources provided by the m hosts in the subset, and E is the efficiency — the fraction of resources in use when the system as a whole begins to become overloaded. Efficiency is a function of the system's ability to properly assign functionality to an appropriate set of sites and of load balancing; better load balancing results in values of E approaching one [8]. In a few systems, such as SETI@home, all available service resources will be used; thus, u approaches d, and it makes sense for m to equal n. However, in most systems each node can provide far more resources than it is likely to consume; thus, $u \ll d$, and given reasonable efficiency, $m \ll n$.

Self-organizing subsets address key problems of scale and adaptation that are inherent in the massively decentralized approach. For example, routing state and hop count for request routing in existing peer-to-peer systems typically grow with $O(\lg n)$ at best. While this may qualify as "scalable," it still imposes significant overhead even for systems of modest size. Using only a subset of the available hosts reduces this overhead. More importantly, massively decentralized structures may be limited to services with little or no mutable state (e.g., immutable file sharing), since coordination of updates quickly becomes intractable. Our recent study of availability [16] shows that services with mutable state may suffer from too much replication: adding replicas may compromise availability rather than improve it. Random distribution of functionality among replicas means that more replicas are required to deliver the same level of performance and availability. Thus, there is an interesting tension between the locality and availability improvements on the one hand and the degradation on the other that comes from replication of mutable data in peer to peer systems. A primary goal of our work is to show the *resource waste* that comes from random distribution, i.e., the inflation in the number of randomly placed replicas required to deliver the same levels of performance and availability as a smaller number of "well-placed" replicas. Finally, massively decentralized approaches are not sufficiently sensitive to the rapid status changes of a large subset of the client population. We propose to restrict service functions to subsets of nodes with significantly better connectivity and availability than the median, leading to improved stability of group membership.

Our approach adapts Marxist ideology—"from each according to his abilities, to each according to his needs"—to peer-to-peer systems. The first challenge is to gather information about the *abilities* of the participants, e.g., network connectivity, available storage and CPU power, and the *needs* of the application, e.g., demand levels, distribution of content popularity, and network location. The second challenge is to apply this information to select a subset of the participants to host service functions, and a network overlay topology allowing them to coordinate. Status monitoring and reconfiguration must occur automatically and in a decentralized fashion.

Thus, we are guided by the following design philosophies in building scalable, highly available peer-to-peer systems:

- It is important to dynamically select subsets of participants to host service functions in a decentralized manner. In the wide area, it is not necessary to make optimal choices; rather, it is sufficient to make good choices in a majority of cases and to avoid poor decisions. For example, current research efforts place functionality randomly and use replication to probabilistically deliver acceptable performance for individual requests. Our approach is to place functionality deterministically and to replicate it as necessary based on network and application characteristics. This requires methods to evaluate expected performance and availability of candidate configurations to determine if they meet the targets.

- A key challenge to coordinating peer-to-peer systems is collecting metadata about system characteristics. Configuring a peer-to-peer system requires tracking the available storage, bandwidth, memory, stability (in terms of uptime and availability), computational power, and network location of a large number of hosts. At first glance, maintaining global state about potentially billions of hosts is intractable. However, good (rather than optimal) choices require only approximate information: aggressive use of hierarchy and aggregation can limit the amount of state that any node must maintain. Once a subset is selected, the system must track only a small set of candidate "replacement" nodes to address failures or evolving system characteristics. Similarly, clients maintain enough system metadata to choose the replica likely to deliver the best quality of service (where QoS is an application-specific measure). Once again, the key is to make appropriate request routing decisions almost all of the time, without global state.
- Service availability is at least as important as average-case performance. Thus, we are designing and building algorithms to replicate data and code in response to changing client access patterns and desired levels of availability. Some important questions include determining the level of replication and placement of replicas needed to achieve a given minimum level of availability as a function of workload and failure characteristics. Once again, a key idea is that a few well-placed replicas will deliver higher availability than a larger number of randomly placed replicas because of the control overhead incurred by coordination among replicas.

The rest of this position paper elaborates on some of the challenges we see in fully distributing service functionality among a large number of nodes. It then describes Opus, a framework we are using to explore the structure and organization of peer-to-peer systems.

2   Challenges to Massive Decentralization

In this section, we further elaborate on our view of why fully distributing functionality among a large number of Internet nodes is the wrong way to build peer-to-peer systems. While a number of techniques have been proposed to minimize per-node control traffic and state requirements, it still remains true that in a fully decentralized system with millions of nodes, the characteristics of all million nodes have to be maintained somewhere in the system. To pick one example, each node in a million-node Pastry system must track the characteristics of 75 (given the suggested representative tuning parameters) individual nodes [8], potentially randomly spread across the Internet. We believe that by choosing an appropriate subset of global hosts (m of n) to provide application functionality and by leveraging hierarchy, the vast majority of nodes will maintain state about a constant (small) number of nearby *agents*. Sets of agents are aggregated to form hierarchies and in turn maintain state about a subset of the m nodes and perhaps approximate information on the full set of m nodes[1]. Thus, to route a request to an appropriate server, nodes forward requests to their agent, which in turn determines the appropriate replica (member of m) to send the request to. In summary, massive decentralization requires each system node to maintain state about $O(\lg n)$ other global nodes. If successful in carefully placing functionality at strategic network points, the vast majority of nodes maintain state about a constant and small number of peers (one or more agents), and each agent maintains state about a subset of the m nodes providing application functionality.

Another issue with massive decentralization is dealing with dynamic group membership. Assuming a heavy-tailed distribution for both host uptime and session length, significant network overhead may be required to address host entry or departure of the large group of hosts that exhibit limited or intermittent connectivity (some evidence for this is presented in [11]). This is especially problematic if $O(\lg n)$ other hosts must be contacted to properly insert or remove a host. In our approach, we advocate focusing on the subset of hosts (again, m of n) that exhibit strong uptime and good connectivity— the tail of the heavy-tailed distribution rather than the head. In this way, we are able to focus our attention on hosts that are likely to remain a part of the system, rather than being in a constant state of instability where connectivity is *always* changing in some region of the network. Of course, nodes will be constantly entering and leaving in our proposed system as well. However, entering nodes must contact only a small constant number of nodes upon joining (their agents) and can often leave silently. In particular, node departure is an entirely local event if it never achieved the level of uptime or performance required to be considered for future promotion to an agent or one of the m nodes that deliver application-level functionality.

Finally, a key approach to massive decentralization is randomly distributing functionality among a large set of nodes. The problem then becomes routing requests to appropriate hosts in a small number of steps (e.g., $O(\lg n)$ hops). Because these systems effectively build random application-layer overlays, it can be difficult to match the topology of the underlying IP network in routing requests. Thus, replication and aggressive caching [4, 7, 9] must be leveraged to achieve acceptable performance relative to routing in the underlying IP network. While this approach results in small inflation in "network stress" relative to IP, application-layer store and forward delays can significantly impact end-to-end latency (even when only $O(\lg n)$ such hops are taken). While such inflation of latency is perhaps not noticeable when performing a lookup to download a multi-megabyte file, it can become the bottleneck for a more general class of applications. With our approach, requests can typically be routed in a small and constant number of steps (depending on the chosen depth of the hierarchy). Further, because we have explicit control over connectivity, hierarchy, and place-

---

[1] For simplicity, this discussion assumes a two-level hierarchy, which should be sufficient for most applications. Our approach extends directly to hierarchies of arbitrary depth.

ment of functionality, we can ensure that requests from end hosts are routed to a nearby agent, and from there to an active replica. The random distribution of functionality in massively decentralized systems makes it more difficult to impose any meaningful hierarchy.

3   An Overlay Peer Utility Service

We are pursuing our agenda of dynamically placing functionality at appropriate points in the network in the context of Opus [2], an overlay peer utility service. While our research targets Opus, our techniques and approach are general to a broad range of peer-to-peer services. As a general compute utility, we envision Opus hosting a large set of nodes across the Internet and dynamically allocating them among competing applications based on changing system characteristics. Individual applications specify their performance, availability, and data quality targets to Opus. One challenge is to develop general definitions for availability [1] and data quality [15] as a basis for specifying these targets. Based on this information, we map applications to individual nodes across the wide area. The initial mapping of applications to available resources is only a starting point. Based on observed access patterns to individual applications, Opus dynamically reallocates global resources to match application requirements. For example, if many accesses are observed for an application in a given network region, Opus may reallocate additional resources close to that location.

One key aspect of our work is the use of Service Level Agreements (*SLAs*) to specify the amount each application is willing to "pay" for a given level of performance. In general, these SLAs provide a continuous space over which per-service allocation decisions can be made, enabling prioritization among competing applications for a given system configuration. Using an estimate of the marginal utility of resources across a set of applications at current levels of global demand, Opus makes allocation and deallocation decisions to maximize the expected relative benefit of a set of target configurations [3].

Many individual components of Opus require information on dynamically changing system characteristics. Opus employs a global *service overlay* to interconnect all available service nodes and to maintain soft state about the current mapping of utility nodes to hosted applications (group membership). The service overlay is key to many individual system components, such as routing requests from individual clients to appropriate replicas, and performing resource allocation among competing applications. Individual services running on Opus employ *per-application overlays* to disseminate their own service data and metadata among individual replica sites.

Clearly, a primary concern is ensuring the scalability and reliability of the service overlay. In an overlay with n nodes, maintaining global knowledge requires $O(n^2)$ network probing overhead and $O(n^2)$ global storage requirements. Such overhead quickly becomes intractable beyond a few dozen nodes. Peer-to-peer systems can reduce this overhead to approximately $O(n \lg n)$ but are often unable to provide any information about global system state, even if approximate. Opus addresses scalability issues through the aggressive use of hierarchy, aggregation, and approximation in creating and maintaining scalable overlay structures. Opus then determines the proper level of hierarchy and aggregation (along with the corresponding degradation of resolution of global system state) necessary to achieve the target network overhead.

Security is another important consideration for any general-purpose utility. Opus allocates resources to applications at the granularity of logical nodes, eliminating a subset of the security and protection issues associated with simultaneously hosting multiple applications in a utility model. We believe that a cost model for consumed node and network resources will motivate application developers to deploy efficient software for a given demand level.

We have initial results addressing a number of the challenges outlined above. For instance, we conducted a study to determine the upper bound of service availability as a function of application workload, network failure characteristics, and desired levels of data consistency [16]. One interesting result here is that for a given workload, faultload, and consistency level, there is an optimal number of replicas for availability. That is, beyond some point, additional replicas of a wide-area service actually reduces service availability rather than improves it. The intuition behind this insight is that there is a tension between the desire to widely replicate a service in the hopes that at least one replica is always available to all clients and the desire to centralize the service to minimize the overhead of consistency maintenance. Building on this work, we have also shown how to optimally place replicas in the face of changing network failure characteristics to maximize service availability [17].

Finally, in the space of building both service and application overlays, we have designed, implemented, and evaluated a fully distributed algorithm, called ACDC (for Adaptive low-Cost, Delay Constrained), for building two-metric overlays [5]. We assume that each edge in a wide-area network has two dynamically changing weights, one describing the cost incurred from using that edge and the second describing an arbitrary performance characteristic (such as delay, bandwidth, or loss rate). The goal of ACDC is then to build the lowest-cost overlay that meets application-specified performance targets. This is an NP-hard problem even with accurate global knowledge [10, 14]. Our challenge then is to approximate the global optimum using approximate and potentially inaccurate information. ACDC is designed to scale to large-scale overlays of ten thousand nodes or more. Thus, a key challenge is for ACDC overlays to self-organize (and also to adapt to changing network conditions) to meet target levels of performance and cost in a scalable manner. This requirement rules out the straightforward technique of maintaining global knowledge and performing global system probing. We have designed a set of distributed algorithms that enables ACDC nodes to maintain no more than $O(\lg n)$ state and to probe no more than $O(\lg n)$ peers. Our performance evaluation indicates that ACDC is able to quickly converge to performance and cost targets, even in the face of rapidly changing network conditions. We intend to use the ideas from ACDC as the basis for building scalable Opus overlays.

4   Conclusions

This paper argues that a principal challenge in peer-to-peer systems is determining where to place functionality in response to changing system characteristics and as a function of application-specified targets for availability, survivability, and performance. Many current peer-to-peer research efforts focus on fully distributing service functionality across all participating hosts, which could potentially number in the billions. The resulting research fundamentally contributes to our understanding of structuring distributed services. However, we argue that a key challenge in peer-to-peer research is to dynamically determine the proper subset m, of n participating nodes, required to deliver target levels of availability, survivability, and performance, where typically $m \ll n$. For many application classes, especially those involving mutable data, increasing m will not necessarily improve service utility. We describe the architecture of Opus, an overlay peer utility service that dynamically allocates resources among competing applications, which we are using as a testbed for experimenting with the ideas presented in this paper.

References

[1] Guillermo A. Alvarez, Mustafa Uysal, and Arif Merchant. Efficient Verification of Performability Guarantees. In *Fifth International Workshop on Performability Modeling of Computer and Communication Systems (PMCCS 5)*, September 2001.

[2] Rebecca Braynard, Dejan Kostić, Adolfo Rodriguez, Jeffrey Chase, and Amin Vahdat. Opus: an Overlay Peer Utility Service. In *Proceedings of the 5th International Conference on Open Architectures and Network Programming (OPENARCH)*, June 2002.

[3] Jeffrey S. Chase, Darrell C. Anderson, Prachi N. Thakar, Amin M. Vahdat, and Ronald P. Doyle. Managing Energy and Server Resources in Hosting Centers. In *Proceedings of the 18th ACM Symposium on Operating System Principles (SOSP)*, October 2001.

[4] Frank Dabek, M. Frans Kaashoek, David Karger, Robert Morris, and Ion Stoica. Wide-area Cooperative Storage with CFS. In *Proceedings of the 18th ACM Symposium on Operating Systems Principles (SOSP'01)*, October 2001.

[5] Dejan Kostić, Adolfo Rodriguez, and Amin Vahdat. Scalability and Adaptivity in Two-Metric Overlays. Technical report, Duke University, May 2002. http://www.cs.duke.edu/~vahdat/ps/acdc-full.pdf.

[6] John Kubiatowicz, David Bindel, Yan Chen, Patrick Eaton, Dennis Geels, Ramakrishna Gummadi, Sean Rhea, Hakim Weatherspoon, Westly Weimer, Christopher Wells, and Ben Zhao. OceanStore: An Architecture for Global-scale Persistent Storage. In *Proceedings of ACM ASPLOS*, November 2000.

[7] Sylvia Ratnasamy, Paul Francis Mark Handley, Richard Karp, and Scott Shenker. A Content Addressable Network. In *Proceedings of SIGCOMM 2001*, August 2001.

[8] Antony Rowstron and Peter Druschel. Pastry: Scalable, Distributed Object Location and Routing for Large-scale Peer-to-Peer Systems. In *Middleware'2001*, November 2001.

[9] Antony Rowstron and Peter Druschel. Storage Management and Caching in PAST, a Large-Scale, Persistent Peer-to-Peer Storage Utility. In *Proceedings of the 18th ACM Symposium on Operating Systems Principles (SOSP'01)*, October 2001.

[10] H. Salama, Y. Viniotis, and D. Reeves. An Efficient Delay Constrained Minimum Spanning Tree Heuristic, 1996.

[11] Stefan Saroiu, P. Krishna Gummadi, and Steven D. Gribble. A Measurement Study of Peer-to-Peer File Sharing Systems. In *Proceedings of Multimedia Computing and Networking 2002 (MMCN'02)*, January 2002.

[12] Ion Stoica, Robert Morris, David Karger, Frans Kaashoek, and Hari Balakrishnan. Chord: A Scalable Peer to Peer Lookup Service for Internet Applications. In *Proceedings of the 2001 SIGCOMM*, August 2001.

[13] Marc Waldman, Aviel D. Rubin, and Lorrie Faith Cranor. Publius: A Robust, Tamper-evident, Censorship-resistant, Web Publishing System. In *Proc. 9th USENIX Security Symposium*, pages 59–72, August 2000.

[14] Zheng Wang and Jon Crowcroft. Quality-of-Service Routing for Supporting Multimedia Applications. *IEEE Journal of Selected Areas in Communications*, 14(7):1228–1234, 1996.

[15] Haifeng Yu and Amin Vahdat. Design and Evaluation of a Continuous Consistency Model for Replicated Services. In *Proceedings of Operating Systems Design and Implementation (OSDI)*, October 2000.

[16] Haifeng Yu and Amin Vahdat. The Costs and Limits of Availability for Replicated Services. In *Proceedings of the 18th ACM Symposium on Operating Systems Principles (SOSP)*, October 2001.

[17] Haifeng Yu and Amin Vahdat. Minimal Replication Cost for Availability. In *Proceedings of the ACM Principles of Distributed Computing*, July 2002.

Mapping the Gnutella Network: Macroscopic Properties of Large-Scale Peer-to-Peer Systems

Matei Ripeanu, Ian Foster

Computer Science Department, The University of Chicago
1100 E. 58th Street, Chicago IL, 60637, USA
{matei, foster}@cs.uchicago.edu

Abstract. Despite recent excitement generated by the peer-to-peer (P2P) paradigm and the surprisingly rapid deployment of some P2P applications, there are few quantitative evaluations of P2P systems behavior. The open architecture, achieved scale, and self-organizing structure of the Gnutella network make it an interesting P2P architecture to study. Like most other P2P applications, Gnutella builds, at the application level, a virtual network with its own routing mechanisms. The topology of this overlay network and the routing mechanisms used have a significant influence on application properties such as performance, reliability, and scalability. We describe techniques to discover and analyze the Gnutella's overlay network topology and evaluate generated network traffic. Our major findings are: (1) although Gnutella is not a pure power-law network, its current configuration has the benefits and drawbacks of a power-law structure, (2) we estimate the aggregated volume of generated traffic, and (3) the Gnutella virtual network topology does not match well the underlying Internet topology, hence leading to ineffective use of the physical networking infrastructure. We believe that our findings as well as our measurement and analysis techniques have broad applicability to P2P systems and provide useful insights into P2P system design tradeoffs.

1 Introduction

Unlike traditional distributed systems, P2P networks aim to aggregate large numbers of computers that join and leave the network frequently. In pure P2P systems, individual computers communicate directly with each other and share information and resources without using dedicated servers. A common characteristic of this new breed of systems is that they build, at the application level, a virtual network with its own routing mechanisms. The topology of this overlay network and the routing mechanisms used have a significant impact on application properties such as performance, reliability, scalability, and, in some cases, anonymity. The topology also determines the communication costs associated with running the P2P application, both at individual hosts and in the aggregate. Note that the decentralized nature of pure P2P systems means that these properties are emergent properties, determined by entirely local decisions made by individual resources, based only on local information: we are dealing with a self-organized network of independent entities.

P. Druschel, F. Kaashoek, and A. Rowstron (Eds.): IPTPS 2002, LNCS 2429, pp. 85-93, 2002.

These considerations motivate us to conduct a macroscopic study of a popular P2P system: Gnutella (described succinctly in Section 2). In this study, we benefit from Gnutella's large existing user base and open architecture, and, in effect, use the public Gnutella network as a large-scale, if uncontrolled, testbed.

Our measurements and analysis of the Gnutella network are driven by two primary questions (Section 4). The first concerns its connectivity structure. Recent research [1] shows that networks as diverse as natural networks formed by molecules in a cell, networks of people in a social group, or the Internet, organize themselves so that most nodes have few links while a tiny number of nodes, called hubs, have a large number of links. [2] finds that networks following this organizational pattern (power-law networks) display an unexpected degree of robustness: the ability of their nodes to communicate is unaffected even by extremely high failure rates. However, random error tolerance comes at a high price: these networks are vulnerable to attacks, i.e., to the selection and removal of a few nodes that provide most of the network's connectivity. We show that, although Gnutella is not a pure power-law network, it preserves good fault tolerance characteristics while being less dependent than a pure power-law network on highly connected nodes that are easy to single out (and attack).

The second question concerns how well (if at all) the Gnutella virtual network topology maps to the physical Internet infrastructure. There are two reasons for analyzing this issue. First, it is a question of crucial importance for Internet Service Providers (ISP): if the virtual topology does not follow the physical infrastructure, then the additional stress on the infrastructure and, consequently, the costs for ISPs, are immense. This point has been raised on various occasions but, as far as we know, we are the first to provide a quantitative evaluation on P2P application and Internet topology (mis)match. Second, the scalability of any P2P application is ultimately determined by its efficient use of underlying resources.

An orthogonal but important issue concerns the data gathering techniques (Section 3). For the analysis we present here we developed a "crawler" to gather complete topology information on the network. This technique however, is invasive and has limited scalability. (Moreover, recent minor changes in the protocol baffle our 'crawler'.) We are currently looking at ways to explore and characterize the network by adding limited number of cooperating 'probes' (modified nodes) that will monitor the traffic and may insert a small number of control messages.

2   Gnutella Protocol Description

The Gnutella protocol [3] is an open, decentralized group membership and search protocol, mainly used for file searching and sharing. The term Gnutella also designates the virtual network of Internet accessible hosts running Gnutella-speaking applications. Gnutella nodes, called *servents* by developers, perform tasks normally associated with both SERVers and cliENTS. They provide client-side interfaces through which users can issue queries and view search results, accept queries from other servents, check for matches against their local data set, and respond with corresponding results. These nodes are also responsible for managing the background traffic that spreads the information used to maintain network integrity.

In order to join the system a new node/servent initially connects to one of several known hosts that are almost always available (e.g., gnutellahosts.com). Once attached to the network (e.g., having one or more open connections with nodes already in the network), nodes send messages to interact with each other. Messages can be broadcasted (i.e., sent to all nodes with which the sender has open tcp connections) or simply back-propagated (i.e., sent on a specific connection on the reverse of the path taken by an initial, broadcasted, message). Several features of the protocol facilitate this broadcast/back-propagation mechanism. First, each message has a randomly generated identifier. Second, each node keeps a short memory of the recently routed messages, used to prevent re-broadcasting and to implement back-propagation. Third, messages are flagged with time-to-live (TTL) and "hops passed" fields.

The messages allowed in the network are:

- *Group Membership* (ping and pong) Messages. A node joining the network initiates a broadcasted ping message to announce its presence. When a node receives a ping message it forwards it to its neighbors and initiates a back-propagated pong message. The pong message also contains information about the node such as its IP address and the number and size of shared files.

- *Search* (query and query response) Messages. Query messages contain a user specified search string that each receiving node matches against locally stored file names. query messages are broadcasted. Query responses are back-propagated replies to query messages and include information necessary to download a file.

- *File Transfer* (get and push) Messages. File downloads are done directly between two peers using get/push messages.

To summarize: to become a member of the network, a servent (node) has to open one or many connections with nodes that are already in the network. In the dynamic environment where Gnutella operates, nodes often join and leave and network connections are unreliable. To cope with this environment, after joining the network, a node periodically pings its neighbors to discover other participating nodes. Using this information, a disconnected node can always reconnect to the network. Nodes decide where to connect in the network based only on local information, and thus forming a dynamic, self-organizing network of independent entities. This virtual, application level network has Gnutella servents at its nodes and open TCP connections as its links.

In this section we described the original Gnutella protocol (v0.4) as, at the time of our experiment, most nodes complied with this protocol version. We should mention however that a number of protocol changes have been adopted. The most significant result is a switch from the initial, flat, unstructured, peer network toward a two-level network organization: ordinary nodes link to SuperPeers that shield them from some of the traffic. SuperPeers in turn organize themselves into a flat, unstructured, network similar to the original one.

3 Data Collection

We have developed a *crawler* that joins the network as a servent and uses the membership protocol (the PING-PONG mechanism) to collect topology information. The crawler starts with a list of nodes, initiates a TCP connection to each node in the list, sends a generic join-in message (PING), and discovers the neighbors of the contacted node based on the PONG messages that it receives in reply. Newly discovered neighbors are added to the list. We started with a short, publicly available list of initial nodes, but over time we have incrementally built our own list with more than 400,000 nodes that have been active at one time or another.

In order to reduce the crawling time, we developed a client/server crawling strategy. The 'server' is responsible with managing the list of nodes to be contacted, assembling the final graph, and assigning work to clients. Given this dynamic behavior of the nodes, it is important to find the appropriate tradeoff between discovery time and invasiveness of our crawler. Increasing the number of parallel crawling tasks reduces discovery time but increases the burden on the application. Obviously, the Gnutella graph our crawler produces is not an exact 'snapshot' of the network. However, we argue that the result we obtain is close to a snapshot in a statistical sense: all properties of the network: size, diameter, average connectivity, and connectivity distribution are preserved.

Still, our crawling technique is invasive and has limited scalability. Moreover, recent minor modifications to the protocol changed the ping-pong mechanism the crawler is based on. These modifications, aimed at reducing the number of messages broadcasted in the network, lead to the widespread deployment of 'pong caches'. We are currently looking at separate ways to explore and characterize the Gnutella network by adding a limited number of cooperating 'probes' (modified nodes) that monitor the traffic and may insert a small number of control messages. While some network properties can only be analyzed using complete graph information, the data gathered by the 'probes' is sufficient to estimate a some interesting characteristics: message drop rates, network diameter or its tolerance to attacks (the number of redundant network paths).

While during late 2000 the largest connected network component we found had 2,063 hosts, this grew to 14,949 hosts in March 2001 and 48,195 hosts in May 2001. Recent measurements (www.limewire.com) show that the network in the range of 80 - 100,000 nodes.

4 Gnutella Network Analysis

We start with a macroscopic analysis of the network and study its connectivity patterns (Section 4.1). We then estimate Gnutella generated traffic volume (Section 4.2), and evaluate the mapping of Gnutella overlay network to the underlying networking infrastructure (Section 4.3).

4.1   Connectivity and Reliability in Gnutella Network. Power-Law Distributions

Recent research [1] shows that many natural networks such as molecules in a cell, species in an ecosystem, and people in a social group organize themselves as so called *power-law networks* (more specifically, in a power-law network the number of nodes with L links is proportional to, L^{-k} where k is a network dependent constant). This structure helps explain why these are generally stable and resilient structures, yet occasional catastrophic collapse does occur [2]. In a power-law network most nodes (molecules, Internet routers, Gnutella servents) have few links, thus a large fraction can be taken away and the network stays connected. But, if just a few highly connected nodes are eliminated, the whole network is broken into pieces. One implication is that these networks are extremely robust when facing random node failures, but vulnerable to well-planned attacks.

Given the diversity of networks that exhibit power-law structure and their properties we were interested to determine whether Gnutella falls into the same category. Figure 1 presents the connectivity distribution in Nov. 2000. Although data are noisy (due to the small size of the networks), we can easily recognize the signature of a power-law distribution: the connectivity distribution appears as a line on a log-log plot. Later measurements (Figure 2) however, show that more recent networks tend to move away from this organization: there are too few nodes with low connectivity to form a pure power-law network. In these networks the power-law distribution is preserved for nodes with more than 10 links while nodes with fewer link follow a quasi-constant distribution.

An interesting issue is the impact of this new, multi-modal distribution on network reliability. We believe that the more uniform connectivity distribution preserves the network's ability to deal with random node failures while reducing the network dependence on highly connected, easy to single out (and attack) nodes.

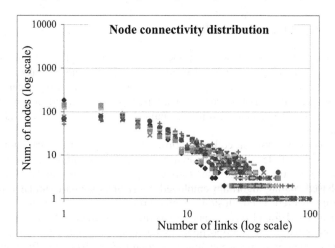

Fig. 1. Connectivity distribution during November 2000. Each series of points represents one Gnutella network topology we discovered at different times during that month. Note the log scale on both axes. Gnutella nodes organized themselves into a power-law network

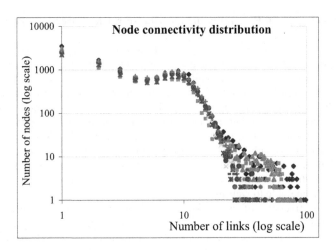

Fig. 2. Connectivity distribution during March 2001. Each series of points represents one Gnutella network topology discovered during March 2001. Note the log scale on both axes. Networks crawled during May/June 2001 show a similar pattern

We speculate that a group of devoted users maintain the small number of Gnutella nodes with the server-like characteristics visible in these power-law distributions. These nodes have a large number of open connections and/or provide much of the content available in the network. Moreover, these server-like nodes have a higher availability: they are about 50% more likely than the average node to be found alive during two successive crawls.

4.2 Estimates of Generated Traffic

We used a modified version of the crawler to eavesdrop the traffic generated by the network. We classified the traffic that goes on randomly chosen links according to message type: 92% QUERY messages, 8% PING messages and insignificant levels of other message types (June 2001). This represents a significant improvement compared to late 2000 when the traffic contained more than 50% overhead messages (PINGs and PONGs).

The topology information collected allows us to analyze the distribution of node-to-node shortest path lengths. Given that 95% of any two nodes are less than 7 hops away (Figure 3), the message time-to-live (TTL=7) preponderantly used, and the flooding-based routing algorithm employed, most links support similar traffic. We verified this theoretical conclusion by measuring the traffic at multiple, randomly chosen, nodes and found 6 Kbps per connection on average. As a result, the total Gnutella generated traffic is proportional to the number of connections in the network. Based on our measurements we estimate the total traffic (excluding file transfers) for a large Gnutella network as 1 Gbps: 170,000 connections (for a 50,000-nodes Gnutella network) times 6 Kbps per connection, or about 330 TB/month. To put this traffic volume into perspective we note that it amounts to about 1.7% of total traffic in

US Internet backbone in December 2000 (as reported in [4]). We infer that the volume of generated traffic is an important obstacle for further growth and that efficient use of underlying network infrastructure is crucial for better scaling and wider deployment.

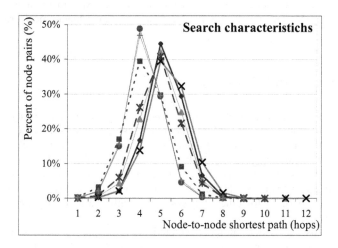

Fig. 3. Distribution of node-to-node shortest paths. Each line represents one network measurement. Note that, although the largest network diameter (the longest node-to-node path) is 12, more than 95% of node pairs are at most 7 hops away

4.3 Internet Infrastructure and Gnutella Network

P2P computing brings an important change in the way we use the Internet: it enables computers sitting at the edges of the network to act as both clients and servers. As a result, P2P applications change radically the amount of bandwidth consumed by the average Internet user. Most Internet Service Providers (ISPs) use flat rates to bill their clients. If P2P applications become ubiquitous, they could break the existing business models of many ISPs and force them to change their pricing scheme.

Given the considerable traffic volume generated by P2P applications, it is crucial from the perspective of both their scalability and their impact on the network that they employ available networking resources efficiently. Gnutella's store-and-forward architecture makes the overlay network topology extremely important: the larger the mismatch between the physical network infrastructure and the overlay's topology, the bigger the "stress" on the infrastructure.

Unfortunately, it is prohibitively expensive to compute exactly the mapping of the Gnutella onto the Internet topology, due both to the inherent difficulty of extracting Internet topology and to the computational scale of the problem. Instead, we proceed with a high-level experiment that highlights the topology mismatch: The Internet is a collection of Autonomous Systems (AS) connected by routers. ASs, in turn, are collections of local area networks under a single technical administration. From an

ISP point of view traffic crossing AS borders is more expensive than local traffic. We found that only 2-5% of Gnutella connections link nodes located within the same AS, although more than 40% of these nodes are located within the top ten ASs (10% in the largest). This result indicates that most Gnutella-generated traffic crosses AS borders, thus increasing costs, unnecessarily. A second, similar experiment showed that the node organization does not follow the DNS domain name hierarchical organization either.

5 Summary and Future Work

Despite recent excitement generated by this paradigm and the surprisingly rapid deployment of some P2P applications, there are few quantitative evaluations of P2P systems behavior. The open architecture, achieved scale, and self-organizing structure of the Gnutella network make it an interesting P2P architecture to study. The social circumstances that have fostered the success of the Gnutella network might change and the network might vanish. However, our measurement and analysis techniques can be used for other P2P systems to enhance general understanding of design tradeoffs.

Our analysis shows that, although Gnutella is not a pure power-law network, it preserves good fault tolerance characteristics while being less dependent than a pure power-law network on highly connected nodes that are easy to single out (and attack).

We have estimated that, as of June 2001, the network generates about 330 TB/month simply to remain connected and to broadcast user queries. This traffic volume represents a significant fraction of the total Internet traffic and makes the future growth of Gnutella network particularly dependent on efficient network usage. We have also documented the topology mismatch between the self-organized, Gnutella network and the underlying physical networking infrastructure. We believe this mismatch has major implications for the scalability of this P2P network or for ISP business models. This problem must be solved if Gnutella or similarly built systems are to reach larger deployment.

We see two other directions for improvement. First, as argued in [6], efficient P2P designs should exploit particular distributions of query values and locality in user interests. Various Gnutella studies show that the distribution of Gnutella queries is similar to the distribution of HTTP requests in the Internet: they both follow Zipf's law (note that, although the Zipf's formulation is widely used, these distributions can also be expressed as power-law distributions).

Other projects [7] try to discover and exploit data sharing patterns emerging at user level for topology optimization and collaborative message filtering.

A second direction of improvement is the replacement of query flooding mechanism with smarter (less expensive in terms of communication costs) routing and/or group communication mechanisms. Several P2P schemes proposed recently fall into the former category: systems like CAN or Tapestry propose a structured application-level topology that allows semantic query routing. We believe, however, that a promising approach is to preserve and benefit from the power-low characteristics that, as shown in this paper, emerge in Gnutella's ad-hoc network topology (and match the underlying, non-uniform resource distribution [5]). A way to preserve the dynamic, adaptive character of the Gnutella network and still decrease resource (network bandwidth) consumption is to use dissemination schemes (e.g., based on epidemic protocols) mixed with random query forwarding.

We have collected a large amount of data on the environment in which Gnutella operates, and plan to use this data in simulation studies of protocol alternatives.

Acknowledgements

This work was supported by the U.S. National Science Foundation under contract ITR-0086044.

References

1. Albert, R., Barabasi, A. L., Statistical mechanics of complex networks, Review of Modern Physics, 74 (47) 2002.
2. Albert, R., Jeong, H., Barabasi, A. L., Attak and tolerance in complex networks, Nature 406(378), 2000.
3. The Gnutella protocol specification v4.0. http://dss.clip2.com/GnutellaProtocol04.pdf.
4. Coffman, K., Odlyzko, A., Internet growth: Is there a "Moore's Law" for data traffic? in Handbook of Massive Data Sets, Abello, J., & all editors., Kluwer Academic Publishers, 2001.
5. Saroiu, S., Gummadi, P., Gribble, S., A Measurement Study of P2P File Sharing Systems, University of Washington Technical Report UW-CSE-01-06-02, July 2001.
6. Sripanidkulchai, K., The popularity of Gnutella queries and its implications on scalability, February 2001.
7. Iamnitchi, A., Ripeanu, M., Foster, I., Locating Data in (Small-World?) P2P Scientific Collaborations, in Proceedings of 1st International Workshop on Peer-to-Peer Systems Cambridge, MA, March 2002.

Can Heterogeneity Make Gnutella Scalable?

Qin Lv[1], Sylvia Ratnasamy[2,3], and Scott Shenker[3]

[1] Princeton University, NJ, USA
[2] University of California at Berkeley, CA, USA
[3] International Computer Science Institute, Berkeley, CA, USA

1   Introduction

Even though recent research has identified many different uses for peer-to-peer (P2P) architectures, file sharing remains the dominant (by far) P2P application on the Internet. Despite various legal problems, the number of users participating in these file-sharing systems, and number of files transferred, continues to grow at a remarkable pace. File-sharing applications are thus becoming an increasingly important feature of the Internet landscape and, as such, the scalability of these P2P systems is of paramount concern. While the peer-to-peer nature of data storage and data transfer in these systems is inherently scalable, the scalability of file location and query resolution is much more problematic.

The earliest P2P file-sharing systems (*e.g.*, Napster, Scour) relied on a *centralized* directory to locate files. While this was sufficient for the early days of P2P, it is clearly not a scalable architecture. These centralized-directory systems were followed by a number of fully *decentralized* systems such as Gnutella and Kazaa. These systems form an overlay network in which each P2P node "connects" to several other nodes. These P2P systems are *unstructured* in that the overlay topology is *ad hoc* and the placement of data is completely unrelated to the overlay topology. Searching on such networks essentially amounts to random search, in which various nodes are probed and asked if they have any files matching the query; one can't do better on such unstructured systems because there is no information about which nodes are likely to have the relevant files. P2P systems differ in how they construct the overlay topology and how they distribute queries. Gnutella, for example, floods all queries and uses a TTL to restrict the scope of the flood. The advantage of such unstructured systems is that they can easily accommodate a highly transient node population. The disadvantage is that it is hard to find the desired files without distributing queries widely.

It seemed clear, at least in the academic research community, that such random search methods were inherently unscalable. As a result, a number of research groups have proposed designs for what we call "highly structured" P2P systems [9, 13, 10, 15]. In these structured systems the overlay topology is tightly controlled and files (or pointers to files) are placed at precisely specified locations.[1] These highly structured systems provide a mapping between the file identifier and location, so that queries can

---

[1] The Freenet system is what we would call a "loosely structured" system; file placement is affected by routing hints that are based on the object's identifiers; these locations are not precisely specified and so not all searches succeed.

P. Druschel, F. Kaashoek, and A. Rowstron (Eds.): IPTPS 2002, LNCS 2429, pp. 94–103, 2002.

be efficiently routed to the node with the desired file (or, again, the pointer to the desired file). These systems thus offer a very scalable solution for "exact-match" queries.[2]

However, structured designs are likely to be less resilient in the face of a very transient user population, precisely because it is hard to maintain the structure (such as neighbor lists, etc.) required for routing to function efficiently when nodes are joining and leaving at a high rate. Moreover, it has yet to be demonstrated that these structured systems, while well-suited for exact-match queries, can scalably implement a full range of partial query techniques, such as keyword searching, which are common on current file-sharing systems.[3]

Despite these issues, we firmly believe in the value of these structured P2P systems, and have actively participated in their design. However, the (sometimes unspoken) motivating assumption behind much of the work on structured P2P systems is that unstructured P2P systems such as Gnutella are inherently not scalable, and therefore should be abandoned. Our goal in this work is to revisit that motivating assumption and ask if Gnutella (and systems like it) could be made more scalable. We do so because, if scalability concerns were removed, these unstructured P2P systems might be the preferred choice for file-sharing and other applications with the following characteristics:

- keyword searching is the common operation,
- most content is typically replicated at a fair fraction of participating sites, and
- the node population is highly transient

The first condition is that one often doesn't know the exact file identifier, or is looking for a set of files all matching a given attribute (*e.g.*, all by the same artist). The second is that one isn't typically searching for extremely *rare* files that are only stored at a few nodes. This would apply to the music-sharing that dominates today's file-sharing systems. The third condition seems to apply to currently popular P2P systems, although it may not apply to smaller community-based systems.

File-sharing is one (widely deployed) example that fits these criteria, distributed search engines [4] might well be another. Unstructured P2P systems may be a suitable choice for these applications because of their overall simplicity and their ability to tolerate transient user populations and comfortably support keyword searches.[4] This all depends, of course, on whether or not such systems can be made scalable, and that is the question we address in this short paper.

Our approach to improving the scalability of these systems is based on recent work which shows the prevalence of heterogeneity in deployed P2P systems and on work improving the efficiency of search in unstructured networks. After reviewing this relevant

---

[2] By an exact-match query, we mean that the complete identifier of the requested data object is known.

[3] There are standard ways to use the exact-match facility of hash tables as a substrate for keyword searches, substring searches, and more fuzzy matches as well [14]. However, it isn't clear how scalable these techniques will be in *distributed* hash tables.

[4] We should add that unstructured P2P systems, whether scalable or not, will never be able match the exact-matching performance of highly structured P2P systems. We expect that these highly structured P2P systems will eventually find many uses; we just aren't convinced that popular-music file-sharing will be one of them.

background in Section 2, we describe, in Section 3, a simple flow control algorithm that takes advantage of heterogeneity and evaluate its performance in Section 4. We offer this algorithm not as a polished design but as a proof-of-concept that heterogeneity can be used to improve the scalability of unstructured P2P systems. Similarly, one should take this position paper not as a "proof" that Gnutella can be made scalable, but as a conjecture in search of more evidence.

2   Background

Our design is based on two ongoing research thrusts: attempts to improve the search facilities in Gnutella [8, 3, 1], and a measurement study [11] that revealed significant heterogeneity in P2P systems.

Improvements to Gnutella: Gnutella uses TTL-limited flooding to distribute its queries. Lv *et al.* [8] investigate several alternative query distribution methods on a range of overlay topologies: random graphs, power-law random graphs,[5] grids and a measured Gnutella topology. Based on their results, they propose the use of multiple parallel random walks instead of flooding. To search for an object using a random walk, a node chooses a neighbor at random and send the query to it. Each neighbor in turn repeats this process until the object is found. Random walks avoid the problem of the exponentially increasing number of query messages that arise in flooding.

Cohen *et al.* [3] study proactive replication algorithms.[6] They find that the optimal replication scheme for random search is to replicate objects proportional to the square-root of their query rate; this replication scheme can be implemented in a distributed fashion by replicating objects a number of times proportional to the length of the search.

These combined approaches, proactive replication plus parallel random walking, was tested via simulation in [8] and was found to significantly improve – by two orders of magnitude in some cases – the performance of flooding and passive replication, as measured by the query resolution time (in terms of number of overlay hops), the per-node query load and the message traffic generated by individual queries. In the case of power-law random graphs it was observed that high degree nodes experienced correspondingly high query loads. For this reason, the authors conclude that P2P network construction algorithms should avoid generating very high degree nodes.

Adamic *et al.*[1] approach the problem from a very different viewpoint. They take power-law random graphs as a given and instead ask how to best search on them. They, too, suggest the use of random walks, but that these random walks should be biased to seek out high-degree nodes. They show how this leads to superior scaling behavior in the limit of large systems. However, their work does not take into account the query load on individual nodes; as shown by [8] the high-degree nodes carry an extremely large share of the query traffic and are likely to be overloaded.

---

[5] Power-law random graphs are graphs where the degree distribution is a power-law and the nodes are connected randomly consistent with the degree distribution. A node's "degree" is its number of neighbors.

[6] With proactive replication, an object may be replicated at a node even though that node has not requested the object. Passive replication, where nodes hold copies only if they've requested the object, is the standard approach in Gnutella-like P2P systems.

Heterogeneity: Saroiu *et al.*, in [11], report on the results of a measurement study of Napster and Gnutella. Their study reveals (amongst other things) a significant amount of heterogeneity in bandwidth across peers. The authors of [11] argue that architectures for P2P systems should take into account the characteristics of participating hosts, including their heterogeneity. However, since the measurement study was the main focus of [11], they did not propose any design strategies.

These two lines of work, the observed heterogeneity and the improvements to Gnutella-like P2P systems, lead us to the following observation. Using random-walk searches on power-law random graphs is an efficient search technique, but results in a very skewed usage pattern with some nodes carrying too large a share of the load. Measurement studies reveal, however, that node capabilities (particularly in terms of bandwidth, but presumably in terms of other resources such as CPU, memory and disk) are also skewed. If we could align these two skews – that is, have the highest capacity nodes carry the heaviest burdens – we might end up with a scalable P2P design.

This is what we attempt to do in the next section.

3 Design

Our design for an unstructured P2P system uses a distributed flow control and topology construction algorithm that (1) restricts the flow of queries into each node so they don't become overloaded and (2) dynamically evolves the overlay topology so that queries flow towards the nodes that have sufficient capacity to handle them.[7] Over this capacity-driven topology we use a biased random walk quite similar to that employed in [1]. The combination of these algorithms results in a system where queries are directed to high capacity nodes, and thus are more likely to find the desired files. Our design results in a quasi-hierarchical organization of nodes with high-capacity nodes at the higher levels in the hierarchy. This "hierarchy" is not, however, explicit; instead it is achieved through a distributed, adaptive and lightweight self-organization algorithm.

Flow control and Topology Adaption: In what follows, we represent the capacity of node i by c_i. For the purpose of this paper, we assume that c_i denotes the maximum number of messages node i is willing/able to process over a given time interval T. A node i is connected, at the application-level, to a set of *neighbor* nodes, denoted $nbr(i)$. For each of $j \in nbr(i)$ a node i maintains the following information:

- $in[j, i]$: the number of incoming messages from node j to i in the last time interval T. Every node i reports its total incoming rate $(in[*, i] = \sum_{j \in nbr(i)} in[j, i])$ to all its neighbors.
- $out[i, j]$: the number of outgoing messages from node i to j in the last time interval T

---

[7] We still consider this an *unstructured* design, even though the topology has become more structured, because the topology adjustment, while it improves scalability, is not required for correctness of operation.

- $outMax[i, j]$: the maximum number of messages node i can send to node j per time interval T. At all times, $out[i, j] <= outMax[i, j]$ and $outMax[i, j] <= out[i, j] + (c_j - in[*, j])$.[8]

When a new node, say i, joins the network, it is connected to a random set of neighbor nodes. For each neighbor node j, i initializes:

- $outMax[i, j] = c_i / d_i$
- $in[j, i] = 0$ and
- $out[i, j] = 0$

Every node i counts the number of messages it receives from and transmits to each neighbor node j ($in[j, i]$ and $out[i, j]$ respectively). Periodically, node i checks whether it is overloaded; *i.e.* whether its total incoming query rate exceeds its capacity. If overloaded, node i attempts to adapt the topology so as to reduce its incoming query load – node i selects a neighbor (say p) with a high incoming query rate (*i.e.* high $in[p, i]$) and redirects it to another neighbor (say q) with high spare capacity. Intuitively, the above topology adaptation rule sets up a direct "link" between a node with high query rate (node p) and one with high capacity (node q) and gets the overloaded node (node i) out of the way. If the overloaded node i cannot find an appropriate neighbor q (as would be the case when all of i's neighbors are running close to capacity), it requests node p to throttle the number of queries it is forwarding to node i (*i.e.* reduce outMax[p,i]). The detailed pseudocode for the above high-level description is shown in Figure 1.

Note that all flow control/topology adjustment decisions made by a node are based on local information (i.e. about the node itself and its neighbors). Further, if a node's capacity was to undergo a sudden change or if the node leaves the system, the topology would automatically re-adapt to accommodate the change.

Our simulations in Section 4 use the algorithm from Figure 1 with parameter values of $\delta = 10$, $\gamma = 1.25$, and $T = 100s$.[9] We defer an exploration of the parameter space to later work. While our initial simulation results appear promising, we expect to incorporate extensions and modifications to the above rules as we continue to study the behavior of our algorithm. For example, a useful measure to factor into the topology adjustment process would be the amount of time a node is expected to remain in the system. We also plan to add a *proactive* component to the above algorithm whereby a node whose capacity is not being utilized can explicitly discover and connect to nodes that are experiencing high query loads.

Search: We use a random-walk based search that combines features from the algorithms in [8, 1]. As in [8] we use TTLs to terminate walks and "state-keeping" to accelerate walks.[10] Similar to [1], a node forwards a query to its neighbor with the

---

[8] $c_j - in[*, j]$ represents node j's current spare capacity. With the above bounds on $out[i, j]$ and $outMax[i, j]$, a node i can at most increase its current outgoing rate to j by an amount equal to node j's current spare capacity.

[9] In practice, δ and γ need not be fixed values. A node can quite easily communicate more precise increase and decrease parameters to its neighboring nodes.

[10] With state-keeping, a search is given a unique identifier. A node then remembers the neighbors to which it has already forwarded queries for a given identifier, and if a query with the same

_checkLoad(node i)
// periodic flow control and topology adaptation at i
If $(\sum_{j \in nbr(i)} in[j, i] > c_i)$ *// i overloaded?*
pick $p \in nbr(i)$ with max $in[j, i]$ over all $j \in nbr(i)$
search for node $q(\neq p) \in nbr(i)$ such that:
$(outMax[i, q] - out[i, q] \geq in[p, i])$ && $(p \notin nbr(q))$

If (q exists) *// adapt topology*
redirect p to q; *// i.e. p disconnects from*
                   // i and connects to q
$outMax[i, q] = outMax[i, q] - in[p, i]$;
$outMax[p, q] = in[p, i]$;
$outMax[q, p] = outMax[i, p]$;

else *// no satisfactory q exists*
node i send p a *slowdown* message
forall $j \in nbr(i)$: $in[j, i] = 0$, $out[i, j] = 0$
forall $j \in nbr(i)$: $outMax[i, j] = outMax[i, j] + \delta$

_recvSlowDown(node j, node i)
// node j receives a slowdown message from i
$outMax[j, i] = outMax[j, i]/\gamma$

Fig. 1. *Pseudo-code for flow control and topology adaptation. Although not explicitly shown in the pseudocode, the upper bound on outMax is enforced in performing the increase operations on outMax.*

maximum value of $outMax$. This causes searches to quickly gravitate towards high capacity nodes. As in Adamic's work, the high degree nodes in our topology do face higher query load. However, unlike Adamic's work, high degree nodes in our system are also high capacity nodes and hence better equipped to handle this load.

In summary, a node i that receives a query, forwards it to its neighbor j with the highest value of $outMax[i, j]$ over all $j \in nbr(i)$ provided (1) $out[i, j] < outMax[i, j]$ and (2) i has not previously forwarded the same query to j.

While we do not explore proactive replication strategies in this paper, any of the solutions described in [8] could be used to further improve the performance of our search algorithm. In its stead, we assume that the number of copies of each object is proportional to the rate at which it is being queried (which would occur under passive replication). Moreover, we assume that these copies are randomly placed proportional to the capacity of nodes.

---

identifier arrives back at the node, it is forwarded to a different neighbor. Using state-keeping thus reduces the likelihood that a random walk will traverse the same path twice. State can be discarded after a short time by a node, so it does not become burdensome.

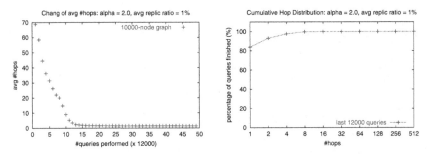

Fig. 2. *Average query resolution time and the distribution of query resolution times for a 10,000 node simulation using a Zipf-like capacity distribution*

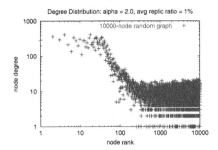

Fig. 3. *Degree distribution for a 10,000 node simulation using a Zipf-like capacity distribution*

4   Evaluation

In this section, we present initial simulation results on the performance of our design outlined in Section 3. We stress that these results are not intended to provide a comprehensive picture of the behavior of our algorithm but merely to serve as a first-order proof of concept.

Simulation Methodology: We use, as our main evaluation metric, the average query resolution time (as measured in terms of the number of application-level hops required to find a particular object). We measure this in simulations in which the object popularity, the rate at which queries are issued for it, follows a Zipf-like distribution where the popularity of the i'th most popular object is proportional to $i^{-\alpha}$. In these simulations we used $\alpha = 1.2$ (based on the Gnutella measurements in [12]). Individual nodes generate queries using a Poisson process with an average query rate of 1.2 queries/minute. Recall, as described above, that we use a proportional replication strategy so that objects are replicated proportional to their query rate. The average degree of replication is 1%[11] – that is, on average objects are replicated on 1% of the nodes, but the more

---

[11] We also experimented with a replication degree of 0.1% and obtained similar results

popular objects are replicated more than the less popular ones. Recall also that objects are assigned randomly to nodes, proportional to their capacity c_i.

To investigate the effect of heterogeneous node capabilities, we use two different capacity distributions. First, we use a Zipf-like distribution where c_i is proportional to $i^{-\beta}$ where $\beta = 2$. Second, we used a distribution based (loosely) on the measured bandwidth distributions of [11]. Here, we assume 5 capacity levels separated by an order of magnitude. The node population is then divided amongst these levels as shown in Table 1. The distribution reflects the observation that a fair fraction of Gnutella clients have dial-up connections to the Internet, the majority are connected via cable-modem or DSL and a small number of participants, via high speed lines.

Table 1. *Gnutella-like node capacity distributions*

Capacity level	Percentage of nodes
x	20%
10x	45%
100x	30%
1000x	4.9%
10000x	0.1%

For each simulation, we start with nodes connected in a uniform random graph topology; for the Zipf-like capacity distribution we use a 10,000 node graph with average degree 9.9, for the Gnutella-like distribution capacity distribution we use a a 5,000 node graph with average degree 7.5. We then assign the capacities and objects to the nodes as described above.

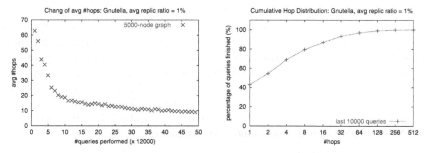

Fig. 4. *Average query resolution time and the distribution of query resolution times for a 5,000 node simulation using a Gnutella-like capacity distribution*

Results: Figures 3 and 5 plot the results of simulations using, respectively, the Zipf and Gnutella-like capacity distribution. The first plots show how, as the topology adjusts,

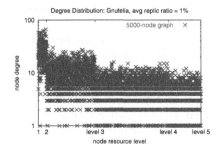

Fig. 5. *Degree distribution for a 5,000 node simulation using a Gnutella-like capacity distribution*

the average query response time rapidly decreases from 70 to ~ 2 in the Zipf case and 9 in the Gnutella case. The middle plots show the distribution of the number of hops for individual queries towards the end of the simulation run. In both cases all queries were successfully resolved. The last plots show the degree distributions of the resulting topology at the end of the simulation run. As one can see, the degree distributions evolve to match the capacity distributions – higher capacity nodes have significantly higher degrees – causing the improved performance.

5   Other Related Work

The Kazaa [5] network appears similar to our work in many respects. Kazaa is a self-organizing, multi-layered network where powerful hosts act as search hubs (called SuperNodes). Details of the Kazaa protocols and code are not publicly available making it hard to draw comparisons to our work.

In [7] the authors propose a cluster-based architecture for P2P systems (CAP). CAP organizes nodes into a two-level hierarchy using a centralized clustering server [6]. Each cluster has a delegate node that acts as directory server for objects stored by nodes within the cluster. Delegate nodes perform intra-cluster node membership registration while a central server tracks existing clusters and their delegates. To some extent, our algorithms achieve the same effect as CAP, with high capacity nodes behaving like delegate nodes. There are however significant differences: unlike CAP our algorithms are completely decentralized and we do not build an *explicit* hierarchy.

File sharing in FreeNet [2] uses hints about the placement of files to improve search scalability. FreeNet does not address node heterogeneity.

6   Discussion

We started this paper with the basic question of whether one could make unstructured P2P systems scalable. Building on the work in [8, 3, 1, 11] we proposed a design that appears to achieve significant scalability. This design is extremely preliminary, and and our evaluation leaves many questions unanswered. We offer it however, merely as support for the conjecture that unstructured P2P systems can be significantly improved, perhaps to the point where their scalability is no longer a barrier.

Our design also raised the more philosophical question of how to deal with heterogeneity. Most of the highly structured P2P designs start with the assumption of a homogeneous node population and then alter their designs to *accommodate* heterogeneity. The design we present here actively *exploits* heterogeneity. One can ask whether the highly structured designs could also be modified to exploit heterogeneity.

References

[1] ADAMIC, L., HUBERMAN, B., LUKOSE, R., AND PUNIYANI, A. Search in power law networks. *Physical Review. E 64* (2001), 46135–46143.

[2] CLARKE, I., SANDBERG, O., WILEY, B., AND HONG, T. Freenet: A Distributed Anonymous Information Storage and Retrieval System. ICSI Workshop on Design Issues in Anonymity and Unobservability, July 2000.

[3] COHEN, E., AND SHENKER, S. Optimal replication in random search networks. preprint, 2001.

[4] INFRASEARCH. http://www.infrasearch.com.

[5] KAZAA. http://www.kazaa.com.

[6] KRISHNAMURTHY, B., AND WANG, J. On network-aware clustering of web clients. In *Proceedings of SIGCOMM '00* (Stockholm, Sweden, Aug. 2000).

[7] KRISHNAMURTHY, B., WANG, J., AND XIE, Y. Early measurements of a cluster-based architecture for p2p systems. In *ACM SIGCOMM Internet Measurement Workshop* (San Francisco, Nov. 2001).

[8] LV, Q., CAO, P., COHEN, E., LI, K., AND SHENKER, S. Search and replication in unstructured peer-to-peer networks. preprint, 2001.

[9] RATNASAMY, S., FRANCIS, P., HANDLEY, M., KARP, R., AND SHENKER, S. A Scalable Content-Addressable Network. In *Proceedings of SIGCOMM 2001* (Aug. 2001).

[10] ROWSTRON, A., AND DRUSCHEL, P. Storage management and caching in PAST, a large-scale, persistent peer-to-peer storage utility. In *Proceedings of the Eighteenth SOSP* (2001), ACM.

[11] SAROIU, S., GUMMADI, K., AND GRIBBLE, S. A measurement study of peer-to-peer file sharing systems. In *Proceedings of Multimedia Conferencing and Networking* (San Jose, Jan. 2002).

[12] SRIPANIDKULCHAI, K. The popularity of gnutella queries and its implications on scalability. In *O'Reilly's www.openp2p.com* (Feb. 2001).

[13] STOICA, I., MORRIS, R., KARGER, D., KAASHOEK, M. F., AND BALAKRISHNAN, H. Chord: A scalable peer-to-peer lookup service for internet applications. In *Proceedings of SIGCOMM 2001* (Aug. 2001).

[14] WITTEN, I. H., MOFFAT, A., AND BELL, T. C. *Managing Gigabytes: Compressing and Indexing Documents and Images*, second ed. Morgan Kaufmann, 1999.

[15] ZHAO, B., KUBIATOWICZ, J., AND JOSEPH, A. Tapestry: An infrastructure for fault-tolerant wide-area location and routing. UCB Technical Report, 2001.

Experiences Deploying a Large-Scale Emergent Network

Bryce Wilcox-O'Hearn

Zooko.Com Software Engineering
zooko@zooko.com

Abstract. "Mojo Nation" was a network for robust, decentralized file storage and transfer. It was first released to the public in July, 2000, and remained in continuous operation until February, 2002. Over 100,000 people downloaded and used the Mojo Nation software. We observe some surprising and problematic behavior of the users as a group. We describe several specific problems in the design of Mojo Nation, some of which appear to be soluble with simple practical improvements, and others of which are not yet addressed in the literature, suggesting opportunities for further research.

Introduction

Mojo Nation[1] was not a file-sharing system (like Gnutella or Napster), but an "emergent file store", in which the storage, transfer and naming of files was performed in a decentralized manner, independent of any individual node. It had much in common with systems like CFS[2], PAST[3] and OceanStore[4], both in goals and in design. Mojo Nation was designed from the start with ambitious goals of attack-resistance and scalability.

The first version of Mojo Nation was released to the public in July of 2000. It had many advanced features, but deployment to large numbers of end users inevitably revealed its architectural deficiencies. During its lifespan, its developers deployed literally hundreds of changes to the protocol in response to observed behavior and in order to take advantage of newly discovered techniques. For example, in August of 2001, shortly after reading a pre-print of [5], we deployed a new version that used consistent hashing to locate a block in a set of servers.

We typically observed more than 10,000 downloads of the Mojo Nation software per month, as shown by statistics published by SourceForge.net[9]. (Note that before August of 2001 downloads were not hosted by SourceForge, although some of the web pages were. The higher number of page views in October of 2000 were a result of Mojo Nation being featured on slashdot.org, the consequences of which will be described below.)

Unfortunately, the vast majority of these users who tried Mojo Nation were not satisfied by the service it offered, as indicated by the fact that they permanently stopped using the network after trying it only briefly. At its largest, Mojo Nation never exceeded 10,000 simultaneously connected nodes, and during the

P. Druschel, F. Kaashoek, and A. Rowstron (Eds.): IPTPS 2002, LNCS 2429, pp. 104–110, 2002.

majority of its 19-month lifespan it had between 100 and 600 persistent nodes. This paper is motivated by the desire to learn from this failure.

There are many important aspects of the Mojo Nation product which we must omit from consideration in this paper. These include Mojo Nation's user interface, marketing, distributed search engine service, lightweight resource accounting scheme, agnostically blindable digital tokens and more. In this paper we will focus on the basics: the individual nodes, connecting them into a network, and building a decentralized file store on that network.

Observed Behavior

Frequent Join / Leave

The most surprising and problematic behavior that users of Mojo Nation displayed was frequent joining and leaving. We observed that the most common behavior was to join the network, stay connected for less than an hour, then leave the network and never return. Measurements taken from two particular 1-month periods (October, 2000 and February, 2001) indicated that between 80% and 84% of the users fell into this group of "one-time, less than one hour" group, and that of the remaining 16% to 20%, a significant fraction stayed connected for less than 24 hours then permanently disconnected.

Even among the remaining persistent nodes (those that recurringly connected to the network over a period of weeks), the typical node remained connected for only a short consecutive time, and only a few times per week. One measurement taken in April 2001 showed that the average node was connected 0.28 of the time, and other, less systematic observations suggest that the distribution was highly skewed, with approximately 1/6 of the nodes connected almost all the time, and the rest connected approximately 3 hours per day.

The overall picture suggested by these observations is that the "network half-life", or the time for replacement of half of the nodes in the network by new arrivals, was usually less than 1 hour, and at times it was much less than 1 hour.

Varying Space Allocation

The default disk allocation per node in Mojo Nation was originally 100 MB. In April of 2001 we raised the default to 500 MB. Users can manually adjust that setting. The Mojo Nation software did not report to us what settings the user chooses, but we do know from support mail and user feedback that no users have complained about the default setting, and that many users are quick to point out that they have raised their limit to a high setting, usually in the range of 10 GB to 60 GB.

Varying Connection Quality

Market research reports (e.g. [7]) typically suggest that around 13% of Internet users have broadband connections, and the rest use relatively slow and intermittent dial-up connections. Anecdotal evidence from Mojo Nation was consistent with this. However, there was an active minority of users with very high quality connections (including academic and corporate networks). These users also tended to be in the minority that stay consistently connected and in the minority that allocate large amounts of disk space.

Routability

Measurements taken at various times over the life of the Mojo Nation network always returned the same answer: 1/3 of Mojo Nation nodes were not directly reachable from the Internet, as observed by the fact that they did not have routeable IP addresses. In addition, some unknown number of users may have had routeable IP addresses, but may have been behind firewalls that did not allow incoming TCP connections.

Which Parts Worked?

Mojo Nation was a complex system and it is difficult to ascribe its successes to individual components. It could be described in general as a file storage and transfer network in which there is a mechanism for global coordination without communication (e.g. consistent hashing to locate nodes and data blocks in a ring), and in which individual nodes use local information to decide how to store, transmit, replicate and cache data. When Mojo Nation worked, it was a demonstration that such a network can be deployed and operated in an environment made up of unmanaged volunteers.

When Mojo Nation failed, its failures can more easily be ascribed to particular design elements.

Which Parts Failed? (Open Problems)

Original Introduction

The only failures which rendered the network completely unusable for all new users (not counting occurences of the authors releasing a new version with fatal bugs), are failures of original introduction. "Original introduction" is the problem of how a node connects to the network for the first time, when it does not yet have any connections to any other nodes in the network. The first version of Mojo Nation used single central introducer. Each new node would contact that introducer and receive in response a list of other nodes.

The Great Slashdotting of October 2000 was a dramatic demonstration of the inherent weakness in this design. In October 2000 an entry was posted on the popular web site slashdot.org headlined: "Forget Napster & Gnutella: Enter Mojo Nation"[8]. The next day our web server reported that downloads of the software had rocketed from 300 copies per day to almost 10,000 copies per day. The central introducer was totally overloaded and was not returning any responses to any users. We struggled for days to make the server operate, but it wasn't until the flash crowd had died down and we took the time to implement a new system of introduction (involving multiple redundant but still centrally managed introducers), that the network became usable again.

The issue of original introduction is largely ignored by the extant literature. There are several solutions to the problem in use on currently deployed networks including redundant centrally-administered introducers (FastTrack, Mojo Nation), bundling a list of original contacts with the download of the software (Limewire, Freenet), asking users to manually configure the original connection (Freenet), and combinations of more than one of these techniques (Limewire).

The scalability, security and attack-resistance trade-offs implicit in these design decisions have not been publically analyzed as far as we know.

Data Availability

Even when the network as a whole was working, a very common failure was that the data that a user sought was unavailable. We ascribe the source of that problem to our design's failure to accomodate the highly unreliable behavior of the nodes. Simultaneously, we believe that the primary reason for frequent join/leave behavior was that the data users sought was unavailable. This constitutes a "chicken-and-egg" problem, which was exaccerbated instead of solved by other elements of our design.

We repeatedly tuned our replication and information dispersal design in order to counteract this problem, but even at the best, data availability was variable, and appeared to depend upon which server nodes were connected at the time an observation was made.

There were two significant mistakes that Mojo Nation made which can be easily avoided.

The first was that it did not discriminate against newly joined nodes. As described in [6], the length of time that a node has been continuously connected to the network is a good predictor of the length of time that it will remain connected into the future. A simple heuristic to favor long-lived nodes, such as that proposed in [6], would have reduced the problems caused by frequent join/leave behavior.

The second was that Mojo Nation used an erasure code to split the data into a set of shares such that any sufficiently large subset of the shares would suffice to rebuild the data. The number of shares required to rebuild the data was equal to $1/2$ of the total number of shares generated. This was intended to increase data robustness, but if the availability of the underlying shares is less than the "required to total" ratio (in this case, less than $1/2$), then such an erasure code has the opposite of the intended effect, dramatically reducing the robustness of the data. We believe that the other problems of data availability were thus compounded by the addition of an erasure code with a "required to total" ratio that was too high for the actual behavior of the nodes.

As noted in the "Future Research" section of [2], the issue of how to manage block storage in the face of servers joining and leaving remains mostly open. More sophisticated caching and replication strategies will hopefully ameliorate this problem. In addition "reputation" or "trust metric" techniques such as described the section on "Attack Resistance" below might help by discriminating against unreliable servers. Mojo Nation deployed software which attempted to do exactly that, but the interaction between this discrimination and other design goals is not well analyzed.

Other Open Problems

Bypassing Firewalls and NAT

The challenge of enabling nodes that live behind firewalls or NAT to act as servers is a challenge that most deployed systems do not yet attempt to address. It is also likely to become more rather than less important in the future as the size of the Internet grows and as application-level connections cross more administrative boundaries.

Mojo Nation used a "relay" technique in which a third node helps two firewalled nodes to communicate with one another, similar in principle to [10].

Attack Resistance / Malicious Nodes / Mutual Distrust / Motivation to Cooperate

Perhaps the most challenging unsolved problem is that of mutual distrust. While a network architect is tempted to assume that all nodes in the system behave as he designed them to behave, this assumption may prove fatal once a network is deployed into multiple disjoint administrative zones.

A fundamentally related issue is that of "motivation to cooperate". Why does a node choose to offer services to the network as well as to make requests of the network? Is there anything preventing a user from altering their copy of the software, or writing their own compatible implementation, which uses the resources of the other nodes but refuses to provide its own resource to them?

Also closely related is the notion of "attack resistance". If a node can use the resources of other nodes without offering them service in return, then it is able to act as a drain on the resources of the network as a whole, possibly constituting a denial-of-service attack on the entire network.

On the other hand, if a node can be coerced into cooperating, perhaps by cutting that node off from the services of the network in retaliation for its lack of cooperation, how can we be sure that the same mechanism cannot be used to attack specific (innocent) nodes, or even to attack the network itself?

Hopefully the research pursued in papers like [11] and [12] will lead to a quantitatively justified method of gaining attack resistance without sacrificing other design goals.

Mojo Nation's experience shows that there are two kinds of attack that are likely to be encountered by any network that is deployed in a large scale on the Internet.

The first attack is when a user alters his client in the attempt to gain more advantage for himself. Several different users made such modifications to their

Mojo Nation software and then helpfully contacted us to describe what they did. Other users have made modifications, but we are aware of those changes only indirectly through observations of anomalous behavior.

The second kind of attack is when an enemy attempts to remove central components of the network through legal means. Legal action was recently initiated[13] against the Fast Track network even though the only centrally administered components are the original introducer service and the design, implementation and distribution of the software.

Conclusion

As an emergent file store, Mojo Nation was partially a success and partially a failure. The parts that failed were a centralized original introduction mechanism and a data storage scheme that proved too fragile when deployed on a network with a surprisingly short half-life.

These two problems can be straightforwardly solved in practice, and they also present possible directions for the advancement of theory.

In addition, we believe that any long term, large scale emergent network will need to address the "other open problems" of attack resistance, malicious nodes, mutual distrust, and motivation to cooperate.

References

1. Web Site: Mojo Nation.
   http://mojonation.net/
2. Dabek, F., Kaashoek, M. F., Karger, D., Morris, R., Stoica, I.: Wide-area cooperative storage with CFS. Proceedings of the 18th ACM Symposium on Operating Systems Principles (SOSP '01) (To appear; Banff, Canada, Oct. 2001).
   http://citeseer.com/dabek01widearea.html
3. Druschel, P., Rowstron, A.: PAST: A large-scale, persistent peer-to-peer storage utility.
   http://citeseer.com/439820.html
4. Kubiatowicz, J., et al.: OceanStore: An Architecture for Global-Scale Persistent Storage. ASPLOS, December 2000.
   http://citeseer.com/kubiatowicz00oceanstore.html
5. Stoica, I., Morris, R., Karger, D., Kaashoek, M. F., Balakrishnan, H. Chord: A scalable peer-to-peer lookup service for Internet applications. Technical Report TR-819, MIT, March 2001.
   http://citeseer.com/stoica01chord.html
6. Maymounkov, P., Mazières, D. Kademlia: A Peer-to-peer Information System Based on the XOR Metric. Proceedings of the 1st International Workshop on Peer-to-Peer Systems (IPTPS02).
   http://scs.cs.nyu.edu/~{}{}dm/kpos.pdf
7. Web Site: Press Release: "U.S. Residential Internet Market Grows in Second Quarter".
   http://www.isp-planet.com/research/2001/us_q2.html

8. Web Site: Slashdot headline: "Forget Napster & Gnutella: Enter Mojo Nation".
   http://slashdot.org/article.pl?sid=00/10/09/1826243
9. Web Site: SourceForge Usage Statistics: Mojo Nation.
   http://sf.net/project/stats/index.php?report=months&group_id=8340
10. Ng, T. S. E., Stoica, I., Zhang, H.: A waypoint service approach to connect het-
    erogeneous internet address spaces. Proceedings of the Usenix Technical Conference
    (June 2001), pp. 319–332.
    http://citeseer.com/ng01waypoint.html
11. Levien, R., Aiken, A. Attack-resistant trust metrics for public key certification. 7th
    USENIX Security Symposium, January 1998.
    http://citeseer.com/levien98attackresistant.html
12. Dingledine, R., Freedman, M.J., Molnar D. The Free Haven project: Distributed
    Anonymous Storage Service. Workshop on Design Issues in Anonymity and Unob-
    servability, July 2000 (LNCS 2009).
    http://www.freehaven.net/doc/berk/freehaven-berk.ps
    http://citeseer.com/dingledine00free.html
13. News Article: "Suit hits popular post-Napster network", CNet News.Com.
    http://news.cnet.com/news/0-1005-200-7389552.html

Anonymizing Censorship Resistant Systems

Andrei Serjantov

University of Cambridge Computer Laboratory,
William Gates Building,
JJ Thomson Avenue,
Cambridge CB3 0FD, UK
Andrei.Serjantov@cl.cam.ac.uk

Abstract. In this paper we propose a new Peer-to-Peer architecture for a censorship resistant system with user, server and active-server document anonymity as well as efficient document retrieval. The retrieval service is layered on top of an existing Peer-to-Peer infrastructure, which should facilitate its implementation. The key idea is to separate the role of document storers from the machines visible to the users, which makes each individual part of the system less prone to attacks, and therefore to censorship.

1 Introduction

Many censorship resistant systems have been proposed recently, yet most still lack one crucial feature – protection of the servers hosting the content.

In the past this was not considered an issue. For instance, in Anderson's eternity service [And96], it was deemed sufficient to guarantee that a document was always available through the system. However, many examples indicate that servers hosting content are vulnerable to censorship, due to "Rubber Hose Cryptanalysis" – various kinds of pressure applied by attackers to shut down servers or remove files. Examples of documents subjected to censorship include DeCSS [DeC], the paper detailing the attack on SDMI [CMW+01], and documents (or quotes from documents) which the Church of Scientology described as their secrets. In cases like this, the server administrators receive "cease and desist" letters when the censor finds the offending document on their server.

Most modern censorship resistant systems, for example Publius [WRC00] and Freenet [CSWH01], have not addressed this problem in a satisfactory way. It is possible for a malicious reader to find out which servers content is stored on, and subsequently try to pressure the server administrators to remove it. With Publius in particular, the situation is slightly more complicated as each document is encrypted and the key is split into n shares, any k of which are recombinable to form the document back. In this case, the attacker needs to remove content from $n - k + 1$ servers, all of which he can easily locate.[1] With the number of servers reasonably small and static, the job of censoring documents becomes easier than one might expect.

---

[1] Indeed, if one server has been pressured into removal, the other server administrators may simply follow the precedent and remove the offending content themselves.

P. Druschel, F. Kaashoek, and A. Rowstron (Eds.): IPTPS 2002, LNCS 2429, pp. 111–120, 2002.
© Springer-Verlag Berlin Heidelberg 2002

An alternative approach to dealing with censorship resistance is taken by systems like Dagster [SW01] and Tangler [WM01] which prevent removal of any single document from the system by entangling documents together. However, in our view, these are not effective enough at dealing with the problem either: if the offending document was entangled with the Declaration of Independence, Das Capital and the Little Red Book, censoring it (thereby removing all of the above from the system) would not be a major problem as all the other documents are readily available from other sources. Given the other documents are *not* available, the censor is unlikely to be discerning enough to want to keep some of them. Furthermore, if the system is run on a single server (eg. Dagster), then the censor may simply try to shut down the entire server.

In our system, we consider the storers of the files valuable entities, and protect them against "Rubber Hose Cryptanalysis". Furthermore, we protect the documents they are storing by providing *active-server document anonymity* (as first introduced in [DFM01]). This is a property which states that the storer should not be able to determine (parts of) which document it is storing, not even by retrieving the document from the system. We now proceed with a description of the system and then give an analysis and critique of it.

2   System Description

Our system consists of many identical peers, each of which can fulfil four different roles:

- Publisher P. The node which has a document and wishes to make it available and censorship resistant.
- Storer s. A node which stores part of a document.
- Forwarder a. A node which has an anonymous pointer to a node storing part of a document.
- Client c. A node which retrieves a document.
- Decrypter l. A node which decrypts part of a document and sends it off to the client.

The system is built on top of an existing Peer-to-Peer document storage service like PAST [DR01]. PAST can be viewed as a network of machines (peers), each with a unique identifier. Neighbouring machines (machines within a certain distance of each other within a logical name space) share state. The only thing required to send a message to a machine is its id, furthermore, PAST guarantees that the message takes no more than $\log N$ hops, where N is the number of nodes in the system. We also assume a public key infrastructure, so any peer is able to learn any other peer's public key. This is further discussed in the next section. Using an existing Peer-to-Peer architecture allows us to abstract from routing, clients leaving and joining the network, and other low level issues. An architecture like PAST also provides robustness, which is further discussed in the next section. We also make use of an anonymous connection system such as Onion Routing [GRS99] which is capable of handling reply blocks.

As usual in censorship resistant systems, the operations available to a node are publishing and retrieval. There is no search facility, therefore we rely on a broadcast mechanism like an anonymous newsgroup to transmit retrieval information to potential readers. We do not support content deletion or modification.

2.1 Publishing

The overall publishing process is illustrated in Figure 1. The main idea is to split the documents into many parts or shares h_i, and store them (encrypted) on machines s_i, while making them accessible through machines a_i which forward requests for the appropriate shares anonymously.

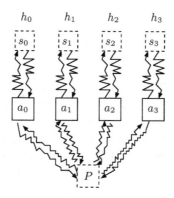

Fig. 1. Publishing

Publisher: To publish a document (see Figure 2), the publisher P splits it into $n + 1$ shares h_i, any $k + 1$ of which can be combined to form the whole document again. This can be done using one of the standard algorithms like Shamir's secret sharing [Sha79]. He then generates $n+1$ keys k_i and encrypts each share with the corresponding key. He now picks $n + 1$ peers $a_0 \ldots a_n$ at random to act as forwarders and constructs onions[2] to send (via the anonymous connections layer) each of them the encrypted share $\{h\}_{k_i}$, the corresponding key k_i and a (large) random integer v_i together with a return address (reply onion)[3] r_P. The publisher can now wait for a confirmation to come back from each of the a_is (via the reply onion) saying whether the publishing has been successful or not. If the operation failed, the publisher should try different a_is.

Forwarder: Each of the forwarders (take a_0 as an example) receives the message, finding an encrypted share $\{h_0\}_{k_0}$, a key k_0 and a random number v_0 and the publisher's return address. He then picks a storer s_0 to store the share and a number v_0' which the storer would associate the share with. He constructs an onion for delivering these to the storer. Thus, he puts the encrypted share, v_0', as well as its own anonymous return address r_{a_0} into the onion as the message and sends it off (see Figure 2a). He remembers

---

[2] This is a technique first described in [Cha81]. A (standard) onion for destination d with message M and peer sequence $a_0 \ldots a_n$ is $\{a_1 \ldots \{d, \{M\}_{k_d}\}_{k_{a_n}} \ldots\}_{k_{a_0}}$ which is sent to a_0. a_i is the address of the server and k_{a_i} is its public key.

[3] A return address is a kind of onion which, if included in an anonymous message, can be used to reply to that message without revealing the original sender (see [Cha81] for details).

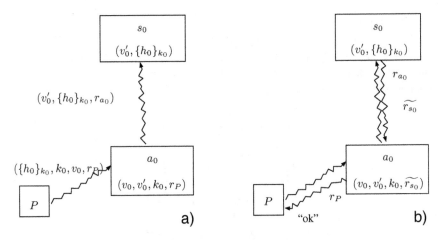

Fig. 2. Inserting share a_0. All communication is done via the anonymous connection system using randomly constructed onions. If the message is sent using a return address, it is displayed at the base of the arrow. Anonymous return addresses are denoted by r, eg. r_{a_0}

Share	h_0
Key	k_0
Storer	s_0
Random numbers	v_0, v_0'
Return addresses	$r_P, \widetilde{r_{s_0}}$

v_0, v_0', k_0 and r_P. If the onion is received by s_0, it stores the share and issues a number of different return addresses $\widetilde{r_{s_0}}$ (to be used for retrieval), sending them back to a_0 via the return address r_{a_0}. Now a_0 associates v_0, v_0' and k_0 with the return addresses $\widetilde{r_{s_0}}$, forgets s_0, and replies "ok" to the publisher via r_P. Once all the shares have been stored, the publisher destroys them and announces the name of the file, together with the $n + 1$ pairs (a_i, v_i) to potential users.

2.2   Retrieval

To retrieve a document (see Figure 3), the client c asks the forwarder a_0 (and each a_i in the same way) to retrieve the share h_0 by sending them an anonymous message with v_0 and their anonymous return address r_c. The forwarder a_0 then picks a random server l to act as a decrypter and sends it k_0, the key it is storing which decrypts the stored share, v_0', r_c, and a return address r_{a_0}, getting back a return address for l. Now a_0 forwards r_l and v_0', which identifies the share, to s_0 via one of the $\widetilde{r_{s_0}}$ (r_{s_0}). Now s_0 looks up the encrypted share corresponding to v_0' and forwards it and v_0' to l, which decrypts the share and sends it to the client via r_c. The process continues until c has accumulated enough shares to reconstruct the document.

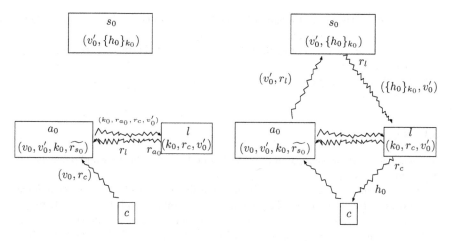

Fig. 3. Retrieval

Share	h_0
Key	k_0
Storer	s_0
Client	c
Decrypter	l
Random numbers	v_0, v_0'
Return addresses	$r_{a_0}, r_{s_0}, r_l, \widetilde{r_{s_0}}, r_c$

We note that when the forwarder starts running out of return addresses for the storer (you can use each one only once), all he needs to do is request some more via one of the return addresses it still has.

The other important detail which we have so far left out of the description of the system is that we use the P2P layer (PAST) to replicate state among neighbouring nodes. This enables requests to be routed to any of the nodes which contain the replicated state. In particular, the forwarder shares $(v_0, v_0', k_0, \widetilde{r_{s_0}})$ with neighbouring nodes which can therefore also answer requests. Similarly, the storer shares $(v_0', \{h_0\}_{k_0})$. The decrypter does not need to share anything as he will only get one request to decrypt the share and will then give up this role.

3 Commentary on the Protocol

In the last section we presented the protocol. Here we will try to explain some of our design decisions and show how they relate to the properties we want our system to satisfy.

There are several novel aspects in the design of our censorship resistant system as compared to existing architectures:

– Replication. The use of a P2P layer like PAST to replicate state in forwarders and storers.

- Forwarders. The use of forwarders to provide an extra layer of indirection and prevent the storers from being visible by clients.
- Encryption of shares. Storing the shares in an encrypted form and keeping the keys at the forwarders.
- Decrypters. The use of separate nodes (as opposed to, for instance, forwarders) to decrypt shares.

3.1 Replication

Replication of state in the system provides fault tolerance, efficiency and prevents several kinds of attacks.

Firstly, it makes denial of service attacks and simple efforts to take down individual forwarders ineffective because there are always a number of other hosts ready to forward requests. Furthermore, even if the attacker succeeds in taking down a particular forwarder, state will be replicated onto a new node which will also start forwarding requests.

More subtly, it reduces the link between any particular forwarder and the share which has been retrieved. This is because the address for a particular forwarder (for instance, a_0) which is published in the anonymous newsgroup denotes a dynamic set of physical hosts rather than a single machine. This is due to the behaviour of the underlying P2P system (PAST): if asked to route a request to a node (a_0), it does not necessarily forward it to that specific node, but instead to any node which shares state with a_0. Therefore, it is not easy to establish precisely which physical machines the address a_0 represents, indeed, this set changes as machines go down and come back up.

However, this introduces a slight complication. Although the forwarders share state, they cannot share private keys (it would be impossible to keep these keys secret because the set of forwarders constantly changes as nodes go up and down). Therefore requests addressed to them must be delivered in plaintext[4]. This turns out not to be problem here as the attacker who watches traffic arriving at the forwarders sees v_0 and r_c. The former is public anyway, and the latter gives away no information about c itself (but enables him to send a fake share to c[5]).

3.2 Forwarders

The use of forwarders serves several purposes. First of all, they help protect the storers against "Rubber Hose Cryptanalysis" by hiding them from the clients. Secondly, they can help provide active-server document anonymity by randomly introducing new dummy requests into the system and dropping some of the valid ones. Thus, it will make it hard for the storer to find out (part of) which document it is storing, even by acting as a client in the system. Finally, we use the forwarders to store keys which decrypt the shares and forward them to the decrypters.

---

[4] This means that the last layer of the onion is not encrypted. Therefore, the message is still anonymous but not secret.

[5] This does not constitute an attack as the adversary would have to perform this active and therefore expensive operation for every share every client requests.

3.3 Encryption of Shares

We have argued that the storers should not be able to see the content they are storing to prevent the possibility of them being pressured into censorship. Therefore, we must make them store the shares in an encrypted form and stop them from getting hold of the keys which would decrypt the shares. Thus these keys cannot be published as this would enable the storers to retrieve all of them and see which one decrypts each of the shares. Hence they are stored by the forwarders.

3.4 Decrypters

Some would argue that the use of decrypters is superfluous. The storer could just send the shares back to the forwarder who would send them back to the client. However, this would expose the forwarder to the risk of being caught red-handed with the share. Furthermore, they might be pressured into installing a filter to censor shares which the attacker does not like. As the forwarder is the publically visible part of the system (and therefore most vulnerable), we decided to delegate the task of decrypting the share to a completely different entity who does not have any information about what it is decrypting.

4 Discussion

In this section we discuss the limitations of our system.

First, one should question the validity of assuming a public key infrastructure on a P2P network. We need each peer to be able to retrieve the public key of any other peer and verify that the key belongs to that particular peer. The simple solution is to use a global repository. However, such a scheme would limit the scalability of the system. We believe that better solutions exist, and are actively working on this problem. A related issue is the fact that working on top of a peer to peer system may result in attacks. For instance, if the attacker is able to arrange requests not to reach a particular set of nodes corresponding to a forwarder at this level (by, for instance, modifying routing tables in some of the nodes of the P2P system), he will effectively censor the corresponding share. We do not address these problems here, but merely point out that this is an active research area and one which we need to pay close attention to. A nice summary of the problems in security of Peer to Peer systems and some solutions to them can be found in [SM02, Dou02].

Secondly, we should consider how likely the forwarders are to suffer from "Rubber Hose Cryptoanalysis" – they are certainly visible to the attacker and contain information which is necessary for share retrieval. However, we argue that they are much less likely to be subjected to such pressure than, for example, Publius servers for the following reasons:

- They are not storing the offending document, not even in an encrypted form, so their connection with it is somewhat indirect.
- They do not store the identities of the s_is, so an attack to try to get it out of them will not succeed.

- The share does not actually go through the forwarders a_i after publication is completed.
- The requests addressed to the forwarder a_i are likely to end up being handled by a number of different physical hosts.

A slight modification to the protocol (which is beyond the scope of this paper) can be introduced to further reduce the role the a_i play in the protocol, and therefore reduce the potential for them to be attacked.

We must also consider the number of compromised peers it takes to remove a document from the system. A possible attack is as follows: each forwarder a_i remembers (rather than forgetting) the corresponding storer s_i at the time of publication in the hope of exposing them later, if the content turns out to be offending. Once the document has been successfully published, a_i notes the correspondence between the random integer (v_i) published with the document and one in its lookup table, and works out the fact that s_i is storing a share of a particular document. It can now pressure s_i into removing the share. However, the chances of the $n - k + 1$ peers picked as a_i being compromised are small and the peers have to be compromised at the time of publication, otherwise the attack fails.

Having stated that we provide active server anonymity, we must pay attention to the amount of information the storer can gain by repeatedly requesting the document and noticing the requests coming in for the share it is storing. However, a suitable number of random requests generated by the forwarders should weaken this attack. Again, the precise details are beyond the scope of this paper.

We also note that we are presenting just one part of a design of a system. Many questions are left unanswered and a few attacks are not addressed. For example, the attacker can simply flood the censorship resistant system with random data so that "real" documents cannot be inserted. This design does not include protection against such an attack.

Neither do we deal with accountability in any systematic way. Consider a scenario where the attacker is powerful enough to insert many nodes into the system. Each of these, when asked to act as a forwarder, replies "ok", but drops the share and fails to answer subsequent requests. In this design, we are relying on the inability of the attacker to insert enough malicious nodes to censor documents in this way. We felt that although standard methods which are summarised in [DFM00] can be used in this system, they are inappropriate in this context or do not provide adequate protection. Therefore, we leave this for future work.

5   Related Work

A variety of censorship resistant systems have been designed, some of which have also been implemented. We have already discussed Publius, Dagster and Tangler but, perhaps the system closest to ours in terms of the aims it tries to achieve is Free Haven [DFM01].

It is built on top of an underlying network of anonymous remailers and deals with reader anonymity, server anonymity and censorship resistance. However, it uses a

Gnutella-like search for retrieval of shares of the document. More specifically, a user request to retrieve a particular document gets broadcast from the user node to all the neighbouring nodes, and so on. When a request arrives at a peer which has a matching share, it gets sent off to the requester via a chain of remailers. This scheme for locating files is rather inefficient, and, as the Gnutella experience has shown, does not scale for large numbers of peers. Furthermore, we note that peers frequently exchange shares with each other. This is costly in terms of network bandwidth and makes it hard to provide guarantees that a document will be located. Our system aims to be more efficient both in terms of bandwidth and share retrieval. Finally, Free Haven has not been implemented, perhaps because it contains complex notions of share trading, reputation, etc. We note, however, that moving the individual shares around in the system is an interesting technique for increasing censorship resistance which may be incorporated into our system. For example, the shares may be moved periodically, and the state in the forwarders updated.

6 Conclusion

We have presented a design of a system which deals with censorship resistance and satisfies strong anonymity requirements. Although (like many other similar systems) it has not yet been implemented, we hope that the fact that it is based on top of an existing infrastructure will make the job easier.

We have argued for building an anonymous censorship resistant system on top of a peer to peer architecture and demonstrated the feasibility of doing so to provide strong anonymity and robustness guarantees.

We have two main future objectives. Firstly, having described an anonymity system and claimed that it satisfies some properties, we consider it worthwhile to formalise those and prove them rigorously.

Finally, we would like to build a prototype implementation on top of a P2P layer like PAST and a suitable anonymity system.

Acknowledgements

I acknowledge support from EPSRC grant GRN24872 *Wide-area Programming* and EU grant PEPITO. A variety of ideas have resulted from conversations with Richard Clayton, George Danezis, Peter Pietzuch, Peter Sewell and Keith Wansbrough and from comments by various members of the Cambridge Security Group.

References

[And96]   R. J. Anderson. The eternity service. In *Pragocrypt*. 1996. http://www.cl.
            cam.ac.uk/users/rja14/eternity/eternity.html.
[Cha81]   D. Chaum. Untraceable electronic mail, return addresses and digital pseudonyms.
            Communications of the A.C.M., 24(2):84–88, 1981.

[CMW+01] S. A. Craver, J. P. McGregor, M. Wu, B. Liu, A. Stubblefield, B. Swartzlander, D. S. Wallach, D. Dean, , and E. W. Felten. Reading between the lines: lessons from the SDMI challenge. In *Information Hiding Workshop*. 2001.

[CSWH01] I. Clarke, O. Sandberg, B. Wiley, and T. W. Hong. Freenet: A distributed anonymous information storage and retrieval system. In Federrath [Fed01], pages 46–66. http://freenet.sourceforge.net.

[DeC] Gallery of CSS descramblers.   http://www-2.cs.cmu.edu/~dst/DeCSS/Gallery/.

[DFM00] R. Dingledine, M. J. Freedman, and D. Molnar. *Peer-to-Peer: Harnessing the Power of Disruptive Technologies*, chapter 16. O'Reilly, 2000.

[DFM01] R. Dingledine, M. J. Freedman, and D. Molnar. The Free Haven Project: Distributed anonymous storage service. In Federrath [Fed01], pages 67–95. http://freehaven.net.

[Dou02] J. R. Douceur. The sybil attack. In *First International Workshop on Peer-to-Peer Systems (IPTPS '02)*. Cambridge, MA, 2002.

[DR01] P. Druschel and A. Rowstron. Past: A large-scale, persistent peer-to-peer storage utility. In *The 8th Workshop on Hot Topics in Operating Systems*. 2001.

[Fed01] H. Federrath, editor. *Designing Privacy Enhancing Technologies: International Workshop on Design Issues in Anonymity and Unobservability*, volume 2009 of *Lecture Notes in Computer Science*. Springer-Verlag, 2001. ISBN 3-540-41724-9.

[GRS99] D. Goldschlag, M. Reed, and P. Syverson. Onion routing for anonymous and private internet connections. *Communications of the ACM (USA)*, 42(2):39–41, 1999.

[Sha79] A. Shamir. How to share a secret. *Communications of the ACM*, 22:612–613, 1979.

[SM02] E. Sit and R. T. Morris. Security considerations for peer-to-peer distributed hash tables. In *First International Workshop on Peer-to-Peer Systems (IPTPS '02)*. Cambridge, MA, 2002.

[SW01] A. Stubblefield and D. Wallach. Dagster: Censorship-resistant publishing without replication. Technical report, Rice University, 2001.

[WM01] M. Waldman and D. Mazieres. Tangler: A censorship resistant publishing system based on document entanglements. In *8th ACM Conference on Computer and Communcation Security (CCS-8)*. 2001.

[WRC00] M. Waldman, A. D. Rubin, and L. F. Cranor. Publius: A robust, tamper-evident, censorship-resistant, web publishing system. In *Proc. 9th USENIX Security Symposium*. 2000.

Introducing Tarzan, a Peer-to-Peer Anonymizing Network Layer

Michael J. Freedman, Emil Sit, Josh Cates, and Robert Morris

MIT Lab for Computer Science
200 Technology Square
Cambridge, MA 02139, USA
{mfreed,sit,cates,rtm}@lcs.mit.edu

Abstract. We introduce Tarzan, a peer-to-peer anonymous network layer that provides generic IP forwarding. Unlike prior anonymizing layers, Tarzan is flexible, transparent, decentralized, and highly scalable. Tarzan achieves these properties by building anonymous IP tunnels between an open-ended set of peers. Tarzan can provide anonymity to existing applications, such as web browsing and file sharing, without change to those applications. Performance tests show that Tarzan imposes minimal overhead over a corresponding non-anonymous overlay route.

1 Introduction

The ultimate goal of Internet anonymization is to allow a host to communicate with a non-participating server in such a manner that *nobody* can determine his identity. Toward this goal, we envision an Internet-wide pool of nodes, numbered in the millions, that relay each others' traffic to gain anonymity. This paper describes a design aimed at realizing that vision. First, however, we discuss why less ambitious approaches are not adequate.

In the simplest alternative to our vision, a host connects to a server through a proxy, such as a Anonymizer.com [1]. This system fails if the proxy reveals a user's identity or an adversary can observe traffic on the proxy's network. Furthermore, servers can block these centralized proxies and adversaries can prevent usage with denial-of-service attacks.

To overcome this single point of failure, a host can connect to a server through a set of mix relays [2]. The anonymous remailer system [5], Onion Routing [11], and Zero-Knowledge's Freedom [7] offer such a model, relying on a small, fixed core set of relays to provide service. Such reliance increases vulnerability to individual node failures, and still provides obvious targets for attacking or blocking. Furthermore, a corrupt relay can perform network-edge traffic analysis on such a system: if the relay receives traffic from a non-core node, that node must be the ultimate origin of the traffic. A corrupt entry relay can conspire with a corrupt exit to determine both source and destination, using timing analysis. An external adversary capable of observing traffic that enters and exits the set of core relays can make the same analysis. This reduces the anonymizing power of long mix paths.

P. Druschel, F. Kaashoek, and A. Rowstron (Eds.): IPTPS 2002, LNCS 2429, pp. 121–129, 2002.

Tarzan, the design presented in this paper, involves sequences of mix relays chosen from a large pool of volunteer participants. All participants are equal peers; they are all potential originators of traffic, as well as potential relays. This design overcomes the edge-analysis weakness: a relay cannot tell if it is the first hop in a mix path. This design is still vulnerable if an adversary can observe traffic throughout the Internet, but this attack seems unlikely.

Tarzan is composed of an open-ended set of participating nodes, with no centralized component; as in other peer-to-peer systems, this lowers the barriers to participation. Tarzan allows client applications on participating hosts to talk to non-participating servers on the Internet. Tarzan is transparent to both client applications and servers, though it must be installed and configured on the client node.

Tarzan routes packets through tunnels involving a randomly chosen sequence of Tarzan peers using mix-style layered encryption. The two ends of a tunnel are a Tarzan node running a client application and a Tarzan node running a network address translator; the latter forwards the client's traffic to the ultimate destination, an ordinary Internet server. These mechanisms provide anonymity in the face of malicious Tarzan participants, inquisitive Internet servers, and observers who can see traffic on a limited number of network links.

The larger purpose of Tarzan is to support a systems-engineering position: anonymity can be built-in as an underlying transport layer, transparent to most systems, trivial to incorporate, and with a tolerable loss of efficiency. The immediate effect of this approach will be to reduce the effort required for application writers to incorporate anonymity into existing designs, and for users to add anonymity without changing applications. In the long term, the ability for a single anonymizing relay to participate in multiple kinds of traffic may make it easier to achieve a critical mass of anonymizing relays.

2   Architecture and Design

This section describes Tarzan's basic tunnel mechanism. Figure 1 shows a simple Tarzan overlay network. All participating nodes run software that 1) discovers other participating nodes, 2) intercepts packets generated by local applications that should be anonymized, 3) manages tunnels through chains of other participants to anonymize these packets, 4) forwards packets to implement other nodes' tunnels, and 5) operates a NAT (network address translator) to forward other participants' packets onto the ordinary Internet.

Typical use proceeds in three stages. First, a node running an application that desires anonymity selects a set of nodes to form a path through the overlay network. Next, this source-routing node establishes a tunnel using these nodes. Finally, it routes data packets through this tunnel. The exit point of the tunnel is a NAT, which forwards the anonymized packets to servers that are not aware of Tarzan.

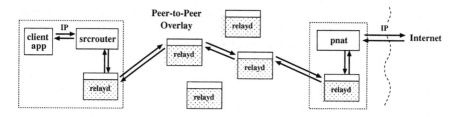

Fig. 1. Tarzan Architecture Overview

Tarzan operates at the IP (Internet Protocol) level and offers a best-effort delivery model. The burden of providing functionality like reliability or authentication is left to the communicating end-hosts.

Tarzan uses layered encryption similar to Chaumian mixes [2]: each leg of the tunnel removes or adds a layer of encryption, depending upon the direction of traversal of the packet. IP headers are sanitized at the tunnel entry-point.

The rest of this section first describes how Tarzan nodes relay packets along existing tunnels; this clarifies the necessary per-node tunnel state. Next, it shows how this state is established during the tunnel setup phase, which includes key distribution. Finally, it describes how IP forwarding happens at the tunnel end-points.

2.1   Packet Relay

A Tarzan tunnel passes two distinct types of messages between nodes: data packets, to be relayed through existing tunnels, and control packets, containing commands and responses that set up and maintain tunnels. Tarzan encapsulates both packet types inside UDP.

A flow tag (similar to MPLS) uniquely tags each hop of each tunnel. A relay rapidly determines how to route a packet based on its tag. Symmetric encryption protects data on a per-hop basis, with separate keys being used in each direction of each hop.

In the forward path, the tunnel entry-point clears each IP packet's source address field, performs a nested encryption per tunnel hop, and encapsulates the result in a UDP packet. More precisely, if the tunnel consists of a sequence of nodes $T = (h_1, h_2, \ldots, h_l)$ and the forward key for each node is k_{h_i}, the originating node produces the encrypted block $\{\{\cdots\{\{p\}_{k_{h_l}}\}_{k_{h_{l-1}}}\cdots\}_{k_{h_2}}\}_{k_{h_1}}$ from the input packet p. The origin tags this block with the first hop's flow identifier and forwards it to h_1. That node's relay will decrypt the data, *i.e.* strip off one layer of encryption, retag the packet, and forward it on to the next hop. This process continues until the packet reaches the last hop, which strips off the innermost layer of encryption, revealing the original IP packet.

On the reverse path, each successive relay performs a single encryption with its appropriate reverse key, re-tags and forwards the packet back towards the origin. This process wraps the packet in layers of encryption, which the origin of the tunnel must unwrap by performing l decryptions. Note that this design places the bulk of the encryption workload on the node seeking anonymity.

2.2   Tunnel Setup

When forming a tunnel, Tarzan selects a series of nodes uniformly at random from existing peers in the network. Each relay publishes a public key that is generated locally the first time it enters the network. We rely on this relay being the only one that knows the corresponding private key. Section 2.4 describes how Tarzan selects tunnel nodes and acquires their public keys.

Tunnels are established on an iterative hop-by-hop basis. The tunnel entry-point is responsible for setting up the entire tunnel, which consists mainly of generating and distributing the symmetric encryption keys.

Each hop is set up using the same procedure. An establish request sent to node h_i is relayed as a normal data packet from h_1 through h_{i-1}. Node h_i cannot distinguish whether the packet originated from node h_{i-1} or from one of that node's predecessors; node h_{i-1} cannot distinguish successive establish requests from ordinary tunneled data. The establish request contains the forward decryption key that h_{i-1} will use when sending packets to h_i and the encryption key that should be used for sending packets received from h_{i+1}. Additionally, it establishes the flow identifiers that will be used to tag packets going in each direction. The initiating node uses the public key of node h_i to encrypt the initial forward session key and then this session key to encrypt the subsequent reverse key, node addresses, and flow identifiers. When h_i has successfully stored the state for this request, it responds to the origin for an end-to-end check of correctness.

For path length l, this algorithm takes $O(l)$ public-key operations and $O(l^2)$ inter-hop messages to complete. This overhead is sufficiently small for realistic choices of l.

2.3   IP Packet Forwarding

Tarzan provides a client IP forwarder and a server-side pseudonymous network address translator (PNAT) to create a generic anonymizing IP tunnel. The IP forwarder diverts certain packets from the client's network stack and ships them over a Tarzan tunnel. The client NATs its own address to a random address assigned by the PNAT from the reserved private address space. The PNAT translates this private address to one of its real addresses. Remote hosts can communicate with PNAT normally, as if it originated the traffic. Correspondingly, response packets are deNAT'ed twice, once at each end of the tunnel.

The IP forwarder only hides Internet Protocol address, and origin port numbers for TCP and UDP packets. For existing application-level protocols that leak information (such as `http`), the client can choose to run an application-level sanitizer.

The pseudonymous NAT can also offer port forwarding to allow ordinary Internet hosts to connect through Tarzan tunnels to servers. This mode of operation provides anonymity to the server. For example, a user can join a file-sharing network such as Napster as an anonymous server by simply registering her PNAT's address. To Napster, the PNAT would appear to be the client. In fact, any two parties can communicate anonymously by each creating a tunnel to a different PNAT; a normal connection between these two PNATs will form a *double-blinded* channel.

2.4   Peer Selection

Tarzan requires that its peer selection mechanism provide three functions: new peer discovery, scalability, and random selection. Additionally, these mechanisms should be robust against adversaries attempting to bias the selection process.

Tarzan uses the Chord lookup algorithm [10] to obtain this functionality, although and peer-to-peer lookup system that provides these functions would be suitable. Chord is a distributed peer-to-peer hash function mapping flat keys to nodes. Each Chord node has a unique 160-bit node identifier (ID) obtained with a cryptographic hash of its IP address.

All Tarzan relays participate in a single Chord ring. New relays join the ring by contacting an existing relay to discover its proper set of overlay neighbors. We assume that an adversary may only impersonate a limited address space, such as the subnetwork from which he is connected. Therefore, he may only join the network with a relatively small number of Chord nodes, as he cannot respond to RPCs sent to other IP addresses.

Keys are mapped in Chord into the 160-bit space by a universal hash function. The *successor* of a key is the node with the smallest ID greater than or equal to that key (with wrap-around), much as in consistent hashing [8]. The Chord operation, $lookup(K)$, discovers the IP address of the successor of K by iteratively sending RPCs to nodes around the Chord ring until reaching the desired successor.

The system is highly scalable, as the expected number of messages involved in a lookup is $O(\log n)$, for network size n nodes. The iterative lookup allows the peer to validate the existence of every intermediate node.

A peer efficiently picks a random peer by generating a random lookup key and finding that key's successor. The successor responds with its IP address and public key.

This lookup would reveal the tunnel initiator's identity if performed immediately before tunnel establishment. Therefore, every node occasionally performs a random key lookup (say, once per minute) and caches this information for later use. The freshness of the key-to-node mapping is not important; the critical mapping is public key to IP address.

3   Policy Issues

Anonymity policy decisions affect tunnel selection and maintenance. While Tarzan provides easy and efficient peer discovery, user concerns may direct the actual choice of peers for a tunnel. For example, tunnel intra-hop latencies have a noticeable impact on end-to-end performance. However, a user more concerned with anonymity may sacrifice some latency to ensure that packets are routed through certain points – for example, nodes outside his government's jurisdiction – or even desire hops that explicitly delay and batch messages.

While each peer responds to lookup queries with a set of its attributes, intra-hop latency is highly dependent upon the underlay network path. Tarzan provides explicit ping meassages to measure per-hop latency through our overlay network. These messages are relayed as normal data packets in the tunnel until reaching the specified node.

4   Security and Anonymity Analysis

This section explains how Tarzan provides adequate security and anonymity against a limited active adversary.

A limited active adversary cannot possibly sniff and perform traffic analysis on all system participants. As all Tarzan users run relays, there are at least as many relays as active participants. We imagine several hundred nodes in early stages of deployment and possibly thousands or more in later stages. Because tunnel setup selects its path randomly at runtime (modulo any policies) from the large set of peers, an adversary cannot consequently predict or target nodes for attack.

Additionally, Tarzan confounds adversaries by funneling all communication through each node's relay, regardless of whether the traffic originates locally or remotely. Tarzan achieves *sender anonymity*: both passive sniffers and malicious participants cannot distinguish whether a node initiates a message or merely relays it.

This argument follows the Crowds analysis [9], but differs in several important ways. First, Tarzan rebuilds individual *links* following failures rather than entire *paths*, minimizing the intersection attack on linked flows. Second, adversaries cannot link flows as easily. Tarzan layer-encrypts data in the tunnel, while all nodes in a Crowds path see plaintext. This encryption additionally protects against message coding attacks (and provides data confidentiality in the tunnel as a second-order effect). Third, Tarzan chooses nodes independently. With c colluding adversaries in a n-node network, the probability of choosing a fully-compromised l-length route is roughly $\left(\frac{c}{n}\right)^l$.

A peer-to-peer system also offers new challenges. An adversary can pseudospoof the system and create a multitude of identities: only the number of IP addresses available limits the number of virtual identities usable by him. However, the user can specify policies to avoid routes with similar IP prefixes.

Pkt size	Latency	Throughput	
(bytes)	(μ-sec)	(pkts/s)	(Mbits/s)
64	244	14000	7.2
512	376	8550	35.0
1024	601	7325	60.0

Table 1. Per-hop latency and forwarding rate for Tarzan relays

Tunnel length	Pre-fetch latency	On-demand latency
1	30.19	29.51
2	46.54	59.77
3	68.37	106.70
4	91.55	146.37

Table 2. Setup latency (msec) for tunnel establishment

5   Implementation

We have implemented Tarzan in C++ on Unix to validate our approach. The core component of Tarzan is a stand-alone relay server that performs the per-hop packet relaying. This server can be run by unprivileged users. Another component, the Tarzan library, communicates with the relay server to establish tunnels, to listen for connections, and to send and receive data. The Tarzan library presents an API modelled after standard Unix sockets, albeit asynchronous, that is executable on a variety of BSD, Linux, and Unix platforms.

Applications such as the IP forwarder and the pseudonymous NAT are built on top of this library. For IP forwarding, we take advantage of FreeBSD's divert sockets.

6   Performance

In this section, we present some preliminary performance measurements running on a 1.2 GHz Athlon PC with 128 MB of RAM running FreeBSD 4.3 connected to a 100 Mbps switched ethernet.

Table 1 shows the average latency as the time needed by the Tarzan relay to read a packet off the network, decrypt and route it, and send it out to the next relay. For large-enough packet, latency scales linearly with packet size. Throughput also scales roughly linearly.

Table 2 shows end-to-end latency required to setup a tunnel. To differentiate Tarzan's overhead from the cost of Chord lookups, we provide two measurements. Clients pre-fetch node information in one and perform lookups on-demand in the second. On average, we incur a setup cost of 20 msec per hop. Therefore, underlay network latency still dominates, even during tunnel setup.

7   Related Work

Prior work in this area falls into two categories: systems that provide application-specific anonymity, and systems that offer a more generic transport framework.

The majority of application-specific anonymous systems focus on email, web browsing, or file sharing. For example, the mix networks proposed by Chaum [2]

were designed to achieve untraceable anonymous email. The Cypherpunk and Mixmaster remailers [5] incorporate these techniques. Web systems range from centralized sanitizers [1] to non-mix peer-to-peer systems [9] that lack self-organization. Systems for anonymous publishing include Publius [13] and Freenet [3]. In contrast to application-layer solutions, Tarzan provides a single general-purpose anonymizer that can be used transparently by many applications. We note that Free Haven [4] and Tangler [12] specifically cite the need for a system like Tarzan for their anonymity requirements.

Few systems attempt anonymity for low-level, real-time communication. The Onion Routing system [11] creates a mix-net over TCP connections. Application integration is achieved using application proxies that create paths in the system by successively encrypting a control packet, or *onion*. Zero-Knowledge Systems developed the first commercial mix-net system, known as the Freedom network [7]. The Freedom network consists of nodes deployed at various ISPs to route traffic between them, using a model similar to Onion Routing. Client-side integration is closely tied to the operating system and Freedom's pseudonym authentication system. Unfortunately, Freedom was shut down in mid-2001 for financial reasons. Both of these systems only provide plausible unlinkability of sender and recipient.

In comparison, Tarzan uses the same basic idea to mix traffic, but achieves IP-level anonymity by generic and transparent packet forwarding. Tarzan offers sender anonymity in addition to the unlinkability of sender and recipient provided by Onion Routing and Freedom. This is derived from its peer-to-peer architecture, which removes any notion of entry-point into the anonymizing layer. Tunnel construction is client-driven, allowing users to find more efficient paths by incrementally building tunnels. Tarzan's IP forwarding architecture also allows servers to interact transparently with anonymous clients by performing dynamic pseudonymous network address translation; Onion Routing's proposed *reply onions* are static and thus more vulnerable to node failure, brute force decryption, and even subpoena attacks. Finally, we hope to provide an anonymization tool that is free to use and plan on making our source code available under the GNU Public Licence.

8   Conclusion

Tarzan provides a flexible, transparent layer for providing anonymity to generic IP connections. Tarzan's peer-to-peer design makes it decentralized, highly scalable, and easy to manage.

We show that Tarzan imposes minimal overhead over a corresponding non-anonymous overlay route. Latency through Tarzan tunnels is completely dominated by transmission speed through the Internet.

Tarzan's ability as a single anonymizing relay to participate in multiple kinds of traffic furthers its usefulness and, hopefully, adoption.

References

[1] The Anonymizer. http://anonymizer.com.

[2] David Chaum. Untraceable electronic mail, return addresses, and digital pseudonyms. *Communications of the ACM*, 4(2), February 1982.

[3] Ian Clarke, Oscar Sandberg, Brandon Wiley, and Theodore W. Hong. Freenet: A distributed anonymous information storage and retrieval system. In Federrath [6], pages 46–66. http://freenet.sourceforge.net.

[4] Roger Dingledine, Michael J. Freedman, and David Molnar. The Free Haven Project: Distributed anonymous storage service. In Federrath [6], pages 67–95. http://freehaven.net.

[5] Electronic Frontiers Georgia (EFGA). Anonymous remailer information. http://anon.efga.org/Remailers/.

[6] Hannes Federrath, editor. *Designing Privacy Enhancing Technologies: International Workshop on Design Issues in Anonymity and Unobservability*, volume 2009 of *Lecture Notes in Computer Science*. Springer-Verlag, 2001.

[7] Ian Goldberg and Adam Shostack. Freedom network 1.0 architecture, November 1999.

[8] D. Karger, E. Lehman, T. Leighton, M. Levine, D. Lewin, and R. Panigrahy. Consistent hashing and random trees: Distributed caching protocols for relieving hot spots on the world wide web. In *Proceedings of the 29th Annual ACM Symposium on Theory of Com puting*, pages 654–663, May 1997.

[9] Michael K. Reiter and Aviel D. Rubin. Crowds: anonymity for Web transactions. *ACM Transactions on Information and System Security*, 1(1):66–92, 1998.

[10] Ion Stoica, Robert Morris, David Karger, M. Frans Kaashoek, and Hari Balakrishnan. Chord: A scalable peer-to-peer lookup service for internet applications. In *Proceedings of the ACM SIGCOMM '01 Conference*, San Diego, California, August 2001.

[11] Paul Syverson, D. M. Goldschlag, and M. G. Reed. Anonymous connections and onion routing. In *Proceedings of the IEEE Symposium on Security and Privacy*, pages 44–54, Oakland, California, May 1997.

[12] Marc Waldman and David Mazières. Tangler: A censorship-resistant publishing system based on document entanglements. In *Proceedings of the 8th ACM Conference on Computer and Communications Security*, Philadelphia, Pennsylvania, November 2001.

[13] Marc Waldman, Aviel D. Rubin, and Lorrie Faith Cranor. Publius: A robust, tamper-evident, censorship-resistant, web publishing system. In *Proceedings of the 9th USENIX Security Symposium*, pages 59–72, August 2000.

Mnemosyne: Peer-to-Peer Steganographic Storage

Steven Hand[1] and Timothy Roscoe[2]

[1] University of Cambridge Computer Laboratory, Cambridge CB3 0FD, UK
steven.hand@cl.cam.ac.uk
[2] Sprint Advanced Technology Lab, 1 Adrian Court, Burlingame, CA 94010, USA
troscoe@sprintlabs.com

Abstract. We present the design of Mnemosyne[1], a peer-to-peer steganographic storage service. Mnemosyne provides a high level of privacy and plausible deniability by using a large amount of shared distributed storage to hide data. Blocks are dispersed by secure hashing, and loss codes used for resiliency. We discuss the design of the system, and the challenges posed by traffic analysis.

1 Introduction and Motivation

A steganographic file system, first presented in [2], has the property that it gives a user strong protection against being compelled to disclose (all) its contents. Attackers not in possession of the secret are unable to acquire the contents of files, and they cannot even gain information about whether a given file is present or not. In effect, the system allows an author to plausibly deny the existence of most files[2] in the system.

A distributed, peer-to-peer steganographic storage system like Mnemosyne has further interesting properties. Firstly, in common with systems like Free-Net [6], storage providers can offer a service without being able to know what is being stored. This property may be attractive to a service provider concerned about liability as it *de facto* confers something akin to common-carrier status on the provider.

Secondly, for a single user desiring to store files securely, a distributed steganographic storage system makes information less susceptible to machine failure or denial-of-service: a local storage medium can always be stolen, but a peer-to-peer system is harder to shut down.

Thirdly, such a system may also be used as a shared-memory communication medium with steganographic properties: this allows interpersonal messaging with a high degree of privacy.

A system with these properties is of great potential use to the modern business traveler.

---

[1] Pronounced *ne moz'nē*.
[2] At least some files must be revealed to justify the existence of the system itself.

P. Druschel, F. Kaashoek, and A. Rowstron (Eds.): IPTPS 2002, LNCS 2429, pp. 130–140, 2002.

Mnemosyne takes advantage of the widespread availability and low cost of network bandwidth and disk space. The system comprises servers that provide unreliable block storage, and clients which write and read blocks to and from the servers. A node can serve the function of server and client simultaneously. The servers collectively comprise a peer-to-peer system: a centralized organisation or authority is neither required nor desirable.

Before describing Mnemosyne itself, we present a description of our *local* steganographic file system. We do this for two reasons. Firstly, many of the principles of local steganographic systems carry over to the distributed case, and discussion of these helps establish context for describing Mnemosyne later. Secondly, our implementation of the local case differs from previous systems (most notably that described in [13]) in ways significant when extending the concept to a full peer-to-peer system.

2   A Local Steganographic File System

Anderson et. al. [2] describe two approaches to the steganographic storage of data. In the first, randomly-filled "cover files" are created, and user files are "written" by altering a subset of the cover files (determined by a passphrase) so that the user file is the XOR of that subset.

The second construction, followed here, assumes a disk which can store X blocks of data. To prepare this for use, we first write random data to every block. Then to store a file we simply encrypt each block and write it to a pseudo-randomly chosen location (e.g. one determined by hashing the filename and block number with a secret key). With a sufficiently good cipher and key, the encrypted blocks will be indistinguishable from the random substrate, and so an attacker cannot even determine the existence of the file. On the other hand, someone privy to the filename and key can reconstruct the pseudo-random sequence, retrieve the encrypted blocks, and decrypt them.

This leads to the problem of *collisions*, where blocks are overwritten on the disk by subsequent files. The well-known "birthday paradox" makes this quite likely with even a small load factor (ratio of file blocks to total blocks on the disk), and so replication is used: each block is written to the disk at n independent locations.

We describe our implementation of this scheme (over Linux) by first describing the process for replicating a block on the disk, and then discussing file structures built over this facility.

Writing and Reading a Single Block

Writing a block to the local steganographic file system requires a user's key K, the block data itself, and two further pieces of information: an *initial hash value* h_0 for the block, and a *validity check* (a way of determining whether the block data has been corrupted or not). The initial hash value and validity check vary according to whether one is storing directory blocks, inodes, or file blocks (see below). To write (or overwrite) a block, the procedure is:

– The user computes a sequence of n hash values:

$$h_0, \ h_1 = H(h_0), h_2 = H(h_1), \ \ldots, \ h_{n-1} = H(h_{n-2})$$

– Replica i $(0 \leq i < n)$ is encrypted under the key $k_i = E_K(h_i)$ and stored at block number $b_i = h_i \bmod X$, where X is the number of blocks on the disk[3].

To read a block given the key K and an initial hash value h_0, we read and decrypt each replica in turn from block b_i until we have a block which passes the validity check. If no blocks pass the check, the block is deemed lost. The use of a per-replica key k_i ensures that replicas are not identical on disk. It also means that K alone is not sufficient to determine the validity of a given block.

In our implementation we use SHA256 as the hash function H and AES as the block cipher for encrypting blocks, choosing a key size of 256 bits to match the size of hash values.

Directories, Inodes, and Files

We build a file system over this basic block facility using *directories*, *inodes*, and *file blocks*.

In Mnemosyne directories are used to aggregate files which share a common key K. A directory block contains a known textual name for the directory itself, and a list of textual file names. The validity check for a directory block is the presence of the name of the directory in the block. The initial hash value used for writing a directory block is obtained by encrypting the directory name with the key, K, and hashing the result. Using K in this way prevents different users from overwriting each others' blocks deterministically when they choose identical directory names.

Each file is represented in the file system by an inode block. The inode block is stored using an initial hash value obtained by concatenating the directory name and file name to produce a pathname, encrypting the pathname with K and hashing the result as before; this is the reason directory blocks need only store filenames. The filename is also stored in the inode block, acting as the validity check. Note that in this scheme directories themselves are completely optional, serving simply as a mnemonic device for a set of file names. Directory *names*, on the other hand, are necessary components of path names.

In addition to this file name, the inode block for a file consists of a list of zero or more {*initval, checkval*} pairs, one for each block in the file. These pairs of 256-bit values are analogous to the block pointers in a conventional file system. *initval*, chosen at random, is the initial hash value for locating the file block replicas. *checkval* is a secure hash of the file block and is used as the validity check for file blocks since, unlike directories and inodes, no redundant information is stored within file blocks.

---

[3] We believe that using subkeys $k_i = E_K(h_i)$ improves over $k_i = K \oplus h_i$, used in an earlier version of this paper.

Discussion

As discussed in [2], the choice of n (the number of replicas) is critical. Intuitively, there is a tension between increasing n to make an individual replica set more resilient and decreasing n to reduce the overall number of blocks written (and hence potentially overwritten). Analytical solutions are difficult to obtain, but initial experiments (see §5) suggest overall replication factors of 2 to 8.

This results in a significant cost in disk space, but the factor is constant (while large) over a conventional file system and so we consider it acceptable since what is offered is a specialised service for certain types of information. The key point is that the service scales well in disk size, not how much disk space is required for a given load.

The systems in [2] and [13] present a hierarchical security model, which can be generalised to a matrix controlling access by a fixed number of users (or principals) to a fixed number of security "levels". We eschew such an approach in favor of a simpler, flat key space: if a user possesses a key and the name of a directory, he or she is able to read and write files in that directory. This has two advantages. Firstly, the indefinite number of keys makes it less likely that all the keys can be extracted from a user under duress. Secondly, and more importantly, when we extend the system to a distributed, peer-to-peer scenario, we cannot know in advance how many users, files, or available blocks there will be. The matrix model implies an authority that at least allocates rows of the matrix to users; the flat key space model is more appropriate for a federated, peer-to-peer world.

Note also that even in this local implementation, users don't have to trust the block store, as long as most of the time it doesn't throw away blocks, and the load factor isn't so great that too many blocks have all their replicas overwritten. This feature is significant when we extend the system to the peer-to-peer case.

Finally, note that the local file system requires no coordinated planning or maintenance: there is no "set up" other than the randomisation of the disk.

3   Distributing the Block Store

We first present here the obvious extension of the local system to the distributed case, and then discuss refinements and modifications of this in §4.

Assume there exists a set of M nodes each of which wishes to contribute N blocks of storage to the collective. We can logically treat this as an array of MN blocks, and proceed to store and retrieve files and directories as described in the previous section. Rather than storing the block replica i at block number (h_i mod X), we need to derive both a node identifier and a block number on that node from the 256-bit hash value.

We can do this by leveraging existing work on peer-to-peer object location and routing schemes. We use Tapestry [22], although any of [15,19,20] could serve. All we require is routing of messages tagged with arbitrary n-bit identifiers to nodes.

In Mnemosyne, even in the local case, blocks read from the disk need not be correct. Instead, the validity of blocks is explicitly checked after they have been retrieved. This allows us to build a distributed block store in which there is little reliance on the integrity of any single node. The only operations a node need implement are:

- **putBlock**(*blockid*, *data*)
- **getBlock**(*blockid*) → *data*

The semantics of these are weak: **putBlock** simply requests that the node store the block *data* in such a way that it may be subsequently retrieved by **getBlock** using an identical *blockid*. However, the node is not required (and may not even be able) to ensure this — that is, the **putBlock** operation has at-most-once semantics.

getBlock requests that the node return whatever data it has associated with the given *blockid*. However the node may ignore the request, or return any block of data it chooses. The client will determine if the information is valid after it has been received.

Using this service we construct a first attempt at a distributed steganographic storage system. We assume a set of Tapestry nodes, each of which exports the same amount of storage space (e.g. 1GB arranged as 2^{20} blocks of 1KB each).

To store a block, we follow the block replication algorithm described in §2, except that we choose the leading 160 bits of h_i as the Tapestry node identifier N_i, and the next (e.g.) 20 bits as the blockid b_i on that node.

To retrieve a block, the client requests blockids b_i from nodes N_i. We note that these requests may proceed in parallel. The client then tries to decrypt and verify each block until a valid one is found. If none is found, the block is deemed lost.

We can build directories and files over this basic system as in the local case. Note that it is not necessary for an individual node to respond "correctly" or even at all. All that the client requires is that at least one of the replicas for a block is still available. This makes it difficult for an attacker without a key to destroy any particular piece of information.

We note that with lookup services having a notion of unique "successor" for a node (such as Chord), a new node joining the system can initialize by duplicating the entire block store of its successor; neither the new nor the existing node need be aware of which blocks are "valid". This duplication means that the new node will immediately respond correctly to any **getBlock** requests made of it. With Plaxton-based systems like Tapestry, there are several nodes analogous to a Chord successor (roughly 4 in Tapestry), but we can still usefully copy fractions of the stores of these nodes.

Discussion

This system has the following useful properties:

Firstly, given the obvious implementation for a "cooperative" node (viz. to reserve 1GB of space and then store and retrieve blocks as requested), the owner of the node can plausibly deny knowledge of any of the contents. Indeed, they will in general be unaware even of which blocks are in use.

Secondly, a node can choose to use a smaller amount of storage by mapping the 20-bit block identifiers down to $k < 20$ bits. This produces a less resilient but still valid store.

Finally, a node can provide more than 2^{20} blocks simply by obtaining more than one node identifier (e.g. as with "virtual servers" in CFS [7]).

In summary, Mnemosyne provides information hiding at two levels: first, data is striped widely across different nodes each of which is unaware of the other nodes holding parts of the file. Second, each individual node embeds encrypted blocks in a random substrate, thus making them indistinguishable from one another (without a valid key).

4   Enhancements

Our first enhancement to this basic scheme is to replace simple replication with the information dispersal algorithm (IDA) [14]. Using this, an author chooses two numbers $m \geq n$ and encodes information to be published into m blocks such that any n of these are sufficient to reassemble the original data. Using the IDA gives us much better resilience for a given "redundancy factor" (m/n).

The IDA requires that we replace our simple redundancy-based validity checks with a cryptographic authenticity check on each dispersed block; our current implementation uses the AES in the new OCB mode [18] to get both privacy and authentication in one pass, although CBC-MAC, XCBC, or IACBC [11] would also suffice.

The combination of the IDA and the MAC also mean that "client blocks" are now smaller than "storage blocks". In our current implementation (see §6) we treat client files as comprising a sequence of n 1000-byte blocks. These are mapped by the IDA to m 1002-byte blocks, padded to 1008-bytes, and finally encrypted under OCB-AES to produce a 1024-byte storage block (which includes a 16-byte MAC suffix).

Readers now independently retrieve m' of the m blocks where $m' \geq n$ is chosen by each user so as to obtain a "reasonable" expectation that at least n blocks will be valid. The publisher chooses m so that $\binom{m}{m'}$ is large enough for likely values of m'. Concurrently, readers retrieve r other blocks chosen at random and discard them on receipt.

This allows us to more efficiently address the problem of traffic analysis whereby an adversary who can snoop packet transfers can infer the existence (and possibly location) of a file. If desired some of the r blocks could represent a known piece of content to provide "deniable encryption" [3] in a manner reminiscent of "chaffing" [17].

We also use the flexible dispersal of the IDA to address the problem that any reader of a file can replace or destroy its contents. To combat hijacking we can simply allow authors to use pseudonymous digital signatures, much as in [8]. To prevent destruction of file content we introduce explicit *location keys*: randomly chosen values which are XORed with a (directory or file) name's hash in order to choose the set of m storage locations. An author can now choose any l different location keys and write a total of lm blocks (assuming no collisions).

Each reader is now provided with the name, the encryption key, a location key, and m. This prevents a single reader from destroying more than a fraction of the total replicas. Furthermore, if l is never disclosed, an author under duress can claim to delete all copies but later recover the information, as in the Eternity Service [1].

Writing of data under Mnemosyne also holds interesting challenges. A per-node rate limiter protects against brute-force denial-of-service attacks, as an alternative to the Hash-Cash scheme in [21]. We note that Mnemosyne is less susceptible to such attacks due to its sparse use of storage space.

Nonetheless, over time more and more of a document $\mathcal{D}$'s replicas will be overwritten until at some point it is no longer accessible. To avoid this we need to periodically refresh $\mathcal{D}$. Choosing a good refresh interval in the absence of global knowledge is difficult, and so we expect users to err on the side of caution (i.e. to rewrite rather frequently).

The refresh of files provides us with another traffic analysis problem. We could attempt to resolve this as before: i.e. arrange for additional writes to occur so that the "real" ones may be concealed. Unfortunately this would result in a large number of additional writes, and hence collisions.

A better scheme is to require that all messages to block stores are encrypted and of the same size. A single bit in a request is used to specify if the accompanying payload is to be written. In all cases, a block of data is returned. This makes it impossible for an eavesdropper to distinguish between reads and writes, making traffic analysis more difficult. If bandwidth is cheap, an obvious extension is for all users to issue an isochronous stream of requests in which "real" requests are occasionally embedded.

5   Simulation

Two of the key parameters in the system are the choices of m and n for a given file since there is a tension between maximizing the capacity of the store, and increasing the resilience of each file. This is further complicated in the decentralized case since users are free to choose m and n independently, and no-one knows how many users there are, or how much traffic they are generating. Nevertheless, to give some idea of the trade-offs involved, we present here some initial simulation results for fixed-size files and uniform coding schemes.

The simulation repeatedly adds files to a store of 4 million blocks and keeps track of how many files are still retrievable: i.e. files for which n blocks have not been overwritten in the store. Starting with an empty store, this number converges to a limit for each m as files are added, and we call this limit the capacity of the store. Figure 1 shows how the capacity changes with choice of m. For low values, the birthday paradox comes into play and capacity is limited. As

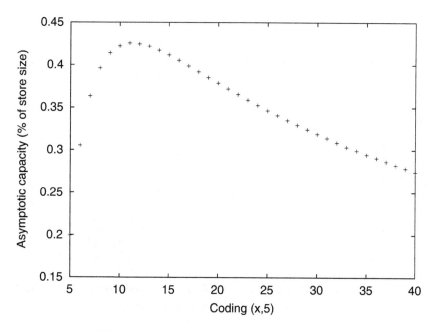

Fig. 1. Capacity of a simulated 4Mblock store

m increases, capacity increases until the large number of writes per file reduce it again.

Of more importance to actual users of the system is the expected lifetime of a file: how long a file lasts before it becomes inaccessible. Figure 2 shows cumulative distributions of file lifetimes (measured as the number of subsequent file writes) for the same coding parameters as before. Of interest to users is where these curves intersect some low probability of file loss, thus giving an idea of how often a file needs to be refreshed.

6   Implementation

We have built a working implementation of Mnemosyne. The client is implemented in C and makes use of freely available implementations of SHA256 and the AES; it provides a command-line interface with operations for creating directories and copying files between Mnemosyne and the Unix filing system.

We use the IDA with polynomials over $GF(2^{16})$ for dispersal, and OCB-AES to provide combined encryption and authenticity. Local performance is plausible: we can copy in at around 64KB/s, and out at circa 375KB/s (for $n = 32$, $m = 96$).

The distributed block storage functionality is implemented as a set of Java classes over Tapestry [22]. The client uses a simple UDP-based protocol to com-

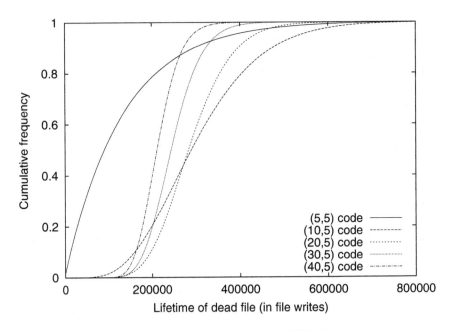

Fig. 2. File lifetimes in a simulated 4Mblock store

municate with a randomly picked Tapestry node. Read and write requests are then routed through Tapestry to the appropriate block store. Responses are returned to the client via the original Tapestry node. In early tests using 3 co-located nodes we can copy in files at around 80KB/s, and copy them out at 160KB/s.

We intend to make the code for Mnemosyne available in the near future.

7   Relation to Existing Work

Some recent systems have used distribution and self-organisation to provide robustness and availability [1,7,9,10,12]. Other systems use their decentralised nature to provide anonymity of access and prevent censorship [4,6,8,21].

Mnemosyne is more aligned with the latter class of system. However it provides in addition plausible deniability for clients, and is more suited to private storage and messaging applications than to the wide-scale publishing of data. Mnemosyne also shares some common ground with private information retrieval systems [5,16].

References

1. Ross Anderson. The Eternity Service. In *Proceedings of the 1st International Conference on the Theory and Applications of Cryptology (PRAGOCRYPT'96)*. CTU Publishing House, Prague, 1996.
2. Ross Anderson, Roger Needham, and Adi Shamir. The Steganographic File System. In *IWIH: International Workshop on Information Hiding*, 1998.
3. R. Canetti, C. Dwork, M. Naor, and R. Ostrovsky. Deniable encryption. *Lecture Notes in Computer Science*, 1294:90–104, 1997.
4. D. Chaum. Untraceable electronic mail, return addresses, and digital pseudonyms. *Communications of the ACM*, 24(2):84–88, February 1981.
5. Benny Chor, Oded Goldreich, Eyal Kushilevitz, and Madhu Sudan. Private Information Retrieval. In *IEEE Symposium on Foundations of Computer Science*, pages 41–50, 1995.
6. Ian Clarke, Oskar Sandberg, Brandon Wiley, and Theodore W. Hong. Freenet: A Distributed Anonymous Information Storage and Retrieval System. In *Workshop on Design Issues in Anonymity and Unobservability*, pages 46–66, July 2000.
7. F. Dabek, M. Kaashoek, D. Karger, R. Morris, and I. Stoica. Wide-area cooperative storage with CFS. In *Proceedings of the 18th ACM Symposium on Operating Systems Principles (SOSP '01), Banff, Canada.*, October 2001.
8. Roger Dingledine, Michael J. Freedman, and David Molnar. The Free Haven Project: Distributed Anonymous Storage Service. In *Workshop on Design Issues in Anonymity and Unobservability*, pages 67–95, July 2000.
9. Peter Druschel and Antony Rowstron. PAST: A large-scale, persistent peer-to-peer storage utility. In *Proceedings of the Eighth Workshop on Hot Topics in Operating Systems (HotOS-VIII). Schloss Elmau, Germany*, May 2001.
10. A. Iyengar, Robert Cahn, Juan A Garay, and Charanjit Jutla. Design and implementation of a secure distributed data repository. In *Proceedings of the 14th IFIP International Information Security Conference (SEC 98), New York, 1998.*, 1998.
11. Charanjit S. Jutla. Encryption modes with almost free message integrity. Cryptology ePrint Archive, Report 2000/039, 2000. http://eprint.iacr.org/.
12. John Kubiatowicz, David Bindel, Yan Chen, Steven Czerwinski, Patrick Eaton, Dennis Geels, Ramakrishna Gummadi, Sean Rhea, Hakim Weatherspoon, Westley Weimer, Chris Wells, and Ben Zhao. OceanStore: An Architecture for Global-Scale Persistent Storage. In *Proceedings of the Ninth international Conference on Architectural Support for Programming Languages and Operating Systems (ASPLOS 2000)*, November 2000.
13. Andrew D. McDonald and Markus G. Kuhn. StegFS: A Steganographic File System for Linux. In *Information Hiding*, number 1768 in LNCS, pages 462–477. Springer Verlag, 1999.
14. M. Rabin. Efficient dispersal of information for security, load balancing, and fault tolerance. *Communications of the ACM*, 36(2):335–348, April 1989.
15. S Ratnasamy, P. Francis, M. Handley, R. Karp, and S. Shenker. A Scalable Content-Addressable Network. In *Proceedings of ACM SIGCOMM 2001, San Diego, California, USA.*, August 2001.
16. Michael K. Reiter and Aviel D. Rubin. Crowds: anonymity for Web transactions. *ACM Transactions on Information and System Security*, 1(1):66–92, 1998.
17. Ronald L. Rivest. Chaffing and winnowing: Confidentiality without encryption. In *CryptoBytes (RSA Laboratories), Vol 4 No 1*, pages 12–17, 1998.

18. Phillip Rogaway, Mihir Bellare, John Black, and Ted Krovetz. OCB: A Block-Cipher Mode of Operation for Efficient Authenticated Encryption. In *Eighth ACM Conference on Computer and Communications Security (CCS-8)*. ACM Press, August 2001.

19. Antony Rowstron and Peter Druschel. Pastry: Scalable, decentralized object location and routing for large-scale peer-to-peer systems. In *Proceedings of the 18th IFIP/ACM Internation Conference on Distributed Systems Platforms (Middleware 2001), Heidelberg, Germany*, November 2001.

20. I. Stoica, R. Morris, D. Karger, F. Kaashoek, and H. Balakrishnan. Chord: A Scalable Peer-to-peer Lookup Service for Internet Applications. In *Proceedings of ACM SIGCOMM 2001, San Diego, California, USA.*, August 2001.

21. Marc Waldman, Aviel D. Rubin, and Lorrie Faith Cranor. Publius: A robust, tamper-evident, censorship-resistant, web publishing system. In *Proceeding of the 9th USENIX Security Symposium*, pages 59–72, August 2000.

22. Ben Y. Zhao, John D. Kubiatowicz, and Anthony D. Joseph. Tapestry: An Infrastructure for Fault-tolerant Wide-area Location and Routing. Technical Report UCB//CSD-01-1141, U. C. Berkeley, April 2000.

ConChord: Cooperative SDSI Certificate Storage and Name Resolution

Sameer Ajmani, Dwaine E. Clarke, Chuang-Hue Moh, and Steven Richman*

MIT Laboratory for Computer Science
200 Technology Square, Cambridge, MA 02139, USA
{ajmani,declarke,chmoh,richman}@lcs.mit.edu

Abstract. We present ConChord, a large-scale certificate distribution system built on a peer-to-peer distributed hash table. ConChord provides load-balanced storage while eliminating many of the administrative difficulties of traditional, hierarchical server architectures.

ConChord is specifically designed to support SDSI, a fully-decentralized public key infrastructure that allows principals to define local names and link their namespaces to delegate trust. We discuss the particular challenges ConChord must address to support SDSI efficiently, and we present novel algorithms and distributed data structures to address them. Experiments show that our techniques are effective and practical for large SDSI name hierarchies.

1 Introduction

SDSI (Simple Distributed Security Infrastructure) [19] is a proposed public key infrastructure that is more powerful and flexible than existing systems like DNS-EXT [7] and X.509 [17]. In SDSI, names are defined in local namespaces, and longer names can link multiple namespaces to delegate trust. This design obviates central certification authorities, allowing principals to declare and modify complex trust relationships.

For example, suppose Acme wants to allow access to their web site only to their partner companies' employees. SDSI allows Acme to define the group "Acme's partners" and delegate trust to each partner to define their own group of "employees." Acme's web server can enforce the access control policy by requiring that each HTTP client prove membership in the group "Acme's partners' employees." A client satisfies this requirement by presenting two certificates: one that shows that she is the "employee" of a company, and another that shows that her company is a "partner" of Acme.

Locating the certificates that a client needs is simple when certificates are stored at a central server, but this defeats the purpose of SDSI's decentralized design and scales poorly. We could distribute certificate storage using a server hierarchy, like DNS. However, unlike DNS, SDSI has no single root, and so requires some non-hierarchical way to locate the server that stores a certificate.

---

* Authors in alphabetical order.

P. Druschel, F. Kaashoek, and A. Rowstron (Eds.): IPTPS 2002, LNCS 2429, pp. 141–154, 2002.

The SPKI/SDSI IETF working group suggests embedding URIs in public keys for this purpose [9], but this seems undesirable, as changes to a key's URI invalidate certificates issued for that key. Also, since one SDSI name can be defined in terms of another, SDSI name resolution is fundamentally more complex than DNS name resolution. Certificates from many different organizations may be required to create a proof, and it is not always clear which organization should store a partial or completed proof.

Server hierarchies also suffer from administrative problems. A large fraction of DNS traffic is caused by "misconfiguration and faulty implementation of the name servers" [5]. Making such systems fault-tolerant requires even more expertise and resources.

To address these challenges, we present ConChord,[1] a distributed SDSI certificate directory built on a peer-to-peer storage system. Peer-to-peer systems [6, 8] configure themselves to provide immense storage capacity, high reliability, balanced load, and efficient lookups. ConChord uses the Chord [22] lookup system, with storage and caching techniques based on the Cooperative File System (CFS) [6]. ConChord locates certificates using relevant information (such as the name a certificate resolves), eliminating the need to embed URIs in public keys.

ConChord supports three operations: inserting a new certificate, resolving a name, and checking whether a name resolves to a specific key. ConChord's prototype implementation supports these operations, but does not yet support replication or recovery from network partitions. ConChord's current design does not handle server failures, restrict access to certificates, enforce storage quotas, or resist malicious attacks; these issues are left for future work.

The rest of this paper is organized as follows: Section 2 describes the capabilities and semantic richness of the SDSI naming system. Section 3 presents ConChord's data structures, algorithms, and storage design, and Section 4 presents a brief evaluation. Section 5 discusses related work, and Section 6 concludes.

2   SDSI Background

The main innovation of SDSI is the use of *local names*. Unlike DNS, in which names must be unique in a global namespace, a SDSI name has meaning relative to the principal defining that name. For instance, Professor X and Professor Y can each define the name "RAs" to refer to their respective research assistants. The two groups of RAs are referred to by the local names "K_{ProfX} RAs" and "K_{ProfY} RAs", where K_P is principal P's public key. In a system that uses SDSI for authorization, Professor X might add "K_{ProfX} RAs" to the access control list for a file, effectively stating that her RAs are the only principals allowed to access that file.

Principals define local names with two kinds of cryptographically-signed certificates: *reducing* and *non-reducing* [4]. A *reducing* certificate binds a local name to a principal. So, if Professor X wants to add Bob and Carol to her group of RAs, she can issue two reducing certificates:

---

[1] Certificates on Chord

$$K_{ProfX} \text{ RAs} \longrightarrow K_{Bob} \tag{1}$$
$$K_{ProfX} \text{ RAs} \longrightarrow K_{Carol}$$

The *value* of a SDSI name is the union of all keys that are bound to it, so here the value of the name "K_{ProfX} RAs" is the set $\{K_{Bob}, K_{Carol}\}$.

We call the operation that returns the value of a name *name resolution*, or simply resolution. We call the operation that verifies that a specified principal is in the set of keys bound to a name *membership checking*. Finally, we call the issuance of a new SDSI certificate *insertion*.

Although certificates can only be issued for local names (which have exactly one string component), resolutions and membership checks can be carried out for longer *extended names*. For instance, if MIT issues the certificate

$$K_{MIT} \text{ faculty} \longrightarrow K_{ProfX} \tag{2}$$

then we can resolve the extended name "K_{MIT} faculty RAs". Semantically, this name denotes all principals that have been designated as RAs by all principals designated as MIT faculty. Given the above certificates, this name resolves to the set $\{K_{Bob}, K_{Carol}\}$. If MIT also issued the certificate "K_{MIT} faculty" $\longrightarrow$ K_{ProfY}, then "K_{MIT} faculty RAs" would also include Professor Y's RAs. Bob can prove that he is a member of "K_{MIT} faculty RAs" by presenting the *sequence* of certificates (2)(1); anyone can verify this proof by checking the signatures on the two certificates.

The second type of certificate is the *non-reducing* certificate, which binds a local name to another (local or extended) name:

$$K_{MIT} \text{ staff} \longrightarrow K_{MIT} \text{ faculty}$$
$$K_{MIT} \text{ staff} \longrightarrow K_{MIT} \text{ faculty RAs} \tag{3}$$
$$K_{MIT} \text{ staff} \longrightarrow K_{HR} \text{ visiting}$$

Notice that the right-hand sides (called *subjects*) of the above non-reducing certificates are names, whereas the subjects of reducing certificates are principals' keys. A non-reducing certificate states that the value of a local name includes the value of the subject. So, given these certificates, we can resolve "K_{MIT} staff" as the union of the values of "K_{MIT} faculty", "K_{MIT} faculty RAs", and "K_{HR} visiting".

Since the name bound by reducing certificate (2), "K_{MIT} faculty", is a prefix of the subject of non-reducing certificate (3), these certificates are called *compatible*, and we can *compose* (3) with (2) to yield a new, derived certificate:

$$K_{MIT} \text{ staff} \longrightarrow K_{ProfX} \text{ RAs} \tag{4}$$

This new certificate doesn't introduce any new trust relationships. Rather, it represents a trust relationship that already exists (we can use the original, signed certificates to prove this fact).

If we repeatedly perform all possible compositions over a certificate set until no more compositions are possible, we eventually have a set of reducing certificates that directly bind each local name to each key in that name's value.

We call such a set *closed* under name-reduction. This closure is important for supporting efficient name resolutions and membership checks.

3   Design

ConChord's key design assumption is that membership checking is by far the most common operation on SDSI names, followed by name resolution. Insertion is comparatively rare. Accordingly, ConChord maintains closure over its certificates on each insertion, thereby reducing the amount of work required for name resolution and membership checking. Users can thus accelerate resolutions and checks for extended names by inserting non-reducing certificates.

ConChord's algorithms use three hash tables: check, value, and compatible (proposed in [10]). These tables are summarized in Table 1.

Membership Checking Every certificate inserted into ConChord is stored in the check table, where the hash key for each certificate c is a function applied to the tuple $\langle c$'s name, c's subject$\rangle$. If multiple certificates that bind the same name to the same subject are inserted into the check table, then the certificate with the latest expiration time overwrites the others.

To check whether a key K is bound to name n, we can resolve n and check whether K is in the resulting set. If n is a local name (like "K_{MIT} staff"), then the closure property guarantees that the binding from n to K (if one exists) is already in the check table, so we can instead fetch $\langle n, K \rangle$ directly from check.

Name Resolution For each local name bound by a certificate, value stores the set of keys bound to that name. The hash key for value is a function of the name.

To resolve a local name, we just look it up in value. To resolve an extended name, we look up the value of the name's *prefix* (the prefix of an extended name "$K\ n_1 \dots n_m$" is the local name "$K\ n_1$"), then we recursively resolve the rest of the name. For instance, to resolve "K_{MIT} staff spouse", we first fetch the value for "K_{MIT} staff". Then, for each staff member K_S, we fetch the value for "K_S spouse" and take the union of those values to compute the result.

Table 1. ConChord Hash Tables

Table	Index	Value
check	name, subject	an entry whose *name* is name and whose *subject* is subject
value	name	a set of entries whose *name* is name and whose *subject* is a public key
compatible	name	a set of entries whose *subject* is a name that starts with name

Insertion The above algorithms rely on two invariants. First, check and value are both up to date with respect to each other (if a name binding is in value, then the corresponding certificate is in check, and vice versa). We maintain this invariant by updating both tables when new certificates are inserted.

Second, closure is always maintained over the certificates. To maintain this invariant, we compose each new certificate with each other compatible certificate in the system. We then recursively insert the resulting derived certificates, since they may trigger further compositions.

When a new non-reducing certificate is inserted, we locate all compatible reducing certificates by looking up the prefix of the new certificate's subject in the value table. When a new reducing certificate is inserted, we must locate all compatible non-reducing certificates. To make this fast, ConChord maintains a third table, compatible, that stores non-reducing certificates, where the hash key of a certificate is a function of the prefix of its subject.

Maintaining Proofs We have said that check stores certificates, value stores keys, and compatible stores non-reducing certificates. In reality, these tables store *entries*, which are proofs of name bindings, and a single proof might consist of a sequence of certificates (if the binding was derived from a composition).

An entry consists of a *name*, a *subject*, and a certificate *sequence* that proves that the *name* is bound to the *subject*. For example, the entry for the derived certificate (4) would be:

$$name = K_{MIT} \text{ staff}$$
$$subject = K_{ProfX} \text{ RAs}$$
$$sequence = (3), (3)_{K_{MIT}^{-1}}, (2), (2)_{K_{MIT}^{-1}}$$

where $X_{K^{-1}}$ represents the digital signature of X using K^{-1}, K's private key.

Like certificates, entries can be composed, in which case their sequences are concatenated. The expiration time of an entry is the earliest expiration of any certificate in its sequence.

Figure 1 presents the complete insertion algorithm using entries.

3.1   Peer-to-Peer Architecture

ConChord locates entries on servers using the Chord [22] distributed lookup system.[2] ConChord distributes its hash tables by mapping each hash key to a *Chord ID*. Clients access the hash tables by calculating the Chord ID for each hash key and contacting the appropriate server.

A problem, however, arises with maintaining our invariants on each insertion. The first invariant (if an entry is in check, it is also in value or compatible, and vice versa) might be violated if a client crashes during an insertion. The second invariant (closure is maintained over the certificate set) might be violated if two compatible certificates are inserted concurrently or if a client crashes before inserting all derived certificates.

---

[2] ConChord could also use CAN [18], Pastry[20], or Tapestry[23].

insert(*certificate c*)
   entry e
   e.name ← $K\ n$ (the name bound by *c*)
   e.subject ← *c's subject*
   e.sequence ← c, c_{K-1}
   insert(*e*)

insert(*entry e*)
  if (check[*e.name, e.subject*] is empty)
     check[*e.name, e.subject*] ← *e*
     if (*e.subject* is a public key)
        value[*e.name*] ← value[*e.name*] ∪ {*e*}
        set ← compatible[*e.name*]
        for each $e' \in set$
           insert(**compose**(e', *e*))
     else
        compatible[**prefix**(*e.subject*)]
          ← compatible[**prefix**(*e.subject*)] ∪ {*e*}
        set ← value[**prefix**(*e.subject*)]
        for each $e' \in set$
           insert(**compose**(*e,e'*))
  else if (check[*e.name, e.subject*] expires before *e*)
     check[*e.name, e.subject*] ← *e*

// requires $e_1.subject = e_2.name \cdot X$
// for some (possibly empty) sequence of strings X
// returns the composed entry e
compose(*entry e_1, entry e_2*)
   e.name ← $e_1.name$
   e.subject ← $e_2.subject \cdot X$
   e.sequence ← $e_1.sequence \cdot e_2.sequence$
   return *e*

Fig. 1. Insertion with closure

We could solve these problems using synchronization to provide transactional consistency for insertions; however, this is slow in the wide area. Instead, we allow the system to temporarily violate our invariants in the rare case that a problem occurs. To restore consistency, each server periodically reinserts the check entries it stores, so all compositions eventually happen. This is an efficient solution because the work of reinsertion is spread among the servers, and reinsertions can be infrequent.

Allowing such temporary inconsistencies safe with respect to security; they can only make some entries temporarily unavailable. Since SPKI/SDSI semantics are monotonic, the inability to locate some certificates cannot grant undeserved authority [12].

Storage Details The value and compatible tables store sets of entries, rather than single entries. A very large set (such as the value of "K_{USA} citizens") might cause load imbalance or even exceed the capacity of a single server. Therefore, ConChord distributes entries in a set among several servers.

We might consider using CFS-style Merkle trees to distribute large data sets [6], but such data structures do not support concurrent modification by multiple clients. Because ConChord periodically reinserts entries and garbage-collects expired entries, sets must support concurrent modification. To do so, ConChord distributes the elements of a set over many servers, but serializes set modifications through a single server.

The members of the set whose Chord ID is s are stored at Chord IDs $hash(s, 1) \ldots hash(s, T)$, where T is the size of the set. The value of T is stored as a *set size record* at ID s. Servers support two atomic operations on set size records: *get-size* and *increment-and-get*.

To fetch the members of a set, a client calls *get-size*, calculates the Chord IDs for all of the set's members, and retrieves them in parallel. To optimize for singleton sets, the client fetches the first entry of a set in parallel with the size.

To add a new entry to set s, a client first calls *increment-and-get*. This increments T and returns the updated size, T'. The client then stores the new entry at ID $hash(s, T')$.

When an entry e in a set expires, the server storing e first looks for updated versions of the expired certificates in the check table. If no new certificates are found, the server storing e tells the set size server that e is no longer valid. The set size server compacts the set by fetching the last element in the set (e'), overwriting e with e', and decrementing the set size. As an optimization, the set size server can instead direct the next set insertion to overwrite e.

Recall that servers periodically reinsert entries; this involves (1) checking that each entry appears in the appropriate value or compatible set, and (2) checking that each entry is composed with all other compatible entries. The first check involves scanning the appropriate set; once done, this check need not be repeated. However, the second check needs to be repeated indefinitely in case new compatible entries are added. In the common case, the set of compatible entries will be unchanged from the previous reinsertion. To make checking for changes fast, we store a version number alongside each set size record. The version number is incremented each time an element is added to the set. Thus, a reinsertion usually only needs to check that the version number is unchanged, which is a single Chord lookup.

Network Partitions If a network partition splits the set of ConChord servers, servers in different partitions may store different values for the same Chord ID. When the partition heals, ConChord automatically resolves such inconsistencies. The server responsible for storing a set entry (in the healed partition) temporarily stores all entries accepted for that ID and lazily diverts all but one entry to the end of the set. Similarly, the server responsible for a set size record temporarily accepts the maximum size value proposed by any server, and lazily corrects the size (if necessary).

Load Imbalance To balance request load for popular entries, ConChord caches entries along lookup paths, as in CFS. Cached entries are expunged from a full cache in LRU order. Cached copies may become out-of-date, so servers assign them time-to-live values. Cached set size records have small TTLs, since insertions and compactions change the actual size values. Cached set entries have relatively larger TTLs, since they only become invalid when a server temporarily stores multiple entries (after a partition heals) or when a set is compacted after an expiration.

Clients and servers can detect and recover from out-of-date set size records by fetching past the expected end of the set until no more entries are found (locations $T+1$, $T+2$, etc.[3]). Set modifications cannot use cached set size records; they must update the original record.

3.2 Accelerating the Operations

Resolution of an extended name requires a value lookup for each part of the name, so resolution latency scales with name length. To reduce the number of lookups, we allow clients to *share resolutions* by caching extended name resolutions in the value and check tables. Then, clients can use cached prefixes when resolving a name. For example, if "K_{MIT} faculty assistant" $\longrightarrow$ $\{K_A, K_B\}$ is cached, resolving "K_{MIT} faculty assistant supervisor" can resolve "K_A supervisor" and "K_B supervisor" directly.

Figure 2 presents a name resolution algorithm that takes advantage of cached resolutions. Calls to **yield** return proofs of the resolution; calls to **insert** mark resolutions that are cached back into ConChord.

Another way to accelerate extended name resolutions is to leverage closure. For example, if we know that we will need to resolve the name "K_{MIT} faculty assistants", we could create an entry whose *name* and *subject* are both that name and whose *sequence* is empty. We call such an entry a *truism*, as it simply states that a name is bound to itself. Since the subject of a truism is a name, it is stored in the compatible table. Closure then causes the values of the name to be stored in the value table, so the name can be resolved in a single lookup!

Li et al. [15] propose that membership checking can adapt between issuer-to-subject and subject-to-issuer searches to avoid large branching factors in the certificate graph. Implementing this algorithm on ConChord simply requires maintaining a subject-to-issuer table for entries and is an area of future work.

4   Evaluation

4.1 DNS Traces

We evaluate the effectiveness of name resolution sharing using a trace of 30,000 DNS requests captured at MIT's Laboratory for Computer Science [14]. We do not propose ConChord as a replacement for DNS; rather, we use the trace to

---

[3] Binary search is possible by fetching $T+2$, $T+4$, etc.

resolve (*name n*)
   entry e
   e.name ← *n*
   e.subject ← *n*
   e.sequence ← ∅
   resolve(*e*)

resolve (*entry e*)
   name n ← *e.subject*
   for *i* ← *n.length* **down to** 1
     set ← value[**prefix**(*n,i*)]
     for each *e'* ∈ *set*
       entry r ← **compose**(*e, e'*)
       if (*r.subject* is a public key)
         yield(*r*)
         insert(*r*)
       else
         resolve(*r*)
         entry p ← **extract**(*r*)
         if(*p* ≠ *e'*)
           insert(*p*)

// *requires r.name* = *N* · *X and r.subject* = *K* · *X*
// *for some name N, public key K,*
// *and (possibly empty) sequence of strings X*
// *returns the extracted entry e*
extract (*entry r*)
   e.name ← *N*
   e.subject ← *K*
   e.sequence ← *r.sequence*
   return *e*

Fig. 2. Name resolution

generate a simple SDSI name hierarchy and a realistic name resolution workload. For each DNS address query of the form "`www.foo.com`", we generate a name resolution request "K_{dns} com foo www" and a set of certificates:

$$K_{dns} \text{ com} \longrightarrow K_{dns.com}$$
$$K_{dns.com} \text{ foo} \longrightarrow K_{dns.com.foo}$$
$$K_{dns.com.foo} \text{ www} \longrightarrow K_{dns.com.foo.www}$$

None of the certificates expire during the trace.

We run the trace between a single client and server and count the number of *sequential* lookups made for each request. While the total number of lookups for a name of length l is $O(l^2)$ (due to fetching prefixes in parallel), the number of

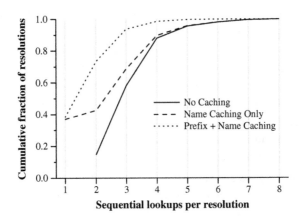

Fig. 3. Cumulative distribution of sequential lookups per DNS name resolution. The *No Caching* distribution is equivalent to the distribution of DNS name lengths.

sequential lookups (thus, latency) is $O(l)$. Since every lookup is for a singleton set in the value table, we expect exactly l sequential lookups per resolution.[4]

To evaluate the effectiveness of name resolution sharing, we run the trace with no caching, with caching of full name resolutions only (like a DNS proxy), and with caching of full names and name prefixes. Each resolution caches all its results (and prefixes) before the next one begins.

Figure 3 plots the cumulative distribution of sequential lookups per name resolution for each trace. With no caching, one lookup is made for each part of the DNS name, so the distribution of lookups is the same as the distribution of name lengths. Full name caching reduces the mean by 23%, but variance is high, since many names are requested only once. Prefix caching reduces the mean by 43%, and 73% of the requests succeed in one or two lookups, suggesting that many requests share a common prefix.

We conclude that prefix caching is quite effective at reducing the latency of name resolutions for this dataset. This is not particularly surprising, as prefix caching is analogous to NS record caching in DNS (shown to be particularly effective in [14]). However, we believe that prefix caching will also benefit other datasets with hierarchical structures.

4.2   Mailing Lists

The DNS dataset is too simple to require closure over its certificates, so we evaluate the overhead of closure using a second dataset based on MIT course mailing

---

[4] We could also have created truisms for each DNS hostname and thus reduced each name resolution to a single lookup. This might be reasonable, since each domain owner knows in advance what hostnames are valid and might be resolved.

lists. Course lists are composed of section lists, which are in turn composed of students, forming a widely-branching hierarchy of large groups. We gathered mailing lists for 27 courses, containing 38 sections and 2,073 students (1,706 distinct) and used the lists to generate a total of 5,624 certificates. For each entry of the form "6.033-students: 6.033-sec9: alice" (6.033 is a course number), we add the following certificates to an insertion trace (suppressing duplicates):

$$K_{mit} \text{ registered} \longrightarrow K_{mit} \text{ courses students}$$
$$K_{mit} \text{ students} \longrightarrow K_{mit} \text{ alice} \qquad K_{mit} \text{ alice} \longrightarrow K_{alice}$$
$$K_{mit} \text{ courses} \longrightarrow K_{mit} \text{ 6.033} \qquad K_{mit} \text{ 6.033} \longrightarrow K_{6.033}$$
$$K_{6.033} \text{ students} \longrightarrow K_{6.033} \text{ secs students} \qquad K_{6.033} \text{ secs} \longrightarrow K_{6.033} \text{ sec9}$$
$$K_{6.033} \text{ sec9} \longrightarrow K_{sec9} \qquad K_{sec9} \text{ students} \longrightarrow K_{mit} \text{ alice}$$

If the course does not have recitation sections, students are added directly to the course's "students" group. This dataset is designed to support a number of useful queries, such as determining whether Alice is registered in 6.033 or enumerating all the students registered in MIT courses.

We count the number of Chord lookups required to insert each certificate and the resulting derived certificates. Figure 4 shows that the number of lookups per insertion is fairly constant. This is because the number of compositions needed to maintain closure after adding a member to a group is proportional to the number of parent groups affected, which is usually small. For example, adding a student to a section only requires closure with the groups that transitively contain that section. The raised parts of the curves correspond to insertions for courses with sections, as these require one addition composition to maintain the section list. Reordering the trace changes the distribution of lookups per insertion, but does not affect the total number of lookups. We conclude that maintaining closure is practical for such datasets.

Since no access trace is available for this dataset, we cannot evaluate the total benefit that closure provides for name resolutions or membership checks. However, specific examples show that the benefit can be substantial: given closure, a membership check for any local name requires a single check lookup. Without closure, a check on a name like "K_{MIT} registered" can require up to eight successive lookups to retrieve the necessary certificates.

5   Related Work

Alternatives to SDSI, such as DNSEXT [7] and X.509 [17], are used almost exclusively for Internet host identification, rather than applications like webs of trust or access control. While X.509 could support richer applications, it is not deployed in any way that facilitates them. PGP [24] supports user-authorized names and webs of trust, but not linked namespaces or named groups. Policy-Maker [3] and Keynote [2] support more general policies than SDSI, but they do not specify a way to locate the certificates needed to satisfy a particular policy.

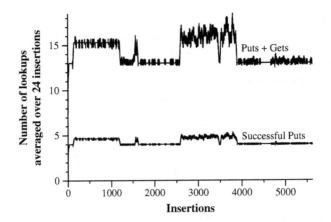

Fig. 4. Number of Chord lookups averaged over 24 insertions for the MIT course mailing list dataset. Each lookup is used either to "put" data in or "get" data from the system. Each insertion triggers a (nearly) constant number of other insertions to maintain closure. Insertions are grouped by course; the raised parts of the curves indicate additional compositions for insertions into courses with sections.

Previous work [1, 15] proposes algorithms for resolving SDSI names using a distributed set of certificates, but does not address the practical challenges of storing and locating those certificates. Nikander and Viljanen [16] describe how to deploy SPKI/SDSI [21] using DNS, but do not support SDSI name resolution.

QCM [11] introduced *policy-directed certificate retrieval* as a general technique for locating the certificates needed to satisfy a given assertion. QCM and its successor, SD3 [13], use authoritative servers to implement distributed resolution of SDSI-like names and rely on embedded URIs or IPs to map principals to servers. ConChord supports policy-directed certificate retrieval to resolve SDSI names and eliminates the need for a mapping between principals and servers. While ConChord loses some of the benefits of authoritative servers, such as online signing and control over certificate dissemination, ConChord gains scalability, self-configuration, and load-balance.

6   Conclusion

We have presented ConChord, a distributed SDSI certificate directory built on a peer-to-peer system. ConChord supports three operations: membership checks, name resolutions, and certificate insertions. To accelerate checks and resolutions, ConChord maintains closure on each insertion and supports name resolution sharing. Experiments show that these techniques are effective and practical.

ConChord provides a novel deployment design that offers a number of practical advantages over traditional, hierarchical server architectures. ConChord eliminates any need to embed location information in certificates and automatically balances load among storage servers. Servers periodically reinsert entries to guarantee eventual consistency and can automatically resolve conflicts that occur due to network partitions.

Our prototype implementation supports the basic features described in this paper. Future work includes implementing replication, supporting SPKI/SDSI authorization certificates and revocation, limiting per-user storage, handling malicious failures, and generalizing ConChord for use with other certificate systems.

Acknowledgments

We thank Barbara Liskov for her valuable advice on the presentation of this paper. We also thank Russ Cox, Robert Morris, Athicha Muthitacharoen, Ronald Rivest, Emil Sit, and the anonymous referees for their helpful comments.

References

[1] S. Ajmani. A Trusted Execution Platform for multiparty computation. Master's thesis, MIT, 2000. App A: Certificate Chain Algorithms.

[2] M. Blaze, J. Feigenbaum, and A. D. Keromytis. Keynote: Trust management for public-key infrastructures (position paper). In *Security Protocols Workshop*, pages 59–63, 1998.

[3] M. Blaze, J. Feigenbaum, and J. Lacy. Decentralized trust management. Technical Report 96-17, 28, 1996.

[4] D. Clarke, J. Elien, C. Ellison, M. Fredette, A. Morcos, and R. L. Rivest. Certificate chain discovery in SPKI/SDSI. *Journal of Computer Security*, 2001.

[5] R. Cox and A. Muthitacharoen. Serving DNS using Chord. In *Proc. IPTPS*, 2002.

[6] F. Dabek, M. F. Kaashoek, D. Karger, R. Morris, and I. Stoica. Wide-area cooperative storage with CFS. In *Proc. ACM SOSP*, Oct. 2001.

[7] DNS extensions (IETF DNSEXT), Mar. 1999. http://www.ietf.org/html.charters/dnsext-charter.html.

[8] P. Druschel and A. Rowstron. PAST a large-scale, persistent peer-to-peer storage utility. In *HotOS VIII*, May 2001.

[9] C. Ellison, B. Frantz, B. Lampson, R. Rivest, B. Thomas, and T. Ylonen. SPKI certificate theory. RFC 2693, Sept. 1999.

[10] C. M. Ellison and D. E. Clarke. High speed TUPLE reduction. Memo, Intel, 1999.

[11] C. A. Gunter and T. Jim. Policy-directed certificate retreival. Technical Report MS-CIS-99-07, U. Penn., Sept. 1998.

[12] J. Y. Halpern and R. van der Meyden. A logic for SDSI's linked local name spaces. *Journal of Computer Security*, 9(1,2):47–74, 2000.

[13] T. Jim. SD3: A trust management system with certified evaluation. In *Proc. 2001 IEEE Symposium on Security and Privacy*, May 2001.

[14] J. Jung, E. Sit, H. Balakrishnan, and R. Morris. DNS performance and the effectiveness of caching. In *Proc. ACM SIGCOMM Internet Measurement Workshop*, 2001.

[15] N. Li, W. H. Winsborough, and J. C. Mitchell. Distributed credential chain discovery in trust management. In *Proc. 8th ACM CCS*, Nov. 2001.

[16] P. Nikander and L. Viljanen. Storing and retrieving internet certificates. In *Proc. 3rd Nordic Workshop on Secure IT Systems*, 1998.

[17] Public-key infrastructure (IETF PKIX), Feb. 2000. http://www.ietf.org/html.charters/pkix-charter.html.

[18] S. Ratnasamy, P. Francis, M. Handley, R. Karp, and S. Shenker. A scalable content-addressable network. In *Proc. ACM SIGCOMM*, 2001.

[19] R. L. Rivest and B. Lampson. SDSI – A simple distributed security infrastructure. Apr. 1996.

[20] A. Rowstron and P. Druschel. Pastry: Scalable, distributed object location and routing for large-scale peer-to-peer systems. In *Proc. IFIP/ACM Middleware*, 2001.

[21] Simple public key infrastructure (IETF SPKI), Feb. 1998. http://www.ietf.org/html.charters/spki-charter.html.

[22] I. Stoica, R. Morris, D. Karger, M. Kaashoek, and H. Balakrishnan. Chord: A scalable peer-to-peer lookup service for Internet applications. In *Proc. ACM SIGCOMM*, Aug. 2001.

[23] B. Y. Zhao, J. Kubiatowicz, and A. Joseph. Tapestry: An infrastructure for fault-tolerant wide-area location and routing. Technical Report UCB/CSD-01-1141, UC Berkeley, Apr. 2001.

[24] P. R. Zimmermann. *The Official PGP User's Guide*. MIT Press, 1995.

Serving DNS Using a Peer-to-Peer Lookup Service

Russ Cox*, Athicha Muthitacharoen, and Robert T. Morris

MIT Laboratory for Computer Science
`rsc,athicha,rtm@lcs.mit.edu`

Abstract. The current domain name system (DNS) couples ownership of domains with the responsibility of serving data for them. The DNS security extensions (DNSSEC) allow verificaton of records obtained by alternate means, opening exploration of alternative storage systems for DNS records. We explore one such alternative using DHash, a peer-to-peer distributed hash table built on top of Chord. Our system inherits Chord's fault-tolerance and load balance properties, at the same time eliminating many administrative problems with the current DNS. Still, our system has significantly higher latencies and other disadvantages in comparison with conventional DNS. We use this comparison to draw conclusions about general issues that still need to be addressed in peer-to-peer systems and distributed hash tables in particular.

1   Introduction and Related Work

In the beginnings of the Internet, host names were kept in a centrally-adminis-tered text file, `hosts.txt`, maintained at the SRI Network Information Center. By the early 1980s the host database had become too large to disseminate in a cost-effective manner. In response, Mockapetris and others began the design and implementation of a distributed database that we now know as the Internet domain name system (DNS) [8, 9].

Looking back at DNS in 1988, Mockapetris and Dunlap [9] listed what they believed to be the surprises, successes, and shortcomings of the system. Of the six successes, three (variable depth hierarchy, organizational structure of names, and mail address cooperation) relate directly to the adoption of an administrative hierarchy for the names. The administrative hierarchy of DNS is reflected in the structure of DNS servers: in typical usage, an entity is responsible not only for maintaining name information about its hosts but also for the serving that information.

The fact that service structure mirrored administrative hierarchy provided a modicum of authentication for the returned data. Unfortunately, IP addresses can be forged and thus it is possible for malicious people to impersonate DNS servers. In response to concerns about this and other attacks, the DNS Security Extensions [5] (DNSSEC) were developed in the late 1990s. DNSSEC provides

---

* Russ Cox was supported by a Hertz Fellowship while carrying out this research.

P. Druschel, F. Kaashoek, and A. Rowstron (Eds.): IPTPS 2002, LNCS 2429, pp. 155–165, 2002.

a stronger mechanism for clients to verify that the records they retrieve are authentic.

DNSSEC effectively separates the authentication of data from the service of that data. This observation enables the exploration of alternate service structures to achieve desirable properties not possible with conventional DNS. In this paper, we explore one alternate service structure based on Chord [10], a peer-to-peer lookup service.

Rethinking the service structure allows us to address some of the current shortcomings in the current DNS. The most obvious one is that it requires significant expertise to administer. In their book on running DNS servers using BIND, Albitz and Liu [1] note that many of the most common name server problems are configuration errors. Name servers are difficult and time-consuming to administer; ordinary people typically rely on ISPs to serve their name data. Our approach solves this problem by separating service from authority — clients can enter their data into the Internet-wide Chord storage ring and not worry about needing an ISP to be online and to have properly configured its name server.

DNS performance studies have confirmed this folklore. In 1992, Danzig *et al.* [4] found that most DNS traffic was caused by misconfiguration and faulty implementation of the name servers. They also found that one third of the DNS traffic that traversed the NSFNet was directed to one of the seven root name servers. In 2000, Jung *et al.* [6] found that approximately 35% of DNS queries never receive an answer or receive a negative answer, and attributed many of these failures to improperly configured name servers or incorrect name server (NS) records. The study also reported that as much as 18% of DNS traffic is destined for the root servers.

Serving DNS data over Chord eliminates the need to have every system administrator be an expert in running name servers. It provides better load balance, since the concept of root server is eliminated completely. Finally, it provides robustness against denial-of-service attacks since disabling even a sizable number of hosts in the Chord network will not prevent data from being served.

Unfortunately, DNS over Chord suffers some notable performance problems as well as significant reductions in functionality.

2   Design and Implementation

We have implemented a prototype of our system, which we call DDNS.

Our system handles lookups at the granularity of resource record sets (RRSets), as in conventional DNS. An RRSet is a list of all the records matching a given domain name and resource type. For example, at the time of writing, *www.nytimes.com* has three address (A) records: 208.48.26.245, 64.94.185.200, and 208.48.26.200. These three answers compose the RRSet for (*www.nytimes.com*, A).

DNSSEC uses public key cryptography to sign resource record sets. When we retrieve an RRSet from an arbitrary server, we need to verify the signature (included as a signature (SIG) record). To find the public key that should have signed the RRSet, we need to execute another DNS lookup, this time for a public key (KEY) RRSet. This RRSet is in turn signed with the public key for the enclosing domain. For example, the (*www.nytimes.com*, A) RRSet should be signed with a key listed in the (*www.nytimes.com*, KEY) RRSet. The latter RRSet should be signed with a key listed in the (*nytimes.com*, KEY) RRSet, and so on to the hierarchy root, which has a well-known public key.

DDNS stores and retrieves resource record sets using DHash [3], a Chord-based distributed hash table. DHash has two properties useful for this discussion: load balance and robustness.

DHash uses consistent hashing to allocate keys to nodes evenly. Further, as each block is retrieved, it is cached along the lookup path. If a particular record is looked up n times in succession starting at random locations in a Chord ring of m nodes, then with high probability each server transfers a given record only $\log m$ times total before every server has the record cached.

DHash is also robust: as servers come and go, DHash automatically moves data so that it is always stored on a fixed number of replicas (typically six). Because the replicas that store a block are chosen in a pseudo-random fashion, a very large number of servers must fail simultaneously before data loss occurs.

To create or update a DDNS RRSet, the owner prepares the RRSet, signs it, and inserts it into DHash. The key for the RRSet is the SHA1 hash of the domain name and the RRSet query type (*e.g.*, SHA1(*www.nytimes.com*, A)). DHash verifies the signature before accepting the data. When a client retrieves the RRSet, it also checks the signature before using the data.

Naively verifying a DNS RRSet for a name with n path elements requires n KEY lookups. We address this problem by allowing the owner to present additional relevant KEYs in the RRSet. To avoid inflating the responses, we can omit KEY RRSets for popular names. For example, the record containing the (*www.nytimes.com*, A) RRSet might also include the (*www.nytimes.com*, KEY) RRSet and the (*nytimes.com*, KEY) RRSet but omit the (*.com*, KEY) RRSet on the assumption that it would be widely cached. The key for the root of the hierarchy is assumed to be known by all DDNS servers, just as the IP addresses of the root servers are known in the current DNS.

To ease transition from conventional DNS to our system, a simple loopback server listening on 127.1 could accept conventional DNS queries, perform the appropriate Chord lookup, and then send a conventional response. Then systems could simply be configured to point at 127.1 as their name server.

3   Evaluation

To evaluate the use of Chord to serve DNS data, we used data from the study by Jung *et al.* on a simulated network with 1000 DDNS nodes. For the test, we turned off replication of records. We did not simulate node failures, so the only effect of replication would be to serve as *a priori* caching. Since the network is a fair amount smaller than the expected size of a real Chord ring, replication would bias the results in our favor because of the increased caching.

We first inserted answers to all the successful queries into the Chord network and then ran a day's worth of successful queries (approximately 260,000 queries), measuring storage balance, load balance while answering queries, and the number of RPCs required to perform each lookup. The distribution of names in the day's queries is heavy tailed.

To simulate failed DNS queries, we started a similar network, did not insert any data, and executed a day's worth of unresolved queries (approximately 220,000 queries), measuring load balance and RPC counts. These queries were distributed similarly to the successful queries.

Finally, to simulate popular entries, we ran what we called the "slashdot test," inserting one record and then fetching it a hundred thousand times.

All three tests began each lookup at a random node in the Chord network.

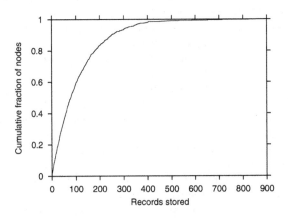

Fig. 1. Storage balance for 120,000 records. The graph shows a cumulative distribution for the number of records stored on each node. Perfect balance would place the same number of records on each node, making the cumulative distribution a vertical line around 120.

For the successful queries test, we inserted approximately 120,000 records to serve as answers to the 260,000 queries. Figure 1 shows that the median number of records stored per node is about 120, as expected. The distribution is exponential in both directions because Chord nodes are randomly placed on a circle and store data in proportion to the distance to their next neighbor clockwise around the circle. Irregularities in the random placement cause some nodes to store more data than others. The two nodes that stored in excess of 800 records (824 and 827) were both responsible for approximately 0.8% of the circle, as compared with an expected responsibility of 0.1%. Even so, this irregularity drops off exponentially in both directions, and can be partially addressed by having servers run multiple nodes in proportion to their storage capacities. We conclude that DDNS does an adequate job of balancing storage among the peers.

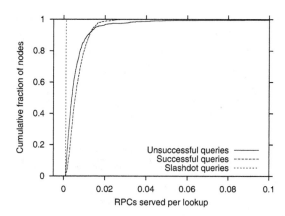

Fig. 2. Load balance. The graph shows a cumulative distribution for the number of RPCs served per lookup by each node during the test. Since there are a thousand nodes, ideal behavior would be involving each node in $n/1000$ RPCs, where n is the median number of RPCs executed per lookup (see Figure 3).

Even though storage is balanced well, some records are orders of magnitude more popular than others. Since records are distributed randomly, we need to make sure that nodes that happen to be responsible for popular records are not required to serve a disproportionate amount of RPCs. DHash's block caching helped provide load balance as measured by RPCs served per lookup per node. As shown in Figure 2, in the successful query test, nodes served RPCs in approximate proportion to the number of records they stored. Specifically, each node serves each of its popular blocks about ten ($\log_2 1000$) times; after that, the block is cached at enough other nodes that queries will find a cached copy instead of reaching the responsible server. A similar argument shows that very quickly every node has a copy of incredibly popular blocks, as evidenced by the Slashdot test: after the first few thousand requests, virtually every node in the system has the record cached, so that subsequent requests never leave the originating node.

For the unsuccessful query test, nodes served RPCs in proportion to the number of queries that expected the desired record to reside on that node. This does a worse job of load balancing since there is no negative caching.

The graphs shows that the loads are similar for both successful and unsuccessful queries, unlike in the current DNS, where unresponsive queries might result in spurious retransmissions of requests.

Figure 3 shows the number of RPCs required by a client for various lookups. Successful queries and unsuccessful queries have the same approximately random distribution of hop counts, except that successful queries usually end earlier due to finding a cached copy of the block. Since the slashdot record got cached everywhere very quickly, virtually all lookups never left the requesting node.

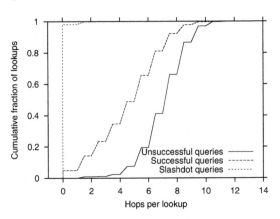

Fig. 3. Client load to perform lookups.

We would like to be able to compare the latency for DNS over Chord with the latency for conventional DNS. This is made difficult by the fact that we do not have an Internet-wide Chord ring serving DNS records. To compensate, we took the latency distribution measured in the Jung. *et al.* study and used it to compute the expected latencies of DNS over Chord in a similar environment.

Figure 4 shows the distribution of latency for successful lookups in the Jung. *et al.* one-day DNS trace. Because we don't have individual latencies for requests that contacted multiple servers, only requests completed in one round trip are plotted. The tail of the graph goes out all the way to sixty seconds; such slow servers would not be used in the Chord network, since timeouts would identify them as having gone off the network. Since such a small fraction of nodes have such long timeouts, we cut the largest 2.5% of latencies from the traces in order to use them for our calculations. To be fair, we also cut the smallest 2.5% of the latencies from the traces, leaving the middle 95%.

Figure 5 shows the correlation between hops per lookup and expected latency. For each hop count h, we randomly selected and summed h latencies from the Jung. *et al.* distribution. Each point in the graph is the average of 1000 random sums; the bars show the standard deviations in each direction.

Figure 6 displays the expected latency in another way. Here, for each query with hop count h, we chose a random latency (the sum of h random latencies from the Jung. *et al.* distribution) and used that as the query latency. The cumulative distribution of these query latencies is plotted.

These experiments show that lookups in DDNS take longer than lookups in conventional DNS: our median response time is 350ms while conventional DNS's is about 43ms. We have increased the median in favor of removing the large tail: lookups taking 60s simply cannot happen in our system.

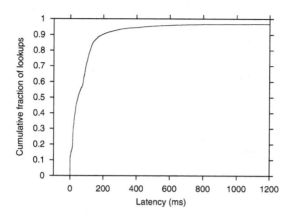

Fig. 4. Lookup latency distribution for successful one-server queries over conventional DNS in the Jung. *et al.* data. Of the 260,000 successful lookups, approximately 220,000 only queried one server.

4   Why (Not) Cooperative DNS?

Serving DNS data using peer-to-peer systems frees the domain owners from having to configure or administer name servers. Anyone who wants to publish a domain only needs the higher-level domain to sign her public key. DDNS takes care of storing, serving, replicating, and caching of her DNS records. Below, we discuss various issues in the current DNS, and the extent to which DDNS solves them.

4.1   DNS Administration

In their 1988 retrospective [9], Mockapetris and Dunlap listed "distribution of control vs. distribution of expertise or responsibility" as one of the shortcomings of their system, lamenting:

> Distributing authority for a database does not distribute corresponding amounts of expertise. Maintainers fix things until they work, rather than until they work well, and want to use, not understand, the systems they are provided. Systems designers should anticipate this, and try to compensate by technical means.

To justify their point, they cited three failures: they did not require proof that administrators had set up truly redundant name servers without a single point of failure; in the documentation, they used hour-long TTLs in the examples but suggested day-long TTLs in the text, with the result that everyone used hour-long TTLs; debugging was made difficult by not requiring servers to identify their server version and type in an automated way. DDNS eliminates much of the need for expertise by automatically providing a routing infrastructure for finding name information, automatically avoiding single points of failure.

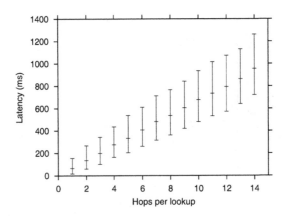

Fig. 5. Hops per lookup in the simulation of the Jung. *et al.* traces over Chord.

In their handbook for the Berkeley DNS server, BIND, Albitz and Liu [1] listed what they believed to be thirteen of the most common problems in configuring a name server. Our system addresses six of them.

Slave server cannot load zone data. DDNS solves this by automatically handling replication via the DHash protocol. There are no slave servers.

Loss of network connectivity. DDNS is robust against server failure or disconnection. Unfortunately, it suffers from network partitions. For example, if a backhoe cuts MIT from the rest of the Internet, even though hosts on the Internet will not see a disruption in any part of the name space (not even MIT's names), hosts at MIT may not even be able to look up their own names! This problem is not new, since a client machine with no knowledge of MIT will require access to the root name servers to get started. Both systems partially avoid this problem with caching: popular DNS data about local machines is likely to be cached and thus available even after the partition. DDNS actually works better in this situation, since the remaining nodes will form a smaller Chord network and pool their caches. If proximity routing is deployed in Chord, then DDNS can use that make some probabilistic guarantee that each record for a domain name gets stored on at least one node close to the record owner.

Missing subdomain delegation. In conventional DNS, a domain is not usable until its parent has created the appropriate NS and glue records and propagated them. DDNS partially eliminates this problem, since there are no NS records. In their place, the domain's parent would have to sign the domain's public key RRSet. At the least, this eliminates the propagation delay: once the parent signs a domain's public key, it is up to the domain's administrator to publish it.

Incorrect subdomain delegation. In conventional DNS, if the parent is not notified when name servers or IP addresses of a domain change, clients will eventually not be able to find the domain's name servers. The analogue in DDNS would be a domain changing its public key but forgetting to get its parent to

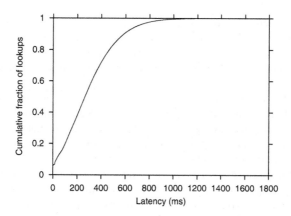

Fig. 6. Imaginary latency to perform DNS lookups over Chord, assuming the latency distribution from the Jung. *et al.* trace plotted in Figure 4

sign the new key. Without getting the signature, though, inserts of records signed with the new key would fail. This would alert the administrator to the problem immediately. (In conventional DNS, the problem can go undetected since the local name server does not check to see whether the parent domain correctly points at it.)

On a similar note, Jung *et al.* [6] reported 23% of DNS lookups failed to elicit any response, partially due to loops in name server resolution. 13% of lookups result in a negative response, many of which are caused by NS records that point to non-existent or inappropriate hosts.

Conventional DNS requires that domain owners manage two types of information: data about hosts (*e.g.*, A records) and data about name service routing (*e.g.*, NS records). The latter requires close coordination among servers in order to maintain consistency; in practice this coordination often does not happen, resulting in broken name service. DDNS completely eliminates the need to maintain name service routing data: routing information is automatically maintained and updated by Chord without any human intervention.

In summary, we believe that using a peer-to-peer system for storing DNS records eliminates many common administrative problems, providing a much simpler way to serve DNS information.

4.2 Dynamically Generated Records

Our system requires that all queries can be anticipated in advance and their answers stored. Since the `hosts.txt` approach required this property and the original DNS papers are silent on the topic, it seems likely that this requirement was never explicitly intended to be relaxed. However, the conventional DNS did relax the requirement: since domains serve their own data, all possible queries need not be anticipated in advance as long as there is some algorithm implemented in the server for responding to queries. For example, to avoid the need to publish internal host names, the name server for *cs.bell-labs.com* will return a valid mail exchanger (MX) record for any host name ending in *.cs.bell-labs.com*, even those that do not exist.

Additionally, responses can be tailored according to factors other than the actual query. For example, it is standard practice to randomly order the results of a query to provide approximate load balancing [2]. As another example, content distribution networks like Akamai use custom DNS responses both for real-time load balancing and to route clients to nearby servers [7].

The system we have described can provide none of these capabilities, which depend on the coupling of the administrative hierarchy and the service structure. If some features were determined to be particularly desirable, they could be implemented by the clients instead of the servers.

4.3   Denial of Service

DDNS has better fault-tolerance due to denial-of-service attacks over the current DNS. Because there is no name server hierarchy, the attacker has to take down a diverse set of servers before data loss becomes apparent.

Another type of denial of service is caused by a domain name owner inserting a large number of DNS records, using up space in the Chord network. We can address this problem by enforcing a quota on how much data each organization can insert depending on how much storage the organization is contributing to DDNS.

5   Conclusions

Separating DNS record verification from the lookup algorithm allows the exploration of alternate lookup algorithms. We presented DDNS, which uses a peer-to-peer distributed hash table to serve DNS records. In our judgement, using DDNS would prove a worse solution for serving DNS data than the current DNS. We believe that the lessons we draw from comparing the two have wider applicability to peer-to-peer systems in general and distributed hash tables in particular.

DDNS eliminates painful name server administration and inherits good load balancing and fault tolerance from the peer-to-peer layer. The self-organizing and adaptive nature of peer-to-peer systems is a definite advantage here, something that conventional manually administered systems cannot easily provide.

DDNS has much higher latencies than conventional DNS. The main problem is that peer-to-peer systems typically require $O(\log_b n)$ RPCs per lookup. Chord uses $b = 2$, requiring 20 RPCs for a million node network. Systems such as Pastry and Kademlia use $b = 16$, requiring only 5 RPCs for a million node network. Our experiments show that using even 5 RPCs results in a significant increase in latency, and of course the problem becomes worse as the peer-to-peer network grows. By contrast, conventional DNS typically needs 2 RPCs; it achieves its very low latency by putting an enormous branching factor at the top of the search tree — the root name servers know about millions of domains. It is easy to hide this problem in the big-O notation, but if peer-to-peer systems are to support low-latency applications, we need to find ways to reduce the number of RPCs per lookup.

DDNS has all the functionality of a distributed `hosts.txt`, but nothing more. Conventional DNS augments this functionality with a number of important features implemented using server-side computation. Distributed hash tables aren't sufficient for serving DNS because they require features to be client-implemented. It is cumbersome to update all clients every time a new feature is desired. At the same time, in a peer-to-peer setting, it is cumbersome to update all servers every time a new feature is desired. This is one case where the enormous size of peer-to-peer networks is not offset by the self-organizing behavior of the network. Perhaps we should be considering "active" peer-to-peer networks, so that new server functionality can be distributed as necessary.

Finally, DDNS requires people publishing names to rely on other people's servers to serve those names. This is a problem for many peer-to-peer systems: there is no incentive to run a peer-to-peer server rather than just use the servers run by others. We need to find models in which people have incentives to run servers rather than just take free rides on others' servers.

References

[1] Paul Albitz and Cricket Liu. *DNS and BIND*. O'Reilly & Associates, 1998.

[2] T. Brisco. DNS support for load balancing. RFC 1794, April 1995.

[3] Frank Dabek, M. Frans Kaashoek, David Karger, Robert Morris, and Ion Stoica. Wide-area cooperative storage with CFS. In *Proceedings of the 18th ACM Symposium on Operating Systems Principles (SOSP '01)*, Chateau Lake Louise, Banff, Canada, October 2001.

[4] P. Danzig, K. Obraczka, and A. Kumar. An analysis of wide-area name server traffic: A study of the internet domain name system. In *Proc ACM SIGCOMM*, pages 281–292, Baltimore, MD, August 1992.

[5] D. Eastlake. Domain name system security extensions. RFC 2535, March 1999.

[6] Jaeyeon Jung, Emil Sit, Hari Balakrishnan, and Robert Morris. Dns performance and the effectiveness of caching. In *Proceedings of the ACM SIGCOMM Internet Measurement Workshop '01*, San Francisco, California, November 2001.

[7] R. Mahajan. How Akamai works.
http://www.cs.washington.edu/homes/ratul/akamai.html.

[8] P. Mockapetris. Domain names - concepts and facilities. RFC 1034, November 1987.

[9] P. Mockapetris and K. Dunlap. Development of the Domain Name System. In *Proc. ACM SIGCOMM*, Stanford, CA, 1988.

[10] Ion Stoica, Robert Morris, David Karger, M. Frans Kaashoek, and Hari Balakrishnan. Chord: A scalable peer-to-peer lookup service for internet applications. In *Proc. ACM SIGCOMM*, San Diego, 2001.

Network Measurement as a Cooperative Enterprise

Sridhar Srinivasan and Ellen Zegura

Networking and Telecommunications Group
College of Computing
Georgia Institute of Technology
Atlanta, GA 30332
{sridhar,ewz}@cc.gatech.edu

Abstract. Real-time network measurements can be used to improve performance of existing Internet services and support the deployment of new services dependent on performance information (e.g., topologically-aware overlay networks). Internet-wide measurement faces numerous scaling-related challenges, including the problem of deploying enough measurement endpoints for wide-spread coverage. We observe that peer-to-peer networks, made up of "volunteer" hosts around the Internet world, have the potential to provide a level of coverage that greatly exceeds that made possible with the tedious human process of negotiating endpoint locations. We therefore propose a distributed peer-to-peer system that can be queried for network performance information. We sketch the architecture and operation of such a system and briefly relate it to alternative proposals for measurement infrastructures. Finally, we list open problems related to the design and realization of such a system.

1 Introduction

Measurements of network performance are valuable for improving performance, assessing utilization, engineering traffic and validating design choices. We are particularly interested in real-time measurements that can be used to improve the performance of existing Internet services and support the deployment of new services dependent on performance information (e.g., topologically-aware overlay networks).

The challenges involved in constructing an Internet-scale measurement infrastructure are considerable. First, there is the difficulty of coverage, that is, obtaining access to a large number of distributed measurement endpoints. Current measurement systems generally involve human-negotiated access to endpoints either with ISPs and/or friends at diverse locations [3, 6]. Second, there is the difficulty of obtaining accurate measurements, given the time-varying nature of network properties of interest (e.g., loss rate, available bandwidth, latency). Third, there is the issue of overhead. Care must be taken to avoid a measurement process that imposes excessive overhead on the overall system. These challenges

P. Druschel, F. Kaashoek, and A. Rowstron (Eds.): IPTPS 2002, LNCS 2429, pp. 166–177, 2002.

have obvious interactions; for example, one can reduce overhead with less accurate measurements or more coarse-grained coverage.

We observe that peer-to-peer networks, made up of "volunteer" hosts around the Internet world, have the potential to provide a level of coverage that greatly exceeds that made possible with the tedious human process of negotiating endpoint locations. *We therefore propose a distributed peer-to-peer system that can be queried for network performance information.* The M-coop (or Measurement co-operative) is a system that answers queries about the path between two arbitrary IP addresses. In addition to performance metric information, the system returns assessments of the metric accuracy and trustworthiness.

Such a system does not, on its own, solve the problem of obtaining accurate measurements. Nor does it solve the problem of measurement overhead. Indeed, because such a system may involve a very large number of end systems, the scaling problem is significant. We will rely on known techniques for dealing with accuracy (e.g., using moving weighted averages); we will introduce mechanisms for reducing the number of end systems that form measurement pairs to help with the scale problem.

Such a system brings with it a number of additional challenges. Well-known are the problems that result from a peer-to-peer system of hosts that may join and leave on a frequent basis [4, 5, 8, 9]. Merely keeping the M-coop system connected can be challenging in this environment. Because the measurement entities are volunteers, and not under any accountable control, we must deal with issues of inaccurate information due to misconfiguration or malicious use. The inclusion of a trustworthiness value recognizes the fact that information quality may vary. We must also consider the question of incentive. What would motivate someone to include their host in an M-coop measurement infrastructure? The limited examples of deployed peer-to-peer systems indicate that people are motivated by self-interest (e.g., Napster, Gnutella) and by a sense of contributing to a larger "good" (e.g., SETI@home). An Internet-scale measurement infrastructure has the potential to tap both sources of motivation.

In the next section, we sketch the design of an M-coop system. In Section3, we describe some details of the architecture and briefly sketch the operation of this system. In Section4, we briefly describe related work and then conclude with a section discussing the open problems in the design and realization of an Internet-wide peer-to-peer measurement system.

2   An M-coop Design

We sketch one possible design of a cooperative measurement system. The system has some features in common with other measurement infrastructures (most notably IDMaps [1] and NIMI [3]). Some similarities and differences are discussed in the Related Work section.

2.1   The Service

The M-coop system answers queries of the form *(IP1, IP2, measurement type)* where *IP1* and *IP2* are IP addresses. The measurement type may be any network quantity measurable by hosts on a network, e.g., delay, bandwidth, jitter. The system returns the answer to the query along with trust and accuracy parameters if available. As a voluntary peer-to-peer system, the possibility of misinformation is high, so a trust value is reported with the information returned. The trust value is an indication of the past reliability (with respect to quality of information) of the node that responded to the query.

The size of the Internet dictates that any measurement will only be an estimate. To keep the system manageable, instead of the measurement being from the requested host, it might be from a "nearby" node on the overlay network. Also, the measurement process may contain some inaccuracies due to changing paths in the Internet, inherent inaccuracies in the measurement process, congestion, etc. The accuracy value tries to quantify the "nearness" of the host to the measurement node as well as the inaccuracies in the measurement process.

We do not address the question of who is allowed to make queries. In the spirit of cooperatives, one might imagine that only participants are allowed access to the community information. This sort of access control (or any other) is orthogonal to the base system design.

2.2   The Architecture

Architecturally, the system consists of two overlay networks, a logical overlay formed as a Distributed Hash Table (DHT), (e.g. Chord [8], SCAN [4], Pastry [5] and Tapestry [9]), and a measurement overlay (m-overlay) based on the AS level graph of the Internet. Each node in M-Coop belongs to both overlays. The DHT is used for storing and looking up information about the nodes that compose the overlays. Nodes connected by edges in the m-overlay form measurement peers. The key of a node in the DHT is generated by its neighbours during the join process while the key of a node in the m-overlay is its IP address. An important issue is the construction of the m-overlay graph to support accurate measurements without undue overhead. Measurements are taken by the endpoints of the edge in two ways, actively, by sending probe packets to each other, and passively, by monitoring the system traffic that traverses this edge of the graph.

For scalability, each node on the network is assigned an "area of responsibility" (AOR), defining a set of addresses for which it can answer queries. The AOR is assigned when the node joins the network. It changes as other nodes join and leave the network.

A query to the system, *(IP1, IP2, measurement type)* is first routed through the DHT to the node which has *IP1* in its AOR. We denote this node *R(IP1)* to indicate it is responsible for *IP1*. If the measurement information is available, node *R(IP1)* will reply, along with the available accuracy and trust information. If the data is not available, it may trigger a measurement or a new query. This new query, called a composition query, will traverse a path on the m-overlay from *R(IP1)* to *R(IP2)* collecting metric data about the links traversed. This data is then returned as the reply to the composition query and finally as a reply to the original query.

A node on the system thus consists of three modules:

1. **Routing.** This module is responsible for maintaining the overlays, communicating with the peer nodes and routing queries and responses through the overlays.
2. **Measurement.** This module performs measurements between itself and its measurement peers, verifies the measurements obtained by other nodes and responds to queries about the node's AOR.
3. **Trust.** This module maintains the trust database, performs trust metric calculations and responds to trust queries.

The next section describes the architecture in more detail.

3 Architecture Details and Operation

3.1 Routing

There are many ongoing research efforts trying to develop better methods of locating data in a distributed system. These efforts are directed towards scalability, reliability, graceful degradation under dynamic conditions, and efficient search [4, 5, 8, 9]. Based on this, we assume that a method of locating data in the DHT is available to us and we will use the generic term *routing* to imply that a packet is using one of the above methods to reach its destination node.

The routing module is responsible for routing data queries and their responses through the DHT overlay based on the AOR of the nodes. Composition queries are routed in the m-overlay. Ideally, the routing on this overlay is the same as the network-layer routing, but this requires knowledge of BGP policies of each AS. To approximate this in the application-layer, a heuristic such as shortest-AS path may be used.

3.2 Measurement

Measurements are taken from the node to its measurement peers. The set of measurement peers is identified when the node joins the system and is updated as needed when peer nodes join or depart. The set of measurement peers is selected to map onto the underlying network topology in the following fashion. If a node in the overlay network is the only one in its autonomous system (AS), then it has one edge for each AS-level neighbor in the underlying network. If there are multiple nodes in the same AS, a clustering protocol is run when new nodes join or depart to ensure that there is only one node which has edges to the AS level neighbors. The remaining nodes are organized to provide redundancy and intra-AS measurement. The AS-level edges join this node to the m-overlay network

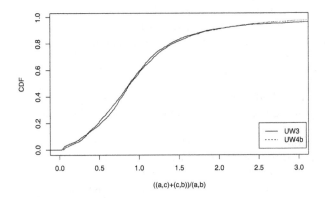

Fig. 1. CDF of composition of AS-level paths

nodes responsible for the IP addresses in the AS level neighbors. The intent is that the measurements obtained by the M-coop system will then better approximate the values seen by a packet on the underlying network. The measurement data obtained is stored with meta-information such as time of measurement, whether it is composed from other measurement data, whether it is cached, the accuracy, trust value, etc. Measurements may be taken periodically, triggered by options in a data query, and/or determined passively by examining packets in the system.

We performed some simple experiments to verify that such an AS-level composition of paths gives reasonable estimates of the original query between two IP addresses. We used data from the UW-3 and UW-4b datasets used for the Detour study[6]. These are end-to-end `traceroute` measurements collected from public `traceroute` servers. Details of the data collection method can be found in the referenced study. For each dataset, we calculate the average latency between two nodes a and b and also compute the AS-level path between them. The composition path is computed by finding a node c which lies on the AS-level path from a to b, and calculating $(a,c)+(c,b)$ for all possible such c's. This simulates our scheme in which the composition packets are forwarded at the AS-level from the source AOR to the destination AOR. The results are shown in Fig. 1 as a CDF of the ratio $((a,c)+(c,b))/(a,b)$. It can be observed that most of the composition values are within a factor of two of the actual path value. These results are similar to those obtained by IDMaps [1] and show that forwarding at the relatively coarse level of ASes can give reasonable estimates.

3.3   Trust

Measurements are of two types, data and verification. Data measurements are performed between a node and its measurement peers and these are reported in response to queries. Verification measurements are performed to verify the

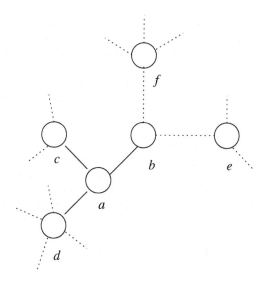

Fig. 2. Verification

responses of the peer nodes. These measurements are part of the process by which the trust module calculates the trust value to be assigned to the peer nodes.

The trust value of a node in the context of a particular link is a measure of its measurement reliability in the past. We assume that past behaviour is an indication of future actions, i.e., if a node has been providing reasonably accurate responses about a link in the past, it is likely to provide a reasonably accurate response when queried now. We will use the term "trust of a node" to implicitly mean the trust of a node with respect to a specific link.

The node checks the operation of its peers in the m-overlay using a *verification process*, which is run regularly. The results of this process are used to calculate the trust value of a node. This trust factors in the time since the last verification process was run, as well as reports from other nodes on the trust of the node in question. In the system, a node reports on the trust of its immediate neighbours. It also gathers information about nodes two and three hops away, which is then reported only if a query about that node's trust passes through.

We now explain the verification process in more detail. In Fig. 2, node a responds to trust queries about the nodes b, c and d. Periodically, a runs the verification process on the nodes b, c and d. The verification process is in two parts: a queries b about the path from b to e, which is b's neighbour; a also performs a measurement from itself to e directly. Since a knows the value of the measurement of the a-b link, it can estimate the b-e link and compare it with the value reported by b. The two values thus obtained are then used to update the trust of node b.

It is important to note that the b-e link measurement estimated by a has higher chances of being inaccurate and so a single value which doesn't tally with the reported value may not be enough to affect the trust value of the node b.

This verification mechanism requires that the point-to-point measurements made by the node be independent of the trustworthiness of the other end point, and hence reliable. This can be partly achieved if at least some of the measurements can be performed without the cooperation of the other node (perhaps at the operating system level). For example, a ping query to measure latency does not reach the application layer of the other end-point and hence is harder to affect maliciously. Given this assumption, a can verify the functioning of b by making a direct measurement to e to estimate the b-e link. Since b is in a's neighbouring AS and similarly, e in b's, it is reasonable to suppose that the direct measurement by a will produce a good estimate. Another solution could be to have a measure directly to other nodes in b's AS, if they exist. These solutions partly address the problem of verification, but do not provide a completely reliable method of verifying a neighboring node's measurements.

Trust Computation Procedure The verification process is used as a building block to compute the trust value of a response to a query using a trust computation procedure. When a query is answered by a node, the querying node has no knowledge of the reliability of the reported result. To determine the trustworthiness of the result, the querying node must ask the neighbours of the responding node about its trustworthiness. To find out the trustworthiness of the neighbours, the node must query their neighbours, and so on. This recursive chain is terminated at the querying node itself. Effectively, the node is building a chain of trust from itself to the responding node. This chain follows the AS-level path from the querying node to the neighbours of the responding node on the m-overlay. Each node on the chain reports the trust it has on the node preceding it (which is its neighbour).

The actual procedure is as follows. A querying node, on receiving a response to its query, can decide to calculate the trustworthiness of the response. It looks up the node information of the responding node on the DHT and gets the list of neighbours. It then selects a set of nodes (potentially all) to send the trust query to. Each query is routed through the m-overlay to its destination (a neighbour of the responding node). The destination node replies with the trust it has on the responding node. As this reply makes its way back to the querying node, each node along the path adds the trust it has on the node from which it received the reply. At the querying node, it applies a function all the information to obtain the trust value of the responding node.

3.4 AOR Assignment and Overlay Construction

The AOR assignment for a node takes place when the node joins the overlay network. The startup procedure for nodes joining the network assumes two things:

a node is capable of finding its AS number;[1] and a node knows the IP address of an existing node on the overlay. (The case of the first node in the overlay is discussed separately.) A further assumption that is useful, but not required, is that a node has access to the list of ASes connected to its AS.[2] The creation and maintenance of the measurement overlay is performed by the following algorithms.

– **Join.** This is executed by a node joining the measurement overlay.
– **Failure recovery.** This is executed by the neighbours of a node that has failed.
– **Maintenance.** This is executed with low frequency to redistribute the AORs across the peers.

The invariant that these algorithms maintain, is that every AS is in the AOR of some node. When a new node joins the overlay, it may be assigned a part of the AS space as its AOR by splitting off portions of a small set of pre-existing nodes' AORs. When a node leaves the overlay, its AOR is merged into the AOR of a neighbour of that node, thus maintaining the invariant. In the following discussion, we restrict our attention to joins and departures in the m-overlay. We do not discuss the DHT join and leave algorithms as they are handled by the DHT schemes.

Join Procedure In Fig. 3, node n is a new node that is attempting to join the network. On startup, the node n contacts a node s that is already a member of M-Coop. The address of the node s is assumed to be known to node n through some out-of-band mechanism such as from a website or an ftp server. In its initial message, n advertises its entire AS as its AOR. Since the overlay already exists, some node b already has this AS in its AOR. Node s queries the DHT to find that node b and returns its address to n. Node n then contacts b with the same advertisement. There are two cases: b and n are in different ASes, or they are in the same AS. We shall consider the cases separately. When the nodes b and n are not in the same AS, n performs the following procedure:

– n uses its list of neighbouring ASes to claim them from the AOR of b. If the ASes appear in b's list of neighbouring ASes as well, then they are assigned to n probabilistically.
– After the basic AOR for n is created, n looks up the nodes responsible for the ASes that are in its neighbour list but not in its AOR or in the AOR of b. These nodes are sent peer requests by n to make them peers in the m-overlay.
– If these nodes have b as a peer, they replace b by n, otherwise they add n as a peer.
– n also peers with b.

---

[1] A repository of AS information is available at www.arin.net/whois.
[2] NLANR maintains such a list at http://moat.nlanr.net/AS/.

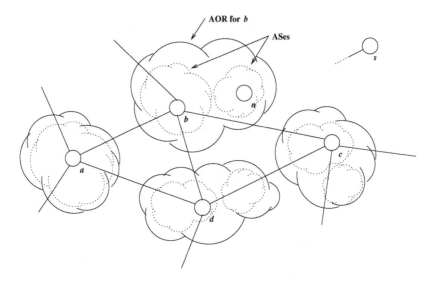

Fig. 3. Overlay before node n joins the network

- n runs the maintenance procedure to resize its AOR and potentially find new peers.

At this point, n is ready to begin making measurements with its peers. A key is generated for n by its peers and n uses this key to join the DHT and publish its peer and AOR information on the DHT. This marks the end of the join algorithm for n.

When the nodes b and n are in the same AS, n joins the cluster of nodes under b for that AS. The process is similar to that of the previous case except that the AORs contain IP prefixes, and the splitting of AORs is based on partitioning the set of IP addresses assigned to each node. Also, the node information of n is published in the intra-AS DHT.

When n is the first node in the overlay, it assigns all AS numbers to its AOR and waits for further nodes to join the network.

Note that the number of nodes that can join the measurement overlay at the intra-AS level is limited. This is to ensure that the AORs remain of reasonable size. Any nodes that join after this limit is reached do not participate in the measurement overlay but remain connected in the DHT. The information published in the DHT is soft state and must be refreshed periodically using the node key generated during the join process.

Failure Recovery The main elements for the failure recovery algorithm, namely the node information and the node key, are put in place during the join process. When a node fails or leaves the M-Coop network, it is the responsibility of its neighbours in the m-overlay to recover from the departure. The periodic measurements taken between the peers in this overlay serve as a heartbeat mechanism and enable the detection of a departure. When a departure is detected, each neighbour of the missing node looks up the missing node information by querying the DHT using the missing node's node key. The neighbour also looks up the node information of the other neighbours to find the neighbour with the smallest AOR. This node incorporates the departed node's AOR into its own AOR, then peers with those nodes from the departed node's peers which are not already peers, runs the maintenance algorithm and then updates the node information published on the DHT.

Maintenance The objective of the maintenance procedure is to distribute the AORs across the peers so that each AS belongs in the AOR of the node closest to it. The maintenance algorithm consists of contacting each of the peers in the m-overlay in turn to negotiate the transfer of ASes based on the contents of the lists of neighbouring ASes. ASes are transferred only if necessary, i.e., if they are in a node's AOR but belong in the peer's neighbour list. This is to minimize the updates of the node information published in the DHT.

We are currently in the process of evaluating these algorithms using simulations.

4 Related Work

There have been several prior projects that concern measurement infrastructures (e.g., IDMaps [1], NIMI [3], SPAND [7] and Remos [2]). Our work is most closely to the IDMaps project, so we limit our related work discussion to that project.

IDMaps [1] is a proposal for a global infrastructure for gathering and distributing Internet host distance information. The goal of the IDMaps project is to provide distance metrics between two hosts on the Internet in an accurate and timely manner. The IDMaps architecture consists of a network of Tracers, which gather Internet distance information, and Clients, which use this information to estimate distances between hosts. The distance estimate between any two IP addresses is calculated from the Address Prefixes (APs) that contain the IP addresses, serving a similar function to our Areas of Responsibility. The calculation is performed by finding the APs to which the IP addresses belong, locating the systems or "boxes" to which the APs are closest and then running a spanning-tree algorithm to to find the shortest distance between the two boxes. This calculation requires that a substantial portion of the box connection topology must be maintained.

The actual box-box topology can be achieved in two ways, the Hop-by-Hop (HbH) and the End-to-End (E2E) models. In the HbH model, every transit backbone router is modeled as a box and the calculation is the sum of inter-AS and intra-AS paths from one AP to the other. The distances on these paths are calculated by the Tracers probing the routers at random intervals. In the E2E model, the Tracers are the boxes and the distances are calculated as the sum of the AP to box distances and the distance between the two boxes.

Our goals for the M-coop system are to provide a generalized metric collection and distribution infrastructure that is simple and rapid to deploy on a large scale. The information returned by the system also contains some indication of how reliable (in terms of accuracy and trustworthiness) the information is. Our approach to the problem is similar to the HbH model proposed in the IDMaps architecture but our method of distance estimation and information dissemination is fundamentally different. We intend to have a little more complexity at the nodes gathering the distance information to avoid the problem of maintaining a global view of the box topology. We also try to address the deployment of the system in the Internet by means of our peer-to-peer design.

5   Open Questions

We have sketched out the design and architecture of an Internet-wide measurement service. Some of the challenges that we are currently working on include:

- **Participation.** Will such a scheme generate enough participation to achieve critical mass, i.e., a level where the query results are a good approximation of the actual values? Related to this issue is the broader question of what will motivate people to participate in peer-to-peer systems. Will people eventually subscribe to peer-to-peer systems, like they subscribe to magazines? Or will they contribute their host to peer-to-peer systems, a la charitable donations? What are the best analogies to the peer-to-peer experience?
- **Generality.** Can a single system be used to satisfy the different measurement requirements of the diverse applications which might want to take advantage of this service? Can such a system be used as a common measurement service for peer-to-peer systems to use for optimizing their operation?
- **Usefulness of Parameters.** Can trust and accuracy be made useful to applications?
- **Composition.** Can composition of measurements from intermediate hops give meaningful values for the actual measurement between two IP addresses?
- **Collusion.** Collusion is a problem in trust systems. Can the amount of collusion required to subvert the system be made large enough to deter attacks?

6   Acknowledgments

The authors would like to acknowledge the helpful suggestions of the anonymous reviewers. We would also like to thank Andy Collins and Stefan Savage for providing the datasets for the experiments.

References

[1] P. Francis, S. Jamin, C. Jin, Y. Jin, D. Raz, Y. Shavitt and L. Zhang. IDMaps: A global Internet host distance estimation service. In *IEEE/ACM Trans. on Networking*, October 2001.

[2] N. Miller and P. Steenkiste. Collecting network status information for network-aware applications. In *Proceedings of Infocom'00*, Tel Aviv, March 2000.

[3] V. Paxson, J. Mahdavi, A. Adams, and M. Mathis. An architecture for large-scale Internet measurement. In *IEEE Communications*, volume 36, pages 48–54, August 1998.

[4] S. Ratnasamy, P. Francis, M. Handley, R. Karp, and S. Shenker. A scalable content-addressable network. In *Proceedings of the ACM SIGCOMM '01*, San Diego, CA, September 2001.

[5] A. Rowstron and P. Druschel. Pastry: Scalable, distributed object location and routing for large-scale peer-to-peer systems. In *Middleware*, 2001.

[6] S. Savage, A. Collins, E. Hoffman, J. Snell, and T. Anderson. The end-to-end effects of Internet path selection. In *Proceedings of the ACM SIGCOMM'99*, Boston, MA, September 1999.

[7] S. Seshan, M. Stemm, and R. H. Katz. A network measurement architecture for adaptive applications. In *Proceedings of Infocom '00*, Tel Aviv, March 2000.

[8] I. Stoica, R. Morris, D. Karger, F. Kaashoek, and H. Balakrishnan. Chord: A peer-to-peer lookup service for Internet applications. In *Proceedings of the ACM SIGCOMM '01*, San Diego, CA, September 2001.

[9] B. Zhao, J. Kubiatowicz, and A. Joseph. Tapestry: An infrastructure for fault-tolerant wide-area location and routing. *UCB Tech. Report UCB/CSD-01-1141*.

The Case for Cooperative Networking*

Venkata N. Padmanabhan[1] and Kunwadee Sripanidkulchai[2]**

[1] Microsoft Research
http://www.research.microsoft.com/~padmanab/
[2] Carnegie Mellon University
http://www.andrew.cmu.edu/~kunwadee/

Abstract. In this paper, we make the case for Cooperative Networking (CoopNet) where end-hosts cooperate to improve network performance perceived by all. In CoopNet, cooperation among peers complements traditional client-server communication rather than replacing it. We focus on the Web flash crowd problem and argue that CoopNet offers an effective solution. We present an evaluation of the CoopNet approach using simulations driven by traffic traces gathered at the MSNBC website during the flash crowd that occurred on September 11, 2001.

1 Introduction

There has been much interest in peer-to-peer computing and communication in recent years. Efforts in this space have included file swapping services (e.g., Napster, Gnutella), serverless file systems (e.g., Farsite [2], PAST [12]), and overlay routing (e.g., Detour [14], RON [1]). Peer-to-peer communication is the dominant mode of communication in these systems and is central to the value provided by the system, be it improved performance, greater robustness, or anonymity.

In this paper, we make the case for Cooperative Networking (CoopNet), where end-hosts cooperate to improve network performance perceived by all. In CoopNet, cooperation among peers complements traditional client-server communication rather than replace it. Specifically, CoopNet addresses the problem cases of client-server communication. It kicks in when needed and gets out of the way when normal client-server communication is working fine. Unlike some of the peer-to-peer systems, CoopNet does not assume that peer nodes remain available and willing to cooperate for an extended length of time. For instance, peer nodes may only be willing to cooperate for a few minutes. Hence, sole dependence on peer-to-peer communication is not an option.

The specific problem case of client-server communication we focus on is *flash crowds* at Web sites. A flash crowd refers to a rapid and dramatic surge in the volume of requests arriving at a server, often resulting in the server being overwhelmed and response times shooting up. For instance, the flash crowds

---

* For more information, please visit the CoopNet project Web page at
  http://www.research.microsoft.com/~padmanab/projects/CoopNet/.
** The author was an intern at Microsoft Research through much of this work.

P. Druschel, F. Kaashoek, and A. Rowstron (Eds.): IPTPS 2002, LNCS 2429, pp. 178–190, 2002.
© Springer-Verlag Berlin Heidelberg 2002

caused by the September 11 terrorist attacks in the U.S. overwhelmed major news sites such as CNN and MSNBC, pushing site availability down close to 0% and response times to over 45 seconds [19]. Flash crowds are typically triggered by events of great interest — whether planned ones such as a sports event or unplanned ones such as an earthquake or a plane crash. However, the trigger need not necessarily be an event of widespread global interest. Depending on the capacity of a server and the size of the files served, even a modest flash crowd can overwhelm the server.

The CoopNet approach to addressing the flash crowd problem is to have clients that have already downloaded content to turn around and serve the content to other clients, thereby relieving the server of this task. This cooperation among clients is only invoked for the duration of the flash crowd. The participation of individual clients could be for an even shorter duration — say just a few minutes. We argue that the CoopNet approach is self-scaling and cost-effective.

The rest of this paper is organized as follows. In Section 2, we present our initial design of CoopNet and discuss several research issues. In Section 3, we analyze the feasibility of CoopNet using traces gathered at MSNBC [21], one of the busiest news sites in the Web, during the flash crowd that occurred on September 11, 2001. We conclude in Section 4 by comparing CoopNet with alternative approaches to addressing the flash crowd problem.

2   Cooperative Networking (CoopNet)

In this section, we present our initial design of CoopNet. We begin by taking a closer look at the impact of a flash crowd on server performance.

2.1   Where Is the Bottleneck?

A key question is what the most constrained resource is during a flash crowd: CPU, disk or network bandwidth at the server, or bandwidth elsewhere in the network. It is unlikely that disk bandwidth is a bottleneck because the set of popular documents during a flash crowd tends to be small, so few requests would require the server to access the disk. For instance, the MSNBC traces from September 11 show that 141 files (0.37%) accounted for 90% of the accesses and 1086 files (2.87%) accounted for 99% of the accesses. It is quite likely that this relatively small number of files would have fit in the server's main memory buffer cache.

The CPU can be a bottleneck if the server is serving dynamically generated content. For instance, Web pages on MSNBC are by default implemented as active server pages (ASPs), which include code that is executed upon each access. (ASPs are used primarily to enable ad rotation and customization of Web pages based on HTTP cookie information.) So when the flash crowd hit in the morning of September 11, the CPU on the server nodes quickly became a bottleneck. For instance, the fraction of server responses with a 500 series HTTP status code (error codes such as "server busy") was 49.4%. However, MSNBC quickly switched to serving static HTML and the percentage of error status codes dropped to 6.7%. Our conversations with the Web site operators have revealed that network bandwidth became the primary constraint at this stage.

Since Web sites typically turn off features such as customization during a flash crowd and only serve static files, it is not surprising that network bandwidth rather than server CPU is the bottleneck. A modern PC can pump out hundreds of megabits of data per second (if not more) over the network. For instance, [4] reports that a single 450 MHz Pentium II Xeon-based system[1] with a highly tuned Web server implementation could sustain a network throughput of well over 1 Gbps when serving static files 32 KB in size.

On the other hand, the network bandwidth of a Web site is typically much lower. In an experiment conducted recently [13], the bottleneck bandwidth between the University of Washington (UW) and a set of 13,656 Web servers drawn from [22] was estimated using the Nettimer tool [7]. The bottleneck bandwidth (server to UW) was less than 1.5 Mbps (T1 speed) for 65% of the servers and less than 10 Mbps for 90% of the servers[2]. So it is clear that in the vast majority of cases network bandwidth will be the constraint during a flash crowd, not server CPU resources.

While it is possible that there may be bottleneck links at multiple locations in the network, it is likely that the links close to the server are worst affected by the flash crowd. So our focus is on alleviating the bandwidth bottleneck at the server.

2.2   Basic Operation of CoopNet

As mentioned in Section 1, the basic idea in CoopNet is to have clients serve content to others clients, thereby alleviating load on the server. Since network bandwidth tends to be the bottleneck rather than server CPU, CoopNet is tailored to drastically reducing bandwidth demands at the server. HTTP requests from clients arrive at the server as usual. During a flash crowd, the server redirects some or all of the requesting clients (depending on how constrained the server's network bandwidth is) to others clients that have downloaded the URL in the recent past. The clients then resend the request to one or more of these peers. Figure 1 illustrates the operation of CoopNet.

Clients indicate their willingness to participate in CoopNet by including a new HTTP `pragma` field in the request header. We call these the "CoopNet clients" and the rest as the "non-CoopNet clients". The server remembers the IP addresses of CoopNet clients that have requested each file in the recent past. For each file, it may be sufficient for the server to remember a relatively small

---

[1] The system had 4 processors, but only one CPU was used for the experiments reported in [4].

[2] Given the good network connectivity of UW, is likely that the bottleneck link in most cases was close to the server. While the bottleneck could have been "in the middle" for some distant servers (e.g., servers overseas), it is still likely to constrain communication between the server and the large number of clients in the U.S.

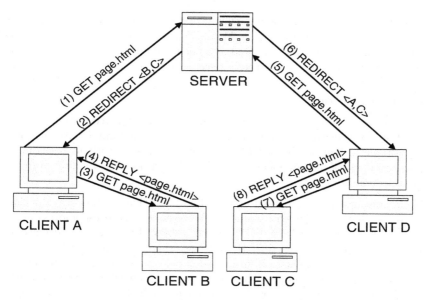

Fig. 1. The basic operation of CoopNet. The numbers in parentheses indicate the ordering of the steps. Note that the list of peers returned by the server is updated as new requests arrive.

number — say a few tens — of client addresses. The server then picks between 5 and 50 addresses at random from this set and includes this in the redirection message. It is quite likely that at least one of these peers is able and willing to serve the requested file. Since the server's list of addresses is constantly being updated as new requests arrive, the redirection procedure would tend to spread load rather evenly across the set of CoopNet clients.

The redirection response, which is a generalization of HTTP redirection, is quite small in size — 200-300 bytes including all protocol headers and the list of peer IP addresses. In contrast, even the slimmed down version of the MSNBC front page during the flash crowd of September 11 was 18-22 KB in size. Thus request redirection saves the server nearly two orders of magnitude in bandwidth. This alone may often be sufficient to help the server tide over the flash crowd problem. Furthermore, server-based redirection often enables a client to locate the desired content within two hops[3] — one to the server and another to a peer. In contrast, a distributed lookup scheme like Chord [16] or Pastry [12] has a lookup cost of $O(log(N))$ hops, where N is the number of nodes in the system. Thus server-based redirection is advantageous in many cases. In some situations, however, it may be desirable to avoid server-based redirection, as we discuss in Section 2.5.

---

[3] We mean end-to-end hops between hosts, not network hops.

We have built a prototype implementation of CoopNet. The server piece is implemented as an extension to the Microsoft IIS server using the ISAPI interface. The client piece is implemented as a client-side proxy that serves requests both from the local browser and from peers.

2.3   Peer Selection

An important question is how a client, upon receiving a redirection message from the server, decides which peer(s) to download a file from. Clearly, it would be desirable to find nearby peers that are well-connected without resorting to expensive network measurements. We employ a multi-pronged approach to the peer selection problem:

1. We use the scheme proposed in [6] to find peers that are topologically close to the client that issued a request. The basic idea is to use address prefix information derived from BGP routing tables. Two peers are deemed to be topologically close if their IP addresses share a common address prefix. The server uses this algorithm to find topologically close peers to include in its redirection response. There exist ways of doing such prefix matching operations very efficiently without imposing much of a burden on the server (e.g., [17]). If it is unable to find any peers with a matching prefix, the server just responds with a random list of peers. However, as we discuss in Section 3.3, the September 11 traces suggest that the server may often be able to find topologically close peers.

2. A match in address prefix does not necessarily mean that two peers are close to one another. For instance, an address prefix may correspond to a large network such as a national or global ISP. Therefore, it may be desirable to have the peers do a quick check to confirm their proximity. Our approach is to have each peer determine its "coordinates" by measuring the network delay to a small number (say 10) of "landmark" hosts. The intuition is that peers that are close to each other would tend to have similar delay coordinates. Similar approaches have been used in a number of contexts recently: network delay estimation [8], geographic location estimation [9], overlay construction [11], and finding nearby hosts [5].

3. For large file transfers, network bandwidth may be a critical metric for peer selection. The last-mile link is often the bottleneck. As in Napster, our approach is to have clients report their bandwidth (suitably quantized — e.g., dialup modem, DSL, T1, etc.) to the server as part of the requests they send. (Clients estimate their last-mile bandwidth by passively monitoring their network traffic in normal course.) The key distinction compared to the Napster approach is that in its redirection messages the server tries to only include peers whose reported bandwidth matches that of the requesting client. The motivation for this is two-fold. First, low-bandwidth clients are anyway constrained by their thin pipes, so they may not gain much from connecting to high-bandwidth peers. Second, clients do not have an incentive to under-report their bandwidth (a problem that afflicts Napster) because that would lead the server to redirect them to peers with a similar low bandwidth.

4. Even after applying the preceding steps, a client may still have a choice of say 2-3 peers to pick from. In such a case, the client could request non-overlapping pieces of data from multiple peers (say using the HTTP byte-range option [3]), determine which connection is the fastest, and then pick the corresponding peer for the remainder of the data transfer. Clearly, this procedure is likely to be worthwhile only in the case of large file transfers. In Section 2.4, we discuss the case of streaming media files where it may be desirable to persist with multiple peers for the entire duration of data transfer.

2.4   Streaming Media Content

Streaming media content presents some interesting issues in the context of a flash crowd. First, due to the large size of streaming media files and the relatively high bandwidth needed for streaming, even a small flash crowd can easily overwhelm the server or its network. For instance, a server behind a T1 link would be able to pump out no more than a dozen 128 Kbps streams simultaneously. Second, unlike static Web content, streaming media content is not normally cached at clients. Third, the burden of serving an entire stream to another client may be too much for a client, which is after all not engineered to be a server.

Our approach is to have clients save a local copy of streams during a flash crowd so that it can be streamed to other clients if needed. Where possible, a group of peers transmits non-overlapping portions of a stream (i.e., a set of "sub-streams") to the requesting client. The client combines these sub-streams on the fly to reconstruct the original stream. Distributed streaming reduces the burden on individual peers and also provides robustness in the face of congestion or packet loss suffered by a subset of the sub-streams. A more detailed discussion of the issues pertaining to streaming media distribution in CoopNet appears in [10].

2.5   Avoiding Server-Based Redirection

In some cases, it may be desirable to avoid having all requests be redirected by the server. First, in an extreme case, the bandwidth and/or processing needed to send the redirection messages may itself overwhelm the server. Second, it may be that only a small fraction of all clients are willing to participate in CoopNet. So cooperation among the CoopNet clients may not help reduce server load noticeably during the flash crowd. While CoopNet clients may still benefit significantly from their mutual cooperation (since they can download most of the bytes from one other instead of from the congested server), even getting the (small) initial redirection message from the congested server may take a long time (because of packet loss and the resulting TCP timeouts). So the total latency for CoopNet clients may remain large.

For these reasons, it may be desirable for CoopNet clients to check with their peers first before turning to the server. How to do this checking efficiently is an interesting and open research question. We present our initial thoughts here.

We term the set of peers among which a client searches for content as its *peer group*. (The peer group could, in principle, include all CoopNet clients.) On the face of it, the problem of searching for content in the peer group is similar to recent work on distributed key searching (e.g., CAN [11], Chord [16], Pastry [12], Tapestry [18]). However, we believe that these schemes may be too heavyweight for the flash crowd problem because (a) individual clients may not participate in the peer-to-peer network for very long, necessitating constant updates of the distributed data structures, (b) as we show in Section 3.1, much of the benefit of cooperation can be obtained even if the peer group size for each client is relatively small (say 30-50 peers), so there is not really the need for a distributed search mechanism that scales to millions of peers, and (c) the search for content in the peer group need not always be successful since there is always the fallback option of going back to the server.

Our approach exploits the observation that the peer group size for each client is relatively small. It may well be feasible for each member of a peer group to know about all other members. For each URL, there would be a designated "root" node within each peer group that would keep track of all copies of the file within the peer group. The assignment of the root node for a URL can be made using a hash function so that any member of the peer group can locate content in just two steps: first finding the root node by hashing on the URL and then finding a node that has the desired content. Redirection via the server can be used both to discover other clients and form a peer group initially, and also as a fallback option in the event that the desired content is not found within the peer group.

2.6   Security Issues

There are two security-related issues to consider: ensuring the integrity of content and ensuring the privacy of peers (i.e., not revealing to a client's peers what content it has accessed).

The integrity of the server's content can easily be ensured by having the server digitally sign the content. A client can obtain the signature either directly from the server (as part of the redirection message) or from a peer. The client can then verify the authenticity of the content it receives from its peers. For the sake of computational efficiency, the server could sign a 160-bit SHA-1 hash of the content rather than the content itself. In any case, since the signature need only be computed once for each version of a file, the burden placed on the server is minimal.

Ensuring privacy is much harder. While there exist proposals for enabling anonymous communication between hosts (e.g., [15]), anonymity comes at the cost of performance. This trade-off may not be appropriate in a flash crowd situation since performance is the key issue. In fact, clients may not care about privacy during a flash crowd because the content served during such times is, in any case, likely to be of widespread interest.

3   Experimental Evaluation

In this section, we evaluate the feasibility and potential performance of end-host cooperation during a flash crowd. The goals of the evaluation are to answer the following questions:

- How often can a client retrieve content from its peer group and avoid accessing the server?
- How much additional load do peers incur by participating in CoopNet?
- How often can cooperating peers be found nearby?
- What is the duration of time for which peers are active?

The cooperation protocol used in our simulations is based on the one described in Sections 2.2 and 2.5. A client who is willing to cooperate initially contacts the server to get IP addresses of other CoopNet clients. The server maintains a fixed size list of the CoopNet clients' IP addresses, and includes the most recent n clients in its redirection message. In our simulations, n ranges from 5 to 50 clients. Once the client has a list, it always contacts peers on the list to ask for content. If content cannot be found at these peers, the client returns to the server to request the full content and an updated peer list.

We use traces collected at the MSNBC website during the flash crowd of September 11, 2001 for our analysis. The flash crowd started at around 6am PDT, and lasted for the rest of the day. The peak load was ten times the typical load. Due to computing limitations, we focus our analysis on the first hour of the flash crowd, between 6:00 am to 7:00 am PDT, containing over 40 million requests.

3.1   Finding Content

In order for cooperation to be effective, clients need to avoid retrieving content from the loaded server to the extent possible. We define two metrics that capture how often content can be retrieved from one's peer group. The first metric is new content hit rate, which is the fraction of requests for new files that can be served by hosts in the peer group. The second metric is fresher content hit rate, which is the fraction of time that a fresher copy of a file can be found within the peer group. Fresher content hit rate only applies to the case when clients are looking for updated versions of files that they had downloaded in the past. If these two hit rates are high, that would indicate that CoopNet is providing an effective mechanism for improving client performance.

Figure 2 depicts the hit rates observed when the number of CoopNet clients is 200 (i.e., only 200 of the many hundreds of thousands of clients are willing to cooperate). The peer list returned by the server, which determines the peer group used by a client, is drawn from this set of 200 CoopNet clients. The peer list size ranges from 5 to 50 clients. The vertical axis in Figure 2 is the observed rate, and the horizontal axis is observation times at every 5 minutes after the beginning of the trace at 6:00am. Each line represents the rate observed for a particular peer list size.

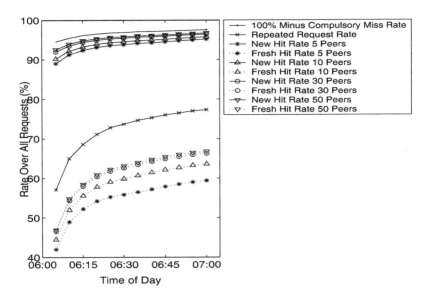

Fig. 2. Average hit rates observed at peers for each peer list size.

We present two analyses — optimistic and pessimistic. In the optimistic analysis, we assume that files are not modified between accesses. So an access is either a repeated request (i.e., a request for a URL that a client has previously accessed) or a request for a new (i.e., previously unseen) URL. The solid line in the middle of Figure 2 is the rate of repeated requests. The solid lines at the top show the sum of the repeated request rate and the hit rate for new content. This sum represents the overall hit rate in the optimistic setting. The upper bound for the overall hit rate is the difference between 100% and the compulsory miss rate (which corresponds to the case when content must be retrieved directly from the server because none of the 200 CoopNet clients has a copy of that content). This upper bound is the line at the top of the figure. We observe that for all peer list sizes, the overall hit rate is close to the upper bound, with less than 5% of requests ending up in a miss. We also observe that hit rates increase with time because of cache warming effects similar to those reported for Web proxies.

In the pessimistic analysis, we assume that a file is updated each second it is retrieved from the server. So in the case of a repeated request, a client would actually look for a fresher copy of the content than it has. The rate for finding fresher content from cooperating peers is represented by the dotted lines in Figure 2. Clearly, the upper bound for finding fresher content is the repeated request rate. After 5 minutes of cooperation, peers find fresher content 46% of the time out of the maximum achievable 56%, using a peer list size of 30. After an hour of cooperation, peers find fresher content 65% of the time out of the

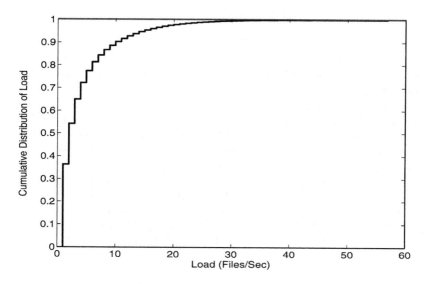

Fig. 3. Load at peers during busy periods.

maximum achievable 77%, using a list size of 30. Increasing the list size from 30 to 50 peers does not significantly improve hit rates.

In summary, we find that cooperation among a small group of peers is effective. Clients need to retrieve content from the server only 15% of the time when using a peer list size of 30.

3.2   Load on Peers

CoopNet clients contribute resources, such as network bandwidth, to the system. To maintain good performance, it is important not to completely exhaust those resources. Here we examine the network bandwidth overhead incurred by clients serving content.

Over 80% of the time, peers are idle and do not serve content. Figure 3 depicts the cumulative distribution of load, measured as the rate at which peers serve files, during the remaining 20% of time for a peer list size of 30. This distribution is representative of the load observed across all simulations of different peer list sizes. For the most part, peers can sustain the bandwidth requirement for serving content. Over half of the time during busy periods, peers serve at most 2 files in a second. However, in a few cases, load may be unevenly distributed, leading to a flash crowd at peers. The load can be as high as 57 files/second. Although the load is much less than that observed at the server, it may be enough to cause an overload at peers. We are presently investigating load distribution and peak bandwidth requirement for peers.

3.3   Finding Nearby Peers

Finding nearby peers can greatly improve the efficiency of peer-to-peer interaction. For example, a peer at CMU can retrieve content more quickly from another peer at CMU than it can from a peer in Europe. In some cases, the peer-to-peer performance could be comparable or better than client-server performance.

We use the following metric to determine network proximity. Peers that are in the same BGP prefix cluster are considered to be "close" to each other. Although this metric does not express closeness of peers that are in different BGP prefix clusters, it provides an approximation to whether or not it is possible to find a nearby peer.

We look at the IP addresses of clients in the trace in the initial 30-minute period. There were 563,284 unique clients, and 69,778 unique BGP prefix clusters. The probability of there being another client in the same prefix cluster during the first 2 minutes of the trace is 80%. The probability grows to 90% for the entire 30-minute period. Therefore, it is likely that peers will cooperate with nearby peers.

3.4   Duration of Activity Period for Peers

The duration of time for which peers are active affects how well CoopNet performs. If peers are active at a website for very short periods of time, peer lists must also be updated frequently.

To determine the period of activity, we consider the interarrival time between requests in the initial 30 minutes of the trace[4]. We treat an interarrival period that is longer than a threshold as representing the end of an activity period (and the start of the next). We consider two values of the threshold — 1 minute and 5 minutes. We find that the average activity period is 1.5 minutes and 4.5 minutes, respectively, in the two cases. This indicates that peer lists may become stale on the order of a few minutes and should be updated frequently.

4   Comparison with Alternative Approaches

We now discuss two alternative approaches to solving the flash crowd problem: proxy caching and infrastructure-based content distribution networks (CDNs). An advantage that both of these approaches have over CoopNet is that they can be deployed transparently to clients.

Proxy caching can help alleviate server load during a flash crowd by filtering out repeated requests from groups of clients that share a proxy cache. However, the effectiveness of proxy caching is limited for a few reasons. First, for them to be really effective in the context of a flash crowd, proxy caches need to be deployed widely. Since this requires substantial infrastructural investments by a large number of organizations, a widespread deployment of proxy caches is only likely if it results in significant performance improvement during "normal" (i.e.,

---

[4] Clearly, the limited length of the trace could bias our results.

non-flash crowd) times as well. However, cache hit rates have remained quite low, and the growing share of dynamic and customized content will only make matters worse.

A second issue is that even a universal deployment of proxy caches may not be sufficient to alleviate a flash crowd in certain situations. For instance, the small Web site for a high school alumni association may be overwhelmed by the flash crowd caused when a link to the video clip of a recent football game is sent out to all members via email. The clients interested in this content are likely to be dispersed across the Internet, so proxy caches at the local or organizational level may not filter out much of the load.

An alternative approach is to depend on an infrastructure-based CDNs (e.g., Akamai [20]) to distribute content. This may be an effective approach for ensuring high availability and good performance both during a flash crowd and in normal times. However, it is unlikely that a small Web site would be in a position to afford the services of a commercial CDN. Moreover, the absolute volume of traffic at such a site even during a flash crowd may not be large enough to be of interest to a commercial CDN.

In summary, we believe that CoopNet offers advantages compared to both proxy caching and infrastructure-based CDNs. CoopNet offers a low-cost but effective solution to the flash crowd problem, which is likely to be especially attractive to small Web sites with limited resources. That said, we do *not* view CoopNet as a replacement for infrastructure-based CDNs. As noted on Section 1, CoopNet's peer-to-peer content distribution kicks in when needed during a flash crowd but lies dormant during normal times. In contrast, an infrastructure-based CDN is engineered to provide a wide range of services (e.g., hit metering, high availability, performance guarantees, etc.) during all times. Thus we believe that there is a role for both CoopNet and infrastructure-based solutions.

Acknowledgements

We are grateful to Jason Bender, Steven Lautenschlager, Perry Stoll, and Ted Thoma for providing us the MSNBC Web logs from September 11. We would like to thank Stefan Saroiu for making his Web server bandwidth measurements available to us. We would also like to thank Lili Qiu for early discussions on CoopNet and the anonymous IPTPS 2002 reviewers for their insightful comments.

References

1. D. G. Andersen, H. Balakrishnan, M. F. Kaashoek and R. Morris. "Resilient Overlay Networks", *ACM SOSP*, October 2001.
2. W. J. Bolosky, J. R. Douceur, D. Ely, and M. Theimer. "Feasibility of a Serverless Distributed File System Deployed on an Existing Set of Desktop PCs", *ACM SIGMETRICS*, June 2000.

3. R. Fielding *et al.* "Hypertext Transfer Protocol – HTTP/1.1", *RFC-2616, IETF*, June 1999.
4. P. Joubert, R. King, R. Neves, M. Russinovich, and J. Tracey. "High-Performance Memory-Based Web Servers: Kernel and User-Space Performance", *Usenix 2001*, June 2001.
5. C. Kommareddy, N. Shankar, and B. Bhattacharjee. "Finding Close Friends on the Internet", *IEEE ICNP*, November 2001.
6. B. Krishnamurthy and J. Wang. "On Network-Aware Clustering of Web Clients", *ACM SIGCOMM*, August 2001.
7. K. Lai and M. Baker. "Nettimer: A Tool for Measuring Bottleneck Link Bandwidth", *USENIX Symposium on Internet Technologies and Systems*, March 2001.
8. T. S. E. Ng and H. Zhang. "Towards Global Network Positioning", *ACM SIGCOMM Internet Measurement Workshop*, November 2001.
9. V. N. Padmanabhan and L. Subramanian. "An Investigation of Geographic Mapping Techniques for Internet Hosts", *ACM SIGCOMM*, August 2001.
10. V. N. Padmanabhan, H. J. Wang, P. A. Chou, and K. Sripanidkulchai. "Distributing Streaming Media Content Using Cooperative Networking", *ACM NOSSDAV*, May 2002.
11. S. Ratnasamy, P. Francis, M. Handley, R. Karp, and S. Shenker. "A Scalable Content-Addressable Network", *ACM SIGCOMM*, August 2001.
12. A. Rowstron and P. Druschel. "Storage Management and Caching in PAST, A Large-scale, Persistent Peer-to-peer Storage Utility", *ACM SOSP*, October 2001.
13. S. Saroiu. "Bottleneck Bandwidths", October 2001.
    http://www.cs.washington.edu/homes/tzoompy/sprobe/webb.htm
14. S. Savage, A. Collins, E. Hoffman, J. Snell, and T. Anderson. "The End-to-End Effects of Internet Path Selection", *ACM SIGCOMM*, August 1999.
15. C. Shields and B. N. Levine. "A Protocol for Anonymous Communication Over the Internet", *ACM Conference on Computer and Communication Security*, November 2000.
16. I. Stoica, R. Morris, D. Karger, F. Kaashoek, and H. Balakrishnan. "Chord: A Scalable Peer-To-Peer Lookup Service for Internet Applications", *ACM SIGCOMM*, August 2001.
17. M. Waldvogel, G. Varghese, J. Turner, and B. Plattner. "Scalable High Speed IP Routing Lookups", *ACM SIGCOMM*, September 1997.
18. B. Zhao, J. Kubiatowicz, and A. Joseph. "Tapestry: An Infrastructure for Fault-Tolerant Wide-Area Location and Routing", *U. C. Berkeley Technical Report UCB//CSD-01-1141*, April 2001.
19. "Web acts as hub for info on attacks", *http://news.cnet.com/news/0-1005-200-7129241.html?tag=rltdnws*, 11 September 2001.
20. Akamai. *http://www.akamai.com/*
21. MSNBC Web site. *http://www.msnbc.com/*
22. List of Web servers. *http://www.icir.org/tbit/daxlist.txt*

Internet Indirection Infrastructure

Ion Stoica[1], Daniel Adkins[1], Sylvia Ratnasamy[1], Scott Shenker[2],
Sonesh Surana[1], and Shelley Zhuang[1]

[1] University of California at Berkeley, Berkeley CA 94720, USA,
{istoica, dadkins, sylviar, sonesh, shelleyz}@cs.berkeley.edu,
[2] The ICSI Center for Internet Research, Berkeley CA 94704, USA,
shenker@cs.icsi.berkeley.edu

Abstract. This paper argues for an Internet Indirection Infrastructure
that replaces the point-to-point communication abstraction of today's
Internet with a rendezvous-based communication abstraction: instead of
explicitly sending a packet to a destination, each packet is associated an
identifier, which is then used by the receiver to get the packet. This level
of indirection decouples the sender and the receiver behaviors, and allows
us to efficiently support basic communication services such as multicast,
anycast and mobility in the Internet. To demonstrate the feasibility of
this approach, we are currently designing and building an overlay network
solution based on the Chord lookup system.

1 Introduction

The Internet is designed around the *unicast* point-to-point communication abstraction. This simple abstraction is one of the main reasons behind its scalability and efficiency. However, as the Internet evolves into a global communication
infrastructure, there is an increasing need to support other communication abstractions such as multicast, anycast, and host mobility. Unfortunately, despite
years of intense research, these services are yet to be deployed in the Internet.
The main difficulty resides in the fact that the point-to-point communication
abstraction—which assumes one sender and one receiver placed at well-known
and fixed network locations—is *not* appropriate for these services. For example,
mobility requires one to remove the assumption that end-hosts are fixed, multicast requires one to remove the assumption that there is only one sender and one
receiver, and anycast requires one to remove the assumption that the receiver's
location is known.

To get around this problem, existing solutions use a simple but powerful
technique: *indirection.* These solutions assume a physical or a logical indirection
point interposed between the sender and the receiver(s) that relays the traffic
between them. By communicating through the indirection point rather than
directly to the end-host, a sender can abstract away the location and the number
of receivers. For instance, mobile IP assumes a home agent that hides the end-
host mobility, while IP multicast assumes a logical indirection point (address)
that hides the number of receivers and their locations.

P. Druschel, F. Kaashoek, and A. Rowstron (Eds.): IPTPS 2002, LNCS 2429, pp. 191–202, 2002.
© Springer-Verlag Berlin Heidelberg 2002

While indirection can enable these services, implementing them at the IP layer in a scalable fashion has proven difficult [4, 9, 14]. Moreover, deploying additional functionality at the IP layer requires a level of community-wide consensus and commitment that is hard to achieve. In short, implementing these more general abstractions at the IP layer poses difficult technical problems and formidable deployment barriers.

In response, many researchers have turned to application-layer solutions (either end-host or overlay mechanisms) to support these abstractions [2, 4, 11, 13]. While these proposals achieve the desired functionality, they do so in a piecemeal fashion; *e.g.*, solutions for mobility do not address multicast, and vice versa. As a result, many similar and largely redundant mechanisms are implemented to achieve these various goals.

The goal of this work is to combine the generality of IP-layer solutions with the deployability of overlay solutions by proposing a general-purpose Internet indirection infrastructure ($i3$). This infrastructure supports indirection by providing a flexible *rendezvous*-based communication abstraction. A variety of communication services, such as multicast, anycast, and mobility, can be easily implemented on top of this communication abstraction. Such an approach will avoid both the technical and deployment challenges inherent in IP-layer solutions and the redundancy and lack of synergy in application-layer approaches.

2   Internet Indirection Infrastructure

The purpose of the Internet Indirection Infrastructure ($i3$) is to provide indirection; that is, it decouples the act of sending from the act of receiving. The $i3$ service model is simple: sources send packets to a logical *identifier*, and receivers express interest in packets sent to an identifier.

This service model is similar to that of IP multicast. The crucial difference is that the $i3$ equivalent of an IP multicast join is more flexible. IP multicast offers a receiver a binary decision of whether or not to receive packets sent to that group (this can be indicated on a per-source basis). It is up to the multicast infrastructure to build efficient delivery trees. The $i3$ equivalent of a join is inserting a *trigger*. It allows receivers to *control* the routing of the packet. This provides two advantages. First, it allows them to create, at the application level, services such as mobility, anycast, and service composition out of this basic service model. Thus, this one simple service model can be used to support a wide variety of application-level communication abstractions, alleviating the need for many parallel and redundant overlay infrastructures. Second, the infrastructure can give responsibility for efficient tree construction to the end-hosts. This allows the infrastructure to remain simple, robust, and scalable.

2.1   Rendezvous-Based Communication

The service model is instantiated as a rendezvous-based communication abstraction. In their simplest form, packets are pairs $(id, data)$ where id is an m-bit identifier[1] and *data* is the payload (typically a normal IP packet payload). Receivers use *triggers* to indicate their interest in packets. In their simplest form, triggers are pairs $(id, addr)$, where id is the trigger identifier, and $addr$ is a node's address, consisting of an IP address and UDP port number. A trigger $(id, addr)$ indicates that all packets sent to identifier id should be forwarded (at the IP layer) by the $i3$ infrastructure to the node with address $addr$. More specifically, the rendezvous-based communication abstraction exports the three primitives shown in Figure 1(a).

$i3$'s Application Programming Interface (API)	
$sendPacket(p)$	send packet
$insertTrigger(t)$	insert trigger
$removeTrigger(t)$	remove trigger

(a)

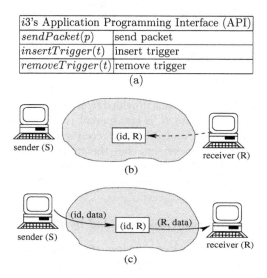

(b)

(c)

Fig. 1. (a) $i3$'s API. Example illustrating communication between two nodes: (b) The receiver R inserts trigger (id, R). (c) The sender sends packet $(id, data)$.

Figure 1(b) illustrates the communication between two nodes, where receiver R wants to receive packets sent to id. The receiver inserts the trigger (id, R) into the network. When a packet is sent to identifier id, the trigger causes it to be forwarded via IP to R. There are two generalization that extend the power and the flexibility of the rendezvous-based communication abstraction. The first generalization allows *inexact* matching between identifiers, and it is used to provide anycast functionality. The second generalization replaces identifiers with a stack of identifiers and is used to provide service composition (see Section 2.3).

---

[1] In our design we use $m = 256$. Such a large value of m allows end hosts to choose trigger identifiers independently since the chance of collision is minimal.

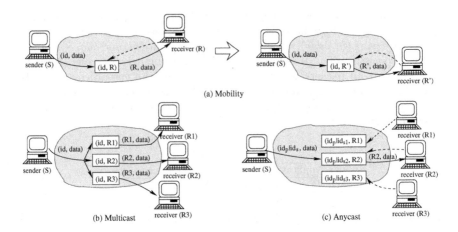

Fig. 2. Communication abstractions provided by $i3$. (a) Mobility: The change of the receiver's address from R to R' is transparent to the sender. (b) Multicast: Every packet $(id, data)$ is forwarded to each receiver R_i that inserts the trigger (id, R_i). (c) Anycast: The packet matches the trigger of receiver R2. $id_p|id_s$ denotes an identifier of size m, where id_p represents the prefix of the k most significant bits, and id_s represents the suffix of the $m - k$ least significant bits.

2.2 Communication Primitives Provided by $i3$

Next, we briefly discuss how $i3$ can achieve the more general communication abstractions of mobility, multicast, and anycast.

Mobility: A mobile host that changes its address from R to R' as a result of moving from one subnetwork to another can preserve the end-to-end connectivity by simply updating each of its existing triggers from (id, R) to (id, R'), as shown in Figure 2(a). The sending hosts need not be aware of the mobile host's current location or address.

Multicast: Creating a multicast group is equivalent to having all members of the group register triggers with the same identifier id. As a result, any packet that matches id is forwarded to all members of the group (see Figure 1(b)). Note that unlike IP multicast, with $i3$ there is no difference between unicast or multicast packets, in either sending or receiving. Such an interface gives maximum flexibility to the application.

Anycast: Anycast ensures that a packet is delivered to at most one receiver in a group. To achieve this, we generalize the exact matching in $i3$ to a longest prefix matching in which the first k bits are required to always match. All hosts in an anycast group maintain triggers which are identical in the k most significant bits. These k bits play the role of the anycast group identifier. To send a packet to an anycast group, a sender uses an identifier whose k-bit prefix matches the anycast group identifier. The packet is then delivered to the member of the group whose trigger identifier best matches the packet identifier according to the longest prefix matching rule (see Figure 1(c)).

2.3 Stack of Identifiers

A second generalization of $i3$ replaces identifiers with identifier *stacks*. An identifier stack is a tuple of identifiers that takes the form $(id_1, id_2, id_3, \ldots, id_k)$ where id_i is either an identifier or an address. Packets p and triggers t are thus of the form: $p = (id_{stack}, data)$, and $t = (id, id_{stack})$, respectively. The generalized form of packets allows a source to send a packet to a series of identifiers, much as in source routing. The generalized form of triggers allows a trigger to send a packet to another identifier rather than to an address.

To illustrate the use of the stack of identifiers, consider the problem of sending an MPEG video stream to one H.263 receiver and one MPEG receiver (see Figure 3). To provide this functionality, we use the ability of the *receiver*, to control the transformations performed on data packets. In particular, the H.263 receiver $R1$ inserts trigger $(id, (id_{MPEG-H.263}, R1))$, and the sender sends packets $(id, data)$. Each packet matches $R1$'s trigger, and as a result the packet's identifier id is replaced by the trigger's stack $(id_{MPEG-H.263}, T)$. Next, the packet is forwarded to the MPEG-H.263 transcoder, and then directly to receiver $R1$. In contrast, an MPEG receiver $R2$ only needs to maintain a trigger $(id, R1)$ in $i3$. This way, receivers with different display capabilities can subscribe to the same multicast group.

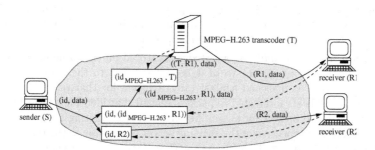

Fig. 3. Heterogeneous multicast: Receiver $R1$ specifies that wants to receive JPEG data, while $R2$ specifies that wants to receive MPEG data. The sender sends MPEG data.

3 $i3$ Implementation

$i3$ is implemented as an overlay network which consists of a set of servers that store triggers and forward packets (using IP) between $i3$ nodes and to end hosts.

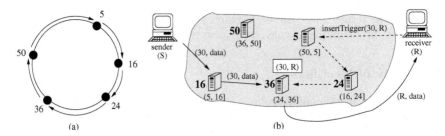

Fig. 4. (a) A Chord identifier circle for $m = 6$, with 5 servers identified by 5, 16, 24, 36, and 50, respectively. Each server is responsible for all identifiers between its identifier and the identifier of the node that precedes it on the circle. (b) Receiver R inserts trigger $(30, R)$, and the trigger is forwarded via $i3$ to server 36 which is responsible for identifier 30. The trigger is stored there (shown in the white box) until explicitly removed or timed out. When sender S sends packet $(30, data)$, it is also forwarded via $i3$ to server 36. Servers identifiers are in bold. The interval of identifiers for which each server is responsible are also shown.

To maintain this overlay network and to route packets in $i3$, we use the Chord lookup protocol [5]. Chord assumes a circular identifier space of integers $[0, 2^m)$, where 0 follows $2^m - 1$. Every $i3$ server has an identifier in this space, and all trigger identifiers belong to the same identifier space. The $i3$ server with identifier n is responsible for all identifiers in the interval $(n_p, n]$, where n_p is the identifier of the node preceding n on the identifier circle. Figure 4(a) shows an identifier circle for $m = 6$. There are five $i3$ servers in the system with identifiers 5, 16, 24, 36, and 50, respectively. All identifiers in the range $(5, 16]$ are mapped on server 16, identifiers in $(17, 24]$ are mapped on server 24, and so on.

When a trigger $(id, addr)$ is inserted, it is stored on the $i3$ node responsible for id. When a packet is sent to id it is routed by $i3$ to the node responsible for its id; there it is matched against (any) triggers for that id and forwarded (using IP) to all hosts interested in packets sent to that identifier. Chord ensures that the server responsible for a identifier is found after visiting at most $O(\log n)$ other $i3$ servers irrespective of the starting server (n represents the total number of servers in the system). To achieve this, Chord requires each node to maintain only $O(\log n)$ routing state. Chord allows servers to leave and join dynamically, and it is highly robust against failures. For more details refer to [5]. Figure 4(b) shows an example in which trigger $(30, R)$ is inserted at node 36 (i.e., the node that maps $(24, 36]$, and thus is responsible for identifier 30). Packet $(30, data)$ is forwarded to server 30, matched against trigger $(30, R)$, and then forwarded via IP to R.

Note that packets are not stored in $i3$; they are only forwarded. $i3$ provides a best-effort service like today's Internet. $i3$ implements neither reliability nor ordered delivery on top of IP. End hosts use periodic refreshing to maintain their triggers in $i3$. Hosts contact an $i3$ node when sending $i3$ packets or inserting triggers. This $i3$ node then forwards these packets or triggers to the $i3$ node responsible for the associated identifiers. Hosts need only know one $i3$ node to use the $i3$ infrastructure. This can be done through a static configuration file, or by a DNS lookup assuming $i3$ is associated with a DNS domain name. In Figure 4(b), the sender knows only server 16, and the receiver knows only server 5.

One important observation is that once the end host finds the server that is responsible for a specific identifier, it can cache that server. As a result the end host can refresh the triggers and send packets with the same identifiers directly to that server. This significantly increases routing efficiency.

4   Security

Unlike IP, where an end-host can only send and receive packets, in $i3$ end-hosts are also responsible for maintaining the routing information through triggers. While this allows flexibility for applications, it also (and unfortunately) creates new opportunities for malicious users. We now discuss several security issues and how $i3$ addresses them.

We emphasize that our main goal here is not to design a bullet proof system. Instead, our goal is to design simple and efficient solutions that make $i3$ not worse and in many cases better than today's Internet. The solutions outlined in this section should be viewed as a starting point towards more sophisticated and better security solutions that we will develop in the future.

An important component for using $i3$ securely is for applications to make a distinction between two types of triggers: *public* and *private*. This distinction is made only at the application level: $i3$ itself doesn't differentiate between private and public triggers.

The identifiers of a public trigger is known by all end-hosts in the system. An example is a web server that maintains a public trigger to allow any client to contact it. A public trigger can be defined as a the hash of the host's DNS name, of a web address, or of the public key associated to a web server. Public triggers are long lived, typically days or months. In contrast, private triggers are chosen by a small number of end-hosts and they are short lived. Typically, private triggers exist only during the duration of a flow. To illustrate the difference between public and private triggers, consider a client accessing a web server. First, the client chooses a private trigger identifier id_c, inserts trigger $(id_s, addr_c)$ into $i3$, and sends id_c to the web server via the server's public trigger. Once contacted, the server selects a private identifier id_s, inserts trigger $(id_s, addr_s)$, and sends id_s to the client via the client's trigger identifier id_c. The client and the server then use the private triggers to communicate. Once the communication terminates, the private triggers are destroyed.

Eavesdropping A user that knows a host's trigger can eavesdrop the traffic towards that host by inserting a trigger with the same identifier and its own address. We consider two cases: (a) private and (b) public triggers.

Private triggers are secretly chosen by the application end-points and are not supposed to be revealed to the outside world. The length of the trigger's

identifier makes it very difficult for a third party to use a brute force attack.[2]
Furthermore, end-points can periodically change the private triggers associated
with a flow. Another alternative would be for the receiver to associate multiple
private triggers to the same flow, and the sender to send packets randomly to one
of these private triggers. The alternative left to a malicious user is to intercept
all private triggers. However this is equivalent to eavesdropping at the IP level
or taking control of the $i3$ server storing the trigger, which makes $i3$ no worse
than IP.

With $i3$, a public trigger is known by all users in the system, and thus anyone
can eavesdrop the traffic to such a trigger. To alleviate this problem, end-hosts
can use the public triggers to choose a pair of private triggers, and then use
these private triggers to exchange the actual data. To keep the private triggers
secret, one can use public key cryptography. To initiate a connection, a host A
chooses a private trigger id_a, encrypts it under the public key of a receiver B,
and then sends it to B via B's public trigger. B decrypts A's private trigger id_a,
then chooses its own private trigger id_b, and sends this trigger back to A over
A's private trigger id_a. Since the sender's trigger is encrypted, a malicious user
cannot impersonate B.[3]

Trigger hijacking A malicious user can isolate a host by removing its public trigger. Similarly, a malicious user in a multicast group can remove other members
from the group by deleting their triggers. While removing a trigger also requires
to specify the IP address of the trigger, this address is, in general, not hard to
obtain.

One possibility to guard against this attack is to add another level of indirection. Consider a server S that wants to advertise a public trigger with identifier
id_p. Instead of inserting the trigger (id_p, S), the server can insert two triggers,
(id_p, x) and (x, S), where x is an identifier known only by S. Since a malicious
user has to know x in order to remove either of the two triggers, this simple
technique provides effective protection against this type of attack. To avoid performance penalties, the receiver can choose x such that both (id_p, x) and (x, S)
are stored at the same server. With the current implementation this can be easily
achieved by having i_p and x share the same k-bit prefix.

DoS Attacks The fact that $i3$ gives end-hosts control on routing opens new
possibilities for DoS attacks. We consider two types of attacks: (a) attacks on
end-hosts, and (b) attacks on the infrastructure. In the former case, a malicious
user can insert a hierarchy of triggers in which all triggers on the last level point

---

[2] While other application constraints such as storing a trigger at a server nearby can
limit the identifier choice, the identifier is long enough (i.e., 256 bits), such that the
application can always reserve a reasonable large number of bits (e.g., 64 bits) that
are randomly chosen.

[3] Note that an attacker can still count the number of connection requests to B. However, this information is of very limited use, if any, to the attacker. If, in the future,
it turns out that this is a problem, then other security mechanisms such as public
trigger authentication will need to be used.

to the victim. Sending a single packet to the trigger at the root of the hierarchy will cause the packet to be replicated and all replicas to be sent to the victim. This way an attacker can mount a large scale DoS attack by simply leveraging the $i3$ infrastructure. In the later case, a malicious user can create trigger loops, for instance by connecting the leaves of a trigger hierarchy to its root. In this case, each packet sent to the root will be exponentially replicated!

To alleviate these attacks, $i3$ uses three techniques:

1. **Challenges** $i3$ assumes implicitly that a trigger that points to an end-host R is inserted by the end-host itself. An $i3$ server can easily verify this assumption by sending a challenge to R the first time the trigger is inserted. The challenge consists of a random nonce that is expected to be returned by the receiver. If the receiver fails to answer the challenge the trigger is removed. As a result an attacker cannot use a hierarchy of triggers to mount a DoS attack (as described above), since the leaf triggers will be removed as soon as the server detects that the victim hasn't inserted them.

2. **Resource allocation** Each server uses Fair Queueing [7] to allocate resources amongst the triggers it stores. This way the damage inflicted by an attacker is only proportional to the number of triggers it maintains. An attacker cannot simply use a hierarchy of triggers with loops to exponentially increase its traffic. As soon as each trigger reaches its fair share the excess packets will be dropped. While this technique doesn't solve the problem, it gives $i3$ time to detect and to eventually break the cycles.

   To increase protection, each server can also put a bound on the number of triggers that can be inserted by a particular end-host. This will preclude a malicious end-host from monopolizing a server's resources.

3. **Loop detection** When a trigger that doesn't point to an IP address is inserted, the server runs a procedure to detect whether the new trigger doesn't create a loop. A simple procedure is to send a special packet with a random nonce. If the packet returns back to the server, the trigger is simply removed. To increase the robustness, the server can invoke this procedure periodically after such a trigger is inserted.

4.1   Anonymity

Point-to-point communication networks such as the Internet provide limited support for anonymity. Packets usually carry the destination and the source addresses, which makes it relatively easy for an eavesdropper to learn the sender and the receiver identities. In contrast, with $i3$, eavesdropping the traffic of a sender will not reveal the identity of the receiver, and eavesdropping the traffic of a receiver will not reveal the sender's identity. The level of anonymity can be further enhanced by using chain of triggers or stack of identifiers to route packets.

5   Related Work

The rendezvous-based communication is similar in spirit to the tuple space work in distributed systems [1, 10, 18]. However, tuple spaces have richer semantics, and they guarantee persistence and atomicity. Providing these properties in very large scale distributed systems is, however, very difficult. In contrast, $i3$ trades these properties for a scalable and efficient implementation.

$i3$ shares many similarities with naming systems. This should come as no surprise, as identifiers can be viewed as semantic-less names. DNS maps hostnames to IP addresses [12]. DNS names are hierarchical while $i3$ identifiers are flat. DNS resolvers form a static overlay hierarchy, while $i3$ servers form a self-organizing overlay. $i3$ integrates identifier resolution with packet forwarding. Active Names (AN) maps a name to a chain of mobile code responsible for locating the remote service [16]. While AN names are used primary to describe services, $i3$ identifiers are used primary to abstract away the end-host location. Intentional Naming System (INS) is a resource discovery and service location system for mobile hosts [17]. $i3$ differs from INS in that from network's point of view an identifier does not carry any semantic. Another difference is that $i3$ allows end-hosts to explicitly control (via triggers) the application-level path followed by the packets.

The rendezvous-based abstraction is similar to the IP multicast abstraction [6]. An IP multicast address identifies the receivers of a multicast group in the same way an $i3$ identifier identifies the multicast receivers. However, unlike IP which allocates a special range of addresses (i.e., class D) to multicast, $i3$ does not put any restrictions on the identifier format. In addition, $i3$ has ability to support multicast groups with heterogeneous receivers.

TRIAD [3] and IPNL [8] have been recently proposed to solve the IPv4 address scarcity problem. Both schemes use DNS names rather than addresses for global identification. One difference between $i3$ and both TRIAD and IPNL is that the path of a packet is determined by end-hosts, instead of being determined during the DNS name resolution by network specific protocols.

6   Status

We have implemented an early prototype of $i3$ based on the Chord lookup service [15]. We use 256 bit identifiers, and the matching procedure requires exact matching on the 128 most significant bits. This choice makes it very unlikely that a packet will erroneously match a trigger, and at the same time gives the applications up to 128 bits to encode application specific information such as the host location. The untuned prototype is able to forward up to 35,000 packets per second, and to handle about 80,000 trigger insertions/refreshes per second on a 700 MHz Pentium III processor. By choosing the set of Chord fingers based on the network proximity, we are able to reduce the routing latency between any two servers in $i3$ within a factor of two (on the average) from the optimal routing in the underlying network.

7   Conclusions

This paper argues for an Internet Indirection Infrastructure based on a rendezvous-based communication abstraction to implement basic communication primitives such as multicast and anycast, and to support end-host mobility and service composition in the Internet. End-hosts no longer communicate by specifying the packet's destination. Instead, they associate with each packet an identifier which is then used by the destination to get the packet. In the best case, we expect this research to open the door to new and exciting research directions in computer networks similar to the way associative memories did in computer architectures, and tuple spaces did in changing the landscape in distributed systems.

References

[1] N. Carriero. *The Implementation of Tuple Space Machines*. PhD thesis, Yale University, 1987.

[2] Yatin Chawathe, Steve McCanne, and Eric Brewer. Rmx: Reliable multicast in heterogeneous networks. In *Proceedings of IEEE INFOCOM 2000*, Tel-Aviv, Israel, March 2000.

[3] D. R. Cheriton and M. Gritter. TRIAD: A new next generation Internet architecture, March 2000. http://www-dsg.stanford.edu/triad/ triad.ps.gz.

[4] Y. Chu, S. G. Rao, and H. Zhang. A case for end system multicast. In *Proceedings of ACM SIGMETRICS'00*, pages 1–12, Santa Clara, CA, June 2000.

[5] Frank Dabek, Frans Kaashoek, David Karger, Robert Morris, and Ion Stoica. Wide-area cooperative storage with cfs. In *Proc. ACM SOSP'01*, pages 202–215, Banff, Canada, 2001.

[6] S. Deering and D. R. Cheriton. Multicast routing in datagram internetworks and extended LANs. *ACM Transactions on Computer Systems*, May 1990.

[7] A. Demers, S. Keshav, and S. Shenker. Analysis and simulation of a fair queueing algorithm. In *Journal of Internetworking Research and Experience*, pages 3–26, October 1990.

[8] P. Francis and R. Gummadi. IPNL: A nat extended internet architecture. In *Proc. ACM SIGCOMM'01*, pages 69–80, San Diego, 2001.

[9] H.W. Holbrook and D.R. Cheriton. IP multicast channels: EXPRESS support for large-scale single-source applications. In *Proceedings of ACM SIGCOMM'99*, Cambridge, Massachusetts, August 1999.

[10] Java Spaces. http://www.javaspaces.homestead.com/.

[11] J. Jannotti, D. K. Gifford, K. L. Johnson, M. F. Kaashoek, and Jr. J. W. O'Toole. Overcast: Reliable multicasting with an overlay network. In *Proceedings of the 4th USENIX Symposium on Operating Systems Design and Implementation (OSDI 2000)*, San Diego, California, October 2000.

[12] P. Mockapetris and K. Dunlap. Development of the Domain Name System. In *Proc. ACM SIGCOMM*, Stanford, CA, 1988.

[13] A. C. Snoeren and H. Balakrishnan. An end-to-end approach to host mobility. In *Proceedings of ACM/IEEE MOBICOM'99*, Cambridge, MA, August 1999.

[14] I. Stoica, T.S.E. Ng, and H. Zhang. REUNITE: A recursive unicast approach to multicast. In *Proceedings of INFOCOM'00*, Tel-Aviv, Israel, March 2000.

[15] Ion Stoica, Robert Morris, David Karger, M. Frans Kaashoek, and Hari Balakrishnan. Chord: A scalable peer-to-peer lookup service for internet applications. In *Proc. ACM SIGCOMM'01*, pages 149–160, San Diego, 2001.

[16] A. Vahdat, M. Dahlin, T. Anderson, and A. Aggarwal. Active names: Flexible location and transport. In *Proceedings of USENIX Symposium on Internet Technologies & Systems*, October 1999.

[17] W. Adjie-Winoto and E. Schwartz and H. Balakrishnan and J. Lilley. The design and implementation of an intentional naming system. In *Proc. ACM Symposium on Operating Systems Principles*, pages 186–201, Kiawah Island, SC, December 1999.

[18] P. Wyckoff, S. W. McLaughry, T. J. Lehman, and D. A. Ford. T Spaces. *IBM System Journal*, 37(3):454–474, 1998.

Peer-to-Peer Caching Schemes to Address Flash Crowds

Tyron Stading, Petros Maniatis, and Mary Baker

Computer Science Department
Stanford University
Stanford, CA 94305, USA
{tstading, maniatis, mgbaker}@cs.stanford.edu
http://identiscape.stanford.edu/

Abstract. Flash crowds can cripple a web site's performance. Since they are infrequent and unpredictable, these floods do not justify the cost of traditional commercial solutions. We describe *Backslash*, a collaborative web mirroring system run by a collective of web sites that wish to protect themselves from flash crowds. Backslash is built on a distributed hash table overlay and uses the structure of the overlay to cache aggressively a resource that experiences an uncharacteristically high request load. By redirecting requests for that resource uniformly to the created caches, Backslash helps alleviate the effects of flash crowds. We explore cache diffusion techniques for use in such a system and find that probabilistic forwarding improves load distribution albeit not dramatically.

1   Introduction

Flash crowds have been the bane of many web masters since the web's explosion in mainstream popularity. The term "flash crowd" is used to describe the unanticipated, massive, rapid increase in the popularity of a resource, such as a web page, that lasts for a short amount of time.

Although their long term effects are hardly noticeable, in the short term, flash crowds incur unbearably high loads on web servers, gateway routers and links. They render the affected resources and any collocated resource unavailable to the rest of the world. Flash crowds are also relatively easy to cause. A mere mention of an interesting web page address in a popular news feed can result in an instant flood that lasts as long as the attention span of the news feed audience. In fact, flash crowds have been commonly referred to as "the Slashdot effect," from the name of the popular news feed, which has caused quite a few floods with its stories.

Although the concept of a malicious flash crowd is certainly within the realm of possibility, the intent behind the effect is usually impossible to distinguish in real time. Therefore in practice, it is important to understand how to adapt efficiently to the changing resource demands so as to distribute the unexpected high load among available resources, regardless of the intent.

P. Druschel, F. Kaashoek, and A. Rowstron (Eds.): IPTPS 2002, LNCS 2429, pp. 203–213, 2002.
© Springer-Verlag Berlin Heidelberg 2002

Commercial solutions have previously addressed this problem for very popular sites, such as large corporations with extensive web presence. Companies such as Akamai earn their income by distributing the load of highly trafficked web sites across a geographically dispersed network in advance. Akamai's solution focuses primarily on using proprietary networks and strategically placed dedicated caching centers to intercept and serve customer requests before they become a flood.

However, for sites such as non-profit organizations, schools and governments, which do not generally *expect* flash crowds, the cost of a high-profile content distribution solution such as Akamai's is not justifiable. Such sites have currently no recourse other than to overprovision or to pay the price of the occasional disastrous flash crowd including unavailability, prolonged recovery, ISP penalties and loss of legitimate, desirable traffic.

The purpose of this paper is to introduce, motivate, describe and begin evaluating *Backslash*, a grassroots web content distribution system based on peer-to-peer overlays. Backslash is a collaborative, scalable web mirroring service run and maintained by a collective of content providers who do not expect consistently heavy traffic to their sites. It relies on a content-addressable overlay [3, 5, 6, 8] for the self-organization of participants, for routing requests and for load balancing.

We use the remainder of this paper to identify the requirements from such a system and to present the overall design in more detail. We focus on the caching aspects of Backslash and limit the scope of the evaluation section to cache diffusion issues. We conclude with a research agenda for further work in this area.

2   System Requirements

In this section we outline the basic requirements for our grassroots web mirroring system.

Backslash is intended as a drop-in replacement for current web servers and reverse proxy caches. The driving requirement for its development is to make deployment completely transparent to the client web browsers.

The setting in which we hope to deploy the system is, for example, a collective of several universities or research institutions (possibly up to several thousand). Each institution dedicates to Backslash a well-connected low-end PC-grade computer; at the time of this writing any Pentium II-class computer with 128 MB of RAM should suffice. Node-to-node links have bandwidths between one and 10 Mbps, and latencies between 10 and 300ms. Client-to-node link characteristics range from 56 Kbps modems to perhaps cable or ADSL home connections. Each node stores a complete copy of the data collection published by its hosting site—that is, the Stanford Backslash node holds the entire web collection of the www.stanford.edu web site— and has enough free storage for caching. We expect the available free space to be a small multiple, say two or three times, of the local collection size.

The objective of Backslash is to offer fair load distribution in the face of flash crowds. Our primary interest is to limit the load on any participating node so as not to overwhelm it, by distributing requests among as many participants as possible. However, we consider the task of identifying and penalizing documents that consistently exhibit disproportionately high popularity to be out of scope for this paper. Similarly, we ignore the security implications of malicious Backslash nodes at this early stage of this work.

Finally, we ignore problems with the mirroring of dynamically generated content. The problem of mirroring static content is, by itself, a formidable one in the grassroots context. Consequently, we tackle it first, before taking on the much harder problem of dynamic content.

3   Design

In the next few sections we describe the design of Backslash at a high level. We first present how Backslash bridges the gap between the resource location subsystem and the traditional browser-server relationship (Section 3.1). Then, we describe the resource location subsystem, which is based on the peer-to-peer *Distributed Hash Table* paradigm (Section 3.2). Finally, we go into cache diffusion in more detail (Section 3.3).

3.1   Redirection

Every Backslash node is primarily a regular web server for the document collection of the hosting site. During its normal mode of operation, that is, as long as the request load perceived by the node is manageable, a Backslash node does little more than what a normal web server does.

When an increased request load is perceived, the Backslash node switches into one of two special modes of operation: the *pre-overload* mode, in which the node sees uncharacteristically high load but is still not overwhelmed, and the *overload* mode, in which the node is nearly overrun with requests. In the pre-overload mode of operation, the node satisfies all requests that arrive, but diverts subsequent requests to associated resources, such as embedded images, away from itself. In the overload mode, the node redirects all requests it receives to surrogate Backslash nodes and otherwise serves no content. Every node has two locally defined *load thresholds* that determine the boundaries of the normal, pre-overload and overload modes.

Backslash nodes diffuse some of the load directed at a flooded document collection via the use of URL rewriting. A node in pre-overload mode overwrites the embedded URLs of the documents it returns so as to divert subsequent follow-up requests. Such requests—for example, embedded images—are directed instead to surrogate Backslash nodes. URL rewriting takes advantage of the two stages of which web requests commonly consist: the DNS lookup and the HTTP request. In fact, every Backslash node runs a simplified DNS server to intercept DNS requests caused by URL rewrites.

Both types of URL rewrites have the same goal: to cause the client browser to look elsewhere for the flooded document. The DNS-based rewrite accomplishes this by directing the DNS lookups for the hostname of the rewritten URL to a Backslash DNS server. For example, the original URL `http://www.backslash.stanford.edu/image.jpg` is rewritten as `http://<hash>˜.backslash.berkeley.edu/www.backslash.stanford.edu/image.jpg`, so as to redirect the requester to a surrogate Backslash node at Berkeley, where `<hash>` denotes the base-32 encoding of a SHA-1 hash of the entire original URL.

Similarly, the HTTP-based rewrite accomplishes the same thing by naming a specific surrogate IP address within the rewritten URL. For example, the original URL `http://www.backslash.stanford.edu/image.jpg` is rewritten as `http://a.b.c.d/www.backslash.stanford.edu/image.jpg`, where `a.b.c.d` is the IP address of a Backslash node at Berkeley.

Although functionally similar, the two rewrite techniques have different performance implications. The DNS-based rewrite can overlap the document location task, triggered by an intercepted DNS request at the surrogate node, with the HTTP/TCP client connection establishment that follows. On the other hand, DNS requests can result in long latencies in high-loss environments because of UDP time-outs, especially for wireless clients. As a result, embedded links served to client browsers coming from "nearby" network locations use DNS rewriting, whereas HTTP rewriting is used for more remote clients, based on local policy.

When URL rewriting is not an option, specifically in the case of the first request to an overloaded node from a particular client, plain redirection is used, again either via DNS or HTTP. For example, the mini DNS server responsible for the `backslash.stanford.edu` domain (which is a Backslash node itself) can return the IP address of a surrogate node when asked for the `A` record of `www`, when the Stanford site is in overload mode. Similarly, HTTP redirection uses `REDIRECT` responses to cause browsers to retry a request at a rewritten URL.

The combination of URL rewriting and redirection allows unaware client browsers to reach an unloaded surrogate Backslash node that will serve their requests. We explain how surrogate Backslash nodes serve requests for content from other sites' collections in the next section.

3.2   Resource Location

Once a surrogate Backslash node has received a request for a document of the afflicted site, it has to act as a gateway between the HTTP client browser and the collaborative mirroring portion of Backslash.

Mirroring is implemented on a peer-to-peer overlay following the *distributed hash table* paradigm. Systems in this category [3, 5, 6, 8] implement a hash table over a large number of self-organized nodes. Each node is responsible for a chunk of the entire hash table. If the hash function used by the table is uniform, then regardless of the distribution of resource names stored, resources are distributed uniformly over the hash space. As long as the chunks of the hash space assigned to participating nodes are of roughly equal size, then each node maintains a roughly equal portion of all resources stored into the distributed hash table, thereby achieving load balancing.

Backslash is specifically implemented on the Content Addressable Network [5], but does not rely on the specifics of CAN for its operation. The overlay used underneath Backslash is mostly interchangeable with any other distributed hash table. In addition to hash table operations, Backslash requires knowledge about the neighborhood of an overlay node, but all such popular systems can be easily modified to export this information through their APIs.

3.3   Caching and Replication

Although using a distributed hash table, such as a CAN, explains how we find a copy of a popular document within the Backslash web mirror, it does not explain how the copy was created or propagated through the system. In this section we explain the basic cache diffusion techniques we explore in the context of Backslash.

Each Backslash node has some available storage for use in caching (a few times the size of its local document collection). This storage is split in two categories: *replica* space and *temporary cache* space. On one hand, a replica is a cached copy of a document that is guaranteed to be where it was placed. Replicas are placed in the overlay by insertion operations of the distributed hash table. A temporary cache, on the other hand, is a cached copy of a document that is placed opportunistically at a node of the overlay to speed up subsequent retrievals. Temporary caches are created in response to retrieval operations of the distributed hash table and are not guaranteed to remain where they are placed. In fact, they might be replaced very soon after they are created if they are the least recently used temporarily cached document of a node. A fixed portion of the available free space of each node is allocated as replica space. Whatever remains unused in the replica space and the remainder of the free space is allocated as temporary cache space.

The Backslash replica space is used exclusively for the first copy of each file in the participating mirrored web collections. Every Backslash node periodically injects the documents in its local document collection into the distributed hash table. The single copy of each such document created at insertion time is a replica. In the cache diffusion schemes we explore in the remainder of this paper we create no other replicas.

The first cache diffusion method we consider is *local diffusion*. In local diffusion, each node serving a document as a replica or temporary cache monitors the rate of requests it receives for that document. When the node determines that the request rate has reached a predetermined *push threshold*, it pushes out a new temporary cache of the document one overlay hop closer to the source of the last request. This technique aims to offload some of the demand by having more nodes in the locality of an observed flood intercept and serve requests. In a sense, a node that observes a local flood creates a "bubble" of temporary caches around itself, diffusing its load over its neighborhood. The diameter of the bubble grows in relation to the intensity of the flood, until no node on the perimeter of the bubble observes high request rates for the document.

The second cache diffusion method we consider is *directory diffusion*. In this method, the distributed hash table stores directories of pointers to document copies instead of the document copies themselves. Replicas and temporary caches are also stored in Backslash nodes, but their location is not related to the hash table structure. When a node receives a newly inserted document, it creates a directory for it, picks a random Backslash node and stores a replica for the document at that node, documenting it in its own directory. When the directory receives a request for the document, it returns as many permuted directory entries pointing to individual copies of the document as it can fit in a single response packet. To create new temporary caches, the directory node monitors the request rate for the document. When the request rate reaches a predetermined threshold, the directory responds to the requester with an invitation to become a new temporary cache along with the list of pointers to copies of the file.

Both cache diffusion techniques require that a node serve a request if it holds a copy of the requested document. We explore a modification of this requirement whereby a node may choose (at random) to forward a request even if it already holds a copy of the requested document. This allows the node to shed probabilistically a fraction of the request load it observes, without creating new temporary caches. We introduce this variation, called *probabilistic forwarding*, to increase the reuse of already existing caches and curtail the creation of new ones. This is especially the case for the local diffusion method, where all requests originating outside the "bubble" are handled by the nodes at the perimeter, leaving caches inside the bubble practically unused. Probabilistic forwarding enables the use of caches in the interior.

4   Evaluation

In our preliminary evaluation efforts, we have focused on the behavior of cache diffusion techniques. The mechanisms responsible for interjecting Backslash into the protocol stream of unaware client browsers (delineated in Section 3.1) or for building simple self-maintained overlays (pointed to in Section 3.2) are available and pose no significant challenges for the purposes of our target application.

Our experimental setup involves 1,000 nodes participating in a single two-dimensional CAN overlay. Each node has twice the size of its own collection in available free space, of which exactly half is allocated to replicas, and the other half to temporary caches. For simplicity, the document collection owned by each node consists of a single document and all documents have exactly the same size. We present a brief preliminary exploration of two particular design choices in our cache diffusion mechanism: diffusion agility and probabilistic forwarding.

Diffusion agility is the speed with which Backslash reacts to a new flash crowd. A highly agile diffusion mechanism spreads out cached copies of the flooded document rapidly, so as to reach a state where the downpour of requests for that document can be served collectively by as many nodes as possible. How-

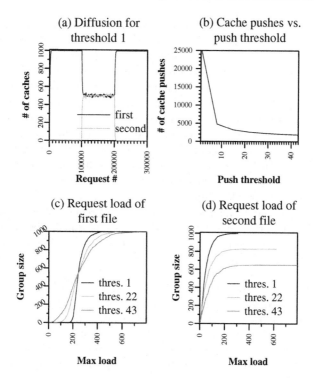

Fig. 1. The simultaneous flash crowd scenario discussed in the text. (a) Cache diffusion during the evolution of the scenario. (b) Number of cache pushes as a function of the push threshold. (c) and (d) Cumulative load distribution of requests served as a function of the number of requests served per node for the first, long flood and the second, short flood, respectively.

ever, high agility also carries an early commitment of heavy resources (storage space, cache diffusion bandwidth) to a flood that might not necessitate them. By controlling agility, we allow the system to moderate the amount of resources it commits to a particular flood.

We represent this agility parameter by a *push threshold*, the number of requests a node must serve before it decides to push to its neighbors a copy of the flooded file. A push threshold of one means that every time a Backslash node receives a request, it also pushes out a copy of the requested file to the neighbor that forwarded it the request for caching. A push threshold of 100 means that the node only pushes out a new cache of a requested file after every 100-th request.

To illustrate the effects of the push threshold, we have simulated a scenario where two floods are handled at the same time by Backslash. The first flood starts alone and manages to saturate the system by causing a copy of the first flooded file to be placed at every node. After saturation, and while the first flood is ongoing, the second flood begins, gradually displacing cached copies of the first file for copies of the second file. Finally, the second flood terminates, allowing the first file to saturate the system again. Figure 1(a) shows how the diffusion evolves in this scenario for a push threshold of one. Note how agility is very high; the system responds very rapidly to changes in offered load.

The benefits of using lower push thresholds are illustrated in Figure 1(d). The figure graphs the cumulative load distribution over the system for the short second flood for three representative thresholds: 1, 22 and 43. On one hand, with a threshold of one, the second flood causes no higher a load than 300 requests to any node. On the other hand, with the highest threshold of 43 only 600 nodes participate in caching the second file at any one time and the maximum per node load reaches almost 800 requests.

However, higher agility makes the satisfaction of requests for the second file more expensive. For the same number of requests satisfied, higher thresholds result in much fewer cache pushes in the face of contention. Figure 1(b) graphs the total number of cache pushes as a function of the push threshold. Threshold one results in almost 25,000 cache pushes, of which 23,000 are mainly due to the oscillations caused by contention between the first and second files. The hysteresis introduced by the highest threshold of 43 mitigates this effect, as indicated by the almost tenfold reduction in cache pushes shown in the graph. Note, however, that the actual point where the benefit of lower threshold justifies the cost is specific to the underlying topology and resource restrictions of each Backslash participant, for whom a temporarily high load might be justified by overall lower traffic.

As described in Section 3.3, in local diffusion cached copies of a flooded document inside a cache "bubble" are only used by requests initiated locally, whereas caches on the perimeter of the bubble are used locally and also by requests initiated outside the bubble. This makes perimeter caches much hotter than internal bubble caches. We explore the use of probabilistic forwarding as a method to spread out the load of the perimeter incurred by requests initiated outside the bubble over all the nodes within the bubble.

We use a probability function that assigns a linearly decreasing forwarding probability to every bubble node on the path from an external request originator to the authority node. In this way, a cache at the perimeter of the bubble has a maximum forwarding probability (60% in our experiments). Subsequent next hop nodes toward the center of the bubble decrease their forwarding probability proportionally as they get closer to the center. We would expect to see a better load distribution among the nodes of a bubble as a result of this technique.

In Figure 2(b) we show the effects of using probabilistic forwarding during a single flood. We have calibrated the push threshold of the probabilistic run so as to achieve similar diffusion patterns between the two runs (see Figure 2(a)). Surprisingly, although we initially expected probabilistic forwarding to even out the distribution of load among nodes with caches of the flooded file, the graph shows only a very small improvement; specifically, there are slightly more lower-load nodes and slightly fewer higher-load nodes. We ascribe this surprising result

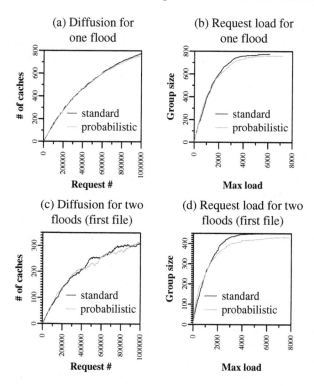

Fig. 2. The effects of probabilistic forwarding in one flood ((a) and (b)), and in two concurrent floods of equal intensity ((c) and (d)). (a) and (c) graph cache diffusion during the evolution of the two experiments, as achieved by calibrating the push threshold in the probabilistic experiments to 220; the standard experiments used a threshold of 500. (b) and (d) show the cumulative load distributions for the two flood scenarios with and without probabilistic forwarding.

to the monolithic fashion in which we measure load in our simulations. We conjecture that although the cumulative load per node in the duration of the experiment seems only a little affected by the use of probabilistic forwarding, it is the distribution of that load *over time* within a single node that improves, that is, becomes less bursty, in this case. Deterministic forwarding creates high bursts of load at the perimeter of the bubble since all nodes must service their requests. Upon reaching its threshold, the bubble expands outward and a new perimeter services incoming requests. Once a node is no longer at the perimeter of the bubble, it does not receive requests from outside and its load drops significantly. Probabilistic forwarding allows interior nodes to continue servicing requests even after they are no longer at the perimeter of the bubble.

The results are similar when two floods of equal intensity compete against each other. Figure 2(d) shows the difference in load distribution with and without probabilistic forwarding for one of the two simultaneous floods. While probabilistic forwarding evens out the load distribution in favor of lower loads, the effect is not quite as significant as we had anticipated. We hope to experiment with different forwarding probability functions to achieve a more pronounced benefit.

This is a very preliminary evaluation of cache diffusion in Backslash. We hope to perform a more thorough analysis in the near future.

5   Related Work

Our work shares many goals with the pioneering work done in the Adaptive Web Caching project [1]. Our local diffusion method is similar to the diffusion method used by AWC. However, AWC offers the benefits of a proxy cache, whereas Backslash replaces a reverse proxy cache.

A lot of work has been done on building self-maintainable overlay networks that follow the distributed hash table paradigm [3, 5, 6, 8]. We use results from that area extensively.

A set of HTTP extensions for the "content addressable web" were proposed recently [2]. Backslash would certainly benefit from the extended HTTP functionality offered by this work.

Other work has explored the use of client Web browser plug-ins to diffuse the effects of flash crowds [4]. However, Backslash is differs by requiring a server-side implementation that is completely transparent to the client.

Finally, Rubenstein and Sahu [7] analyze theoretically a simple peer-to-peer protocol for flash crowd document retrieval based on random walks in an unstructured overlay, similar to Gnutella. We hope to compare the latency characteristics of the two designs in the near future.

6   Conclusions

There exists a need for a cost effective method to combat flash crowds. Backslash addresses this problem and, given preliminary results, is a promising method of mitigating flash crowd effects.

The next steps in this research involve a deeper exploration of different forwarding probability functions and their interactions with the other aspects of cache diffusion, the development of a hybrid local/directory diffusion method to exploit the benefits of both methods, closer cooperation with a cache invalidation scheme, and a higher-fidelity simulation and trial deployment plan.

7   Acknowledgments

This work started out as a summer internship project at the ICSI Center for Internet Research, under the guidance of Mark Handley, Scott Shenker and Sylvia Ratnasamy. We are grateful to them for the unfaltering motivation, guidance and feedback they provided during the formulation of this research.

We are also thankful for the generous funding we have received from the Stanford Networking Research Center, DARPA (contract N66001-00-C-8015) and Sonera Corporation.

References

[1] The Adaptive Web Caching Project. http://irl.cs.ucla.edu/AWC/.

[2] The Content-Addressable Web. http://onionnetworks.com/caw/.

[3] KUBIATOWICZ, J., BINDEL, D., CHEN, Y., CZERWINSKI, S., EATON, P., GEELS, D., GUMMADI, R., RHEA, S., WEATHERSPOON, H., WEIMER, W., WELLS, C., AND ZHAO, B. OceanStore: An Architecture for Global-Scale Persistent Storage. In *Proceedings of ASPLOS 2000* (Cambridge, MA, USA, Nov. 2000), pp. 190–201.

[4] PADMANABHAN, V., SRIPANIDKULCHAI, K. The Case for Cooperative Networking. In *1st International Workshop on Peer-to-Peer Systems (IPTPS 2002) 2002* (Cambridge, MA, USA, March 2002).

[5] RATNASAMY, S., FRANCIS, P., HANDLEY, M., KARP, R., AND SHENKER, S. A Scalable Content-Addressable Network. In *Proceedings of SIGCOMM 2001* (San Diego, CA, U.S.A., Aug. 2001), ACM SIGCOMM, pp. 161–172.

[6] ROWSTRON, A., AND DRUSCHEL, P. Pastry: Scalable, distributed object location and routing for large-scale peer-to-peer systems. In *Proceedings of IFIP/ACM Middleware 2001* (Heidelberg, Germany, Nov. 2001).

[7] RUBENSTEIN, R. AND SAHU, S. An Analysis of a Simple P2P Protocol for Flash Crowd Document Retrieval. Available as *Columbia University Technical Report EE011109-1* (Columbia University, New York, NY, USA, Nov. 2001).

[8] STOICA, I., MORRIS, R., KARGER, D., KAASHOEK, M. F., AND BALAKRISHNAN, H. Chord: A scalable peer-to-peer lookup service for internet applications. In *Proceedings of SIGCOMM 2001* (San Diego, CA, U.S.A., Aug. 2001), ACM SIGCOMM, pp. 149–160.

Exploring the Design Space of Distributed and Peer-to-Peer Systems: Comparing the Web, TRIAD, and Chord/CFS

Stefan Saroiu, P. Krishna Gummadi, and Steven D. Gribble

Department of Computer Science & Engineering
University of Washington
Seattle, WA, 98195-2350, USA
{tzoompy,gummadi,gribble}@cs.washington.edu

Abstract Despite the existence of many peer-to-peer systems, some of their design choices and implications are not well understood. This paper compares several distributed and peer-to-peer systems by evaluating a key set of architectural decisions: naming, addressing, routing, topology, and name lookup. Using the World Wide Web, Triad, and Chord/CFS as examples, we illustrate how different architectural choices impact availability, redundancy, security, and fault-tolerance.

1   Introduction

Peer-to-peer systems are the latest addition to a family of distributed systems whose goal is to share resources across their participants. Previous members of this family include the World Wide Web, distributed file systems, and even the telephony network. To compare these systems, one can decompose them along the following design axes, which are an extension of those proposed by Shoch [9] and Saltzer [7]:

Content name: A name describes *what* a user is looking for, such as a file name in a file system.

Host address: An address describes *where* a resource is, for example, an IP address describes where a host resides in the Internet.

Routing mechanism: A route describes *how* to get to a destination. A routing mechanism (such as BGP across Internet autonomous systems, or ASs) is used to discover or disseminate routes.

Network topology: Topology describes the set of physical or logical *links* between hosts.

Lookup: *Bindings* between names and addresses are registered in the system. Participants use a lookup mechanism that *resolves* a name into an address, based on the registered bindings.

These design axes represent one possible framework to reason about the architecture of a system. Although this framework is clear in the abstract, in practice

P. Druschel, F. Kaashoek, and A. Rowstron (Eds.): IPTPS 2002, LNCS 2429, pp. 214–224, 2002.

real systems often blur the distinction between some of these axes. For example, NAT blurs routing and lookup by introducing a name translation mechanism so that non-routable IP addresses can be "bridged" to the routable Internet. Additionally, IP addresses are converted into MAC Ethernet addresses in a manner similar to name translation. However, we believe that our decomposition is useful both when designing and analyzing a system, and that, by mapping design choices along these axes, we can learn about the trade-offs made by each system.

In this paper, we compare the designs of three different distributed architectures: the World Wide Web, TRIAD [5], and Chord/CFS [2] (as a representative of recently proposed peer-to-peer architectures [3, 6, 10]). We then derive several performance, security, and robustness implications that result from their design choices.

1.1   The World Wide Web

The World Wide Web (WWW) is perhaps the most ubiquitous, popular, and successful distributed system. Fundamentally, the goal of WWW is to enable clients to retrieve hyperlinked content.

Names: Web content names are drawn from an infinite space of globally unique Uniform Resource Locators (URLs), which are structured as a fully qualified domain name (FQDN) combined with a locally unique relative URL [1, 4]. The right to bind an FQDN to an IP address is controlled by hierarchical delegation, and the right to bind relative URLs is controlled by local policy.

Addresses: WWW addresses are globally unique, hierarchically organized IP addresses of Internet hosts (servers, clients, caches, or intermediate routers). There is a finite but large number of IP addresses; addresses are allocated in ranges from a centralized authority, and address assignment rights are delegated locally within these ranges.

Routing: Routing in the WWW is a combination of Internet routing protocols, including BGP, IS-IS, and OSPF. Routing decisions are driven by business policy and performance. The ability to route to an IP address is the result of advertising that address on a routing protocol. There is little control over the right to advertise, as there is typically a lack of authentication and access control in routing protocols.

Topology: The WWW topology is based on the physical topology of Internet hosts. This topology is roughly hierarchical, consisting of interconnected autonomous systems and subnetworks within them.

Lookup: FQDNs within URLs are resolved to IP addresses through the domain name system (DNS); relative URLs are resolved and bound locally by servers. DNS itself is another distributed system; however, the WWW could replace the DNS lookup mechanism with no semantic loss, as is proposed by TRIAD.

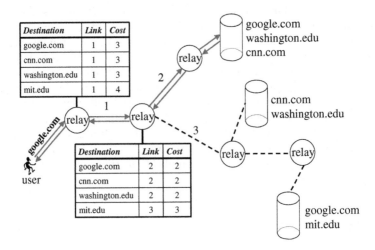

Fig. 1. The TRIAD architecture. A name request (google.com) is forwarded by intermediate relays toward the "best" content replica.

1.2   TRIAD

TRIAD defines a content layer that replaces the Web's address-based routing with a *name-based* routing protocol. An individual piece of content is advertised by each server replica, so that lookup requests are directed from clients along intermediate routers (relay nodes) to servers, and back along the same path. Each relay node maintains a set of name-to-next-hop mappings, just as an IP router maps address prefixes to next hops. When a request for a content name arrives, a relay looks up the name and forwards the request toward the "best" server replica. Once the request reaches a relay responsible for a server replica, that relay sends back a response containing the server's address (Figure 1).

Names: Similar to the Web, TRIAD resources are objects spread across servers. Although TRIAD's content namespace does not require a specific structure, content names are routing table entries, and therefore need to aggregate. Because URLs are hierarchical, TRIAD suggests using the Web's URL namespace for content naming.

Addresses: TRIAD's addresses are a composition of two namespaces: globally unique IP addresses of AS, and locally unique IP addresses within each AS. This results in a finite but very large address space. While the inter-AS address space is controlled by a centralized authority, each intra-AS address space is managed locally by its AS.

Routing: TRIAD uses a name-based, BGP-like routing protocol called NBRP, which distributes name suffix reachability messages. The ability to route to a name is the result of advertising that name across the NBRP protocol.

Topology: TRIAD's topology can be arbitrary, consisting of logical links between relay nodes over which NBRP messages flow. For performance reasons, it is suggested that TRIAD's topology should reflect the physical Internet topology.

Lookup: TRIAD unifies lookup and routing: resolving a name into an address is achieved by routing the name to its destination. Once the destination address is found, the lookup reply is routed back to the source on the same path.

1.3   Chord/CFS

All hosts in Chord/CFS-style peer-to-peer systems serve three roles; they act as servers, clients, and intermediate routers. This symmetry of roles has several design implications.

Names: In Chord, each piece of content is named with a Chord identifier, obtained by hashing the content into 160 bits. The content namespace is flat, large, and uniformly populated.

Addresses: Chord's address namespace is structurally identical to the content namespace. Addresses are obtained by concatenating a host's IP address with a small *virtual host* number, and hashing the result into a 160 bit address. Because addresses have 160 bits, they are probabilistically globally unique in the system.

Routing: Since the content and address namespaces are equivalent, routing can be thought of both as address-based, like the Web, or name-based, like TRIAD. Unlike the Web or TRIAD, any participating Chord host acts as an intermediate router. Routing in Chord is simple: each host directs queries to the neighbor whose address is closest to the name according to a pre-determined lexicographic order.

Topology: Chord's overlay topology is a deterministic function of participating peers' addresses. Each peer has a successor and predecessor based on a total ordering of addresses, and each peer maintains logarithmically-sized "finger table" of connections to other peers. Although stale finger table entries are tolerable, they act as shortcut routes and their freshness ensures efficient routing.

Lookup: Like TRIAD, Chord unifies lookup and routing. Since the content and address namespaces are equivalent, the identity function is sufficient to bind names to addresses: a name is bound to the address which *is* the content name. Name resolution is done by routing the name through the network.

1.4   Summary

A summary of the systems' designs is given in Figure 2. This decomposition has served to illustrate important differences between these systems. For instance, in Chord, binding of names to addresses is done by the identity function and the topology is a deterministic function of host addresses. In both Chord and TRIAD, lookup is unified with routing. We now discuss the implications of these design differences.

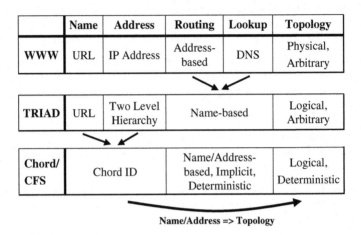

Fig. 2. Routing and lookup are unified in TRIAD and Chord. Chord's name and address spaces are identical, and its topology is a deterministic function of names/addresses.

2   Names and Addresses

In this section, we explore how content names and host addresses are created and bound to each other in our three systems. Unlike the Web and TRIAD, the Chord/CFS name and address spaces are unified, allowing for deterministic bindings of content names to addresses. However, this determinism can easily be exploited to launch various targeted attacks on the system.

2.1   The World Wide Web

Binding names to IP addresses is controlled by the use of DNS as a lookup mechanism. The hierarchical nature of DNS imposes structure on content names, but it also serves to delegate binding rights to authorities that own subtrees of the namespace. While any Web server can create an infinite number of URLs, binding them to IP addresses is restricted by the ability to register FQDNs within a particular DNS subtree. As a result, a malicious Web server cannot pollute the global namespace by registering a large number of dummy or otherwise harmful names, nor can it attack another Web server's content by duplicating its URLs.

There is a similar hierarchical delegation of IP address assignment rights within the Web. An individual may be able to control the assignment of a large but finite number of IP addresses; for instance, MIT owns a class A subnet, allowing it to assign 2^{24} addresses, but all within a fixed range. However, it is difficult for a malicious host to hijack an IP address outside of its allocated range, as this involves injecting false routing advertisements into the network. Although they both are hierarchical, the name and address namespaces in the

Web are completely independent: control over one does not grant control over the other.

2.2 TRIAD

In TRIAD, lookup and routing are unified. As a result, the binding of a name to an address is accomplished by advertising a route across the TRIAD network, and lookup is performed by routing a name to its destination. Similar to the Web, the ability to create a name in TRIAD is unrestricted. Restrictions on binding rights must be enforced by the routing infrastructure; to date, this issue remains unresolved.

Addresses in TRIAD are the composition of globally routable IP addresses assigned to ASs, and locally routable IP addresses within ASs. The authority to assign routable addresses is therefore split across two levels. Similar to the Web, a centralized naming authority delegates globally visible IP ranges to ASs, and ASs enforce local policies for address assignment. Therefore, individual hosts in TRIAD typically cannot affect globally visible IP assignment.

Because name routing in TRIAD involves routing advertisements and routing table formation, the ability to aggregate names is important for scalability. As a result, the content namespace must be hierarchical (or otherwise aggregatable) in practice. For this reason, TRIAD content names are modeled after URLs.

2.3 Chord/CFS-style Peer-to-Peer Systems

Since the peer-to-peer content namespace is flat, the responsibility for managing content is randomly distributed across the address namespace. The insertion of a name-to-address binding (i.e. publishing content) into the system causes some host to accept the responsibility and incur the cost of managing that content. Thus, unless the right to insert a name-to-address binding is controlled, any host can cause unbounded amounts of effort and storage to be expended across the system. Furthermore, attacks on specific victims are also possible. For example, an attacker could overwhelm a targeted victim address with content, or even cause the targeted host to store undesirable or illicit content. In contrast, in the Web and TRIAD, binding a name to an address does not cause the host to store the content, making such attacks impossible.

The set of content names associated with an address is also deterministic. If hosts are allowed to select their own addresses, they can use this deterministic mapping to control access to specific content names. In Chord/CFS, the ability to create an address is restricted by limiting hosts to using hashes of their IP addresses concatenated with a small "virtual host" number. An attacker who has assignment rights over *O(number of Chord nodes / max virtual hosts)* IP addresses can control arbitrary content in the system.

3   Routing, Lookup, and Topology

In this section, we discuss the consequences of unifying routing and lookup in TRIAD and Chord/CFS, in contrast to the Web. Furthermore, because Chord's topology is a function of its address space, several unexpected implications emerge affecting the system's redundancy, availability, fault-tolerance and security.

3.1   The World Wide Web

The structure of the WWW is mapped directly onto the Internet's physical topology: Web servers and clients are addressed by their IP addresses, and the routing of data between them is performed using IP routing protocols such as BGP, IS-IS, and OSPF. Infrastructure such as content-delivery networks and caching hierarchies extend the name-to-address lookup mechanism, but the result of a lookup is still an IP address of the host that will serve the data.

In the WWW, routing policy can be selected independent of both physical topology and content. This flexibility allows policy to be driven by efficiency, business rules, or even local physical characteristics. Routing policy can be altered without affecting naming, binding, or content placement. It is common for ISP operators to adjust policy to achieve a financial or traffic balancing goal; however these adjustments are functionally transparent to the rest of the system.

It is possible to engineer redundancy (and hence higher availability) at two levels in the Web. At the routing level, redundant physical routes provide alternate paths for data transport between clients and servers. At the name binding level, binding the same name to multiple addresses allows clients or middleware to fail over to an alternate address if one destination becomes unavailable.

The endpoints of a Web transfer (servers and clients) are, in general, physically distinct from routers. This physical separation of roles has several benefits. Different degrees of trust can be associated with different roles; for example, core Internet routers are more protected and trustworthy than Web clients or servers. Hosts can be provisioned and optimized for their specific roles; a high-speed router needs different hardware, OS, and software support than a Web server or client. Finally, side-effects of host failures are isolated with respect to the role they play. A Web server failure does not affect the routability of IP addresses, and a router failure doesn't affect content availability (unless the failure partitions the network).

3.2   TRIAD

In TRIAD, because routing *is* the content name-to-address lookup mechanism, routing policy can no longer be selected independently of content. If TRIAD's network topology mirrors the physical topology of the Internet, as suggested by the authors, then an efficient routing policy is enough to enable clients to route requests to their topologically "nearest" content replica. This only works because

TRIAD can route on arbitrary topologies, unlike Chord/CFS-style peer-to-peer systems, as discussed below.

TRIAD also supports two levels of redundancy. Multiple name-to-address bindings are attainable by replicating content on additional routable destinations; if one destination fails, the content is available at the replicas' addresses. This replication technique is possible because TRIAD routing uses content names rather than host addresses. In addition, from a given source, there may be multiple routes to a destination name. Thus, the failure of a link in TRIAD does not necessarily cause content to become unavailable.

Because TRIAD supports routing over arbitrary topologies, it is possible to construct a topology in which content servers are never intermediate nodes in a route, and therefore servers do not need to participate in the routing of requests. Thus, content hosting and routing are still separable roles, enabling the same separations of trust, provisioning, and failure as the Web.

3.3   Chord/CFS-style Peer-to-Peer Systems

The topology of Chord/CFS-style peer-to-peer systems is a deterministic function of the set of participating addresses. As a side-effect, routing tables need not be advertised across the system, eliminating one cause of overhead. Routing tables, approximated by finger tables in Chord, are constructed by each peer upon its entry to the system, and lazily updated to reflect the departure of its neighbors. If finger tables are kept up-to-date, the carefully chosen topology bounds lookup route lengths by $\log(\# \text{ peers})$.

The content name and address namespaces in Chord/CFS are unified, which allows binding to be the identity function: the content name *is* the address towards which a peer routes requests. When combined with Chord's deterministic topology, this implies that all peers are expected to serve both as routers and content destinations. These roles are inseparable: a peer cannot choose an address that will relieve it of routing responsibilities, and the topology cannot be engineered to relieve content destinations of routing responsibilities. However, roles no longer need to be explicitly assigned, and the topology need not be explicitly constructed; they are determined as peers join and leave the system, vastly simplifying and decentralizing the administration of the system.

Redundancy in Chord/CFS can occur at multiple levels. Because binding is the identity function, it is impossible to bind the same content name to multiple addresses. However, a naming convention can assign aliases to any given content name; unlike TRIAD or the WWW, redundancy at this level is not transparent to the user, since it is exposed in the content namespace. A second level of redundancy exists within the overlay itself. There are on the order of $\log(\# \text{ peers})$ mutually disjoint routes between any two given addresses. As long as routes fail independently, this provides a high degree of availability to the system.

The Chord/CFS network is an overlay that maps down to a physical IP network. Redundancy can be added to the physical network, but since the overlay topology is a function of Chord addresses that involves a randomly distributed hash function, physical locality is diffused throughout the overlay. Accordingly, it is difficult to predict the effect of physical network redundancy on the overlay network. For example, the failure of a network link will manifest itself as multiple, randomly scattered link failures in the overlay.

The diffusion of physical links across the logical Chord network tends to amplify the bad properties of a system, but not its good properties. If any link within a lookup path has low bandwidth, high latency, or low availability, the entire path suffers. Conversely, all links within a path must share the same good property for the path to benefit from it. Thus, a single bad physical link can "infect" many routes. As was measured in [8], 20% of the hosts in popular file-sharing peer-to-peer systems connect to the Internet over modems. Since Chord overlay paths traverse essentially random physical links, a simple calculation reveals that for a network of 10,000 peers with similar characteristics to those in [8], there is a 79% probability that a lookup request encounters at least one modem.

As another example, it is possible for a single physical link failure in the Internet to cause a large network partition. Consider a worst-case failure that separates an AS from the rest of the Internet: as long as all Web content within that AS is replicated outside of it, all content is available to all non-partitioned clients. However, in Chord, the number of failed routes that this single link failure will cause is proportional to the number of Chord addresses hosted within the partitioned AS, and these failed routes will be randomly distributed across both peers and content.

A final implication of the deterministic nature of routes in Chord/CFS-style systems is that it is possible for an attacker to construct a set of addresses that, if inserted into the system, will intercept all lookup requests coming from a particular member of the system. Even though mechanisms exist to prevent a peer from selecting arbitrary addresses, if a peer can insert enough addresses, it can (probabilistically) surround or at least become a neighbor of any other peer.

4   Summary and Conclusions

We presented a design decomposition of the World Wide Web, TRIAD [5] and Chord/CFS [2] (as representative of recent peer-to-peer architectures [3, 6, 10]). This decomposition allowed us to describe fundamental system design differences: (1) in Chord/CFS, the content and address namespaces are equivalent, as opposed to WWW and TRIAD; (2) Chord's network topology is a deterministic function of its content and address namespace and (3) unlike the WWW, in both TRIAD and Chord, lookup and routing are unified. To summarize, these design differences impact fundamental properties of these systems, such as:

Access Control:

Because name and address spaces are hierarhical in WWW, local subtrees serve as limits to the registration and maintanance of name-to-address bindings. Instead, TRIAD and Chord/CFS can have flat name and address spaces with global bindings. As a result, single hosts force others to do work by registering bindings in the system.

In WWW and TRIAD, the name and address spaces are independent: control over one does not grant control over the other. They are unified in Chord/CFS, and, therefore, namespace control is equivalent to address-space control.

Content Replication:

In all these systems, content replication is achieved through multiple bindings. However, because a name is an address, Chord/CFS cannot bind the same content name to multiple addresses, unlike WWW and TRIAD.

Path Redundancy:

Unlike WWW and TRIAD, Chord/CFS has many ($\log(\#$ peers$)$) mutually disjoint routes between any two given addresses.

Chord/CFS uses hashing to randomly map its logical overlay to the physical network. One side-effect of hashing is that local properties are randomly diffused throughout the system. As a result, one cannot locally provision the network to ensure the availability of own content.

Security:

Participants in WWW and TRIAD have different roles: some hosts are content servers, others are routers. This decoupling allows for different levels of trust. Typically, in WWW, core Internet routers are more trustworthy than content servers. In Chord/CFS, roles are symmetric: every peer is both a router and a server. In consequence, only a single level of trust can exist across the system.

Failures:

In WWW and TRIAD, multiple roles allow for isolation of failures: a router failure does not affect content availability and, similarly, a server failure does not affect routing. Because servers are routers, in Chord/CFS, a server failure is a router failure, and vice-versa.

Because in Chord/CFS locality is diffused throughout the overlay, local link failures can impact network reachability in unpredictable ways. Instead, in WWW and TRIAD, most local failures only have local effects.

In conclusion, despite the existence of several peer-to-peer architectures, some of their design choices and implications are not well understood. Our decomposition helped us illustrate the side-effects of fundamental trade-offs made by these peer-to-peer systems as opposed to the World Wide Web and TRIAD. Most of these side-effects have resulted in the loss of properties that are crucial to these systems' availability, redundancy, security and fault-tolerance.

References

[1] Tim Berners-Lee, Larry Masinter, and Mark McCahill. RFC 1738 - Uniform Resource Locators (URL), December 1994.

[2] Frank Dabek, M. Frans Kaashoek, David Karger, Robert Morris, and Ion Stoica. Wide-area cooperative storage with CFS. In *Proceedings of the 18th ACM Symposium on Operating Systems Principles (SOSP 2001)*, Lake Louise, AB, Canada, October 2001.

[3] Peter Druschel and Antony Rowstron. Storage management and caching in PAST, a large-scale, persistent peer-to-peer storage utility. In *Proceedings of the 18th ACM Symposium on Operating Systems Principles (SOSP 2001)*, Lake Louise, AB, Canada, October 2001.

[4] Roy Fielding. RFC 1808 - Relative Uniform Resource Locators, June 1995.

[5] Mark Gritter and David R. Cheriton. An Architecture for Content Routing Support in the Internet. In *Proceedings of the 3rd Usenix Symposium on Internet Technologies and Systems (USITS)*, San Francisco, CA, USA, March 2001.

[6] Sylvia Ratnasamy, Paul Francis, Mark Handley, Richard Karp, and Scott Shenker. A Scalable Content-Addressable Network. In *Proceedings of the ACM SIGCOMM 2001 Technical Conference*, San Diego, CA, USA, August 2001.

[7] Jerome Saltzer. RFC 1498 - On the Naming and Binding of Network Destinations, August 1993.

[8] Stefan Saroiu, P. Krishna Gummadi, and Steven D. Gribble. A Measurement Study of Peer-to-Peer File Sharing Systems. In *Proceedings of the Multimedia Computing and Networking Conference (MMCN)*, San Jose, CA, USA, January 2002.

[9] John F. Shoch. Inter-Network Naming, Addressing, and Routing. In *Proceedings of IEEE COMPCON*, pages 72–79, Washington, DC, USA, December 1978. Also in K. Thurber (ed.), Tutorial: Distributed Processor Communication Architecture, IEEE Publ. #EHO 152-9, 1979, pp. 280–287.

[10] Ben Y. Zhao, John Kubiatowicz, and Anthony D. Joseph. Tapestry: An Infrastructure for Fault-Resilient Wide-Area Location and Routing. Technical Report UCB//CSD-01-1141, University of California at Berkeley Technical Report, April 2001.

Are Virtualized Overlay Networks
Too Much of a Good Thing?

Pete Keleher, Bobby Bhattacharjee, and Bujor Silaghi

Department of Computer Science
University of Maryland
College Park, MD 20742, USA
{keleher,bobby,bujor}@cs.umd.edu

Abstract. The majority of recent high-profile work in peer-to-peer networks has approached the problem of location by abstracting over object lookup services. Namespace virtualization in the overlay layer provides load balance and provable bounds on latency at low costs.

We contend that namespace virtualization comes at a significant cost for applications that naturally describe their data sets in a hierarchical manner. Opportunities for enhancing browsing, prefetching and efficient attribute-based searches are lost. A hierarchy exposes relationships between items near to each other in the topology; virtualization of the namespace discards this information even if present at client, higher-level protocols.

We advocate encoding application hierarchies directly into the structure of the overlay network, and revisit this argument through a newly proposed distributed directory service.

1   Introduction

Peer-to-peer (P2P) networks have recently become one of the hottest topics in OS research [1, 2, 3, 4, 5, 6, 7, 8]. Starting with the explosion of popularity of Napster, researchers have become interested because of the unparalleled chance to do relevant research (people might actually use it!), and the naive approach of many of the first P2P protocols deployed.

The majority of most recent high-profile work has described middleware that performs a single task: distributed object lookup. This seemingly simple function can be used as the basic building block of more complex systems that perform a variety of sophisticated functions (file systems, event notification system, etc.).

P2P networks differ from more conventional infrastructures in that the load (whether expressed as CPU cycles, data storage costs or packet forwarding bandwidth) is distributed across participating peers. This load should ideally be balanced, as the load is in some sense the "payment" for participating in the network. Overloading some peers while letting others off without performing any work towards the common good is clearly unfair.

The approach many recent systems [6, 4, 7, 5] have taken towards ensuring load balance is to virtualize data item keys by creating the keys from one-way

P. Druschel, F. Kaashoek, and A. Rowstron (Eds.): IPTPS 2002, LNCS 2429, pp. 225–231, 2002.

hashes (SHA-1, etc.) of the item labels. The peer node ID's are similarly encoded, and data items are mapped to "closest" nodes by comparing keys and hashed node ID's. We refer to this as *virtualization* of the namespace. By contrast, a "non-virtualized" system is one where data items are served by the same nodes that export them.

A virtualized approach helps load balance because data items from one very popular site will be served by different nodes; they will be distributed randomly among participating peers. Similarly, routing load is distributed because paths to items exported by the same site are usually quite different.

Just as importantly, virtualization of the namespace provides a clean, elegant abstraction of routing, with provable bounds on routing latency.

The contention of this position paper is that this virtualization comes at a significant cost, as described below:

1. *Virtualization destroys locality* - By virtualizing keys, data items from a single site are not usually co-located, meaning that opportunities for enhancing browsing, prefetching, and efficient searching are lost.
2. *Virtualization discards useful application-specific information* - The data used by many applications (file systems, auctions, resource discovery) is naturally described using hierarchies. A hierarchy exposes relationships between items near to each other in the hierarchy; virtualization of the namespace discards this information.

The rest of this paper elaborates on these points and outlines an alternative approach.

To be clear, the environment assumed in this paper is that of a set of co-operating, widely-separated peers, running as user-level processes on ordinary PC's or workstations. Peers "export" data, and keys are "mapped" onto overlay servers. The set of peer nodes that export data is also the set of overlay servers. A "node" is a process participating in the system.

2   Locality Is a *Good* Thing

The first form of locality with which we are concerned is spatial locality. Users who access d_i are more likely to also access d_{i+1} than some arbitrary d_j. Consider web browsing: a given page might require many nearby items to be accessed, and the next page accessed by a user is likely to be on the same site as well.

Virtualization of this process loses several opportunities for performance improvement. First, the route to the exporting site only has to be discovered once in a non-virtualized system. Subsequent accesses to a second data item in logical proximity can follow the same route and shortcuts would be employed. In a virtualized system, there will likely be nothing in common between the two routes. Second, an exporting site (or the access initiator) might choose to prefetch nearby data in a non-virtualized system. Prefetching, when it works, enables the system to hide the latency of remote accesses. While prefetching can be made to work with a virtualized namespace, it is much more difficult. For example,

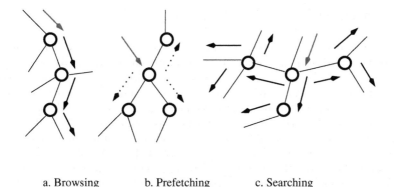

a. Browsing   b. Prefetching   c. Searching

Fig. 1. Browsing, prefetching and searching are examples of operations where object relationships can be exploited by integrating them with the routing procedure in the overlay network layer. This allows subsequent steps, indicated by arrows, to be performed in $O(1)$ as opposed to $O(\log N)$, etc. for virtualized namespace schemes.

CFS [9], a cooperative file system built on top of Chord [6], can prefetch file blocks. A peer receiving a request for block i of a file can prefetch the second by locally reconstructing the virtualized name for block $i+1$ and sending a prefetch message to the site serving it. However, each such prefetch requires an object lookup which translates to additional messages in the network. Worse, prefetching blocks is relatively easy because the name of the nearby object (block $i+1$) is easy to predict. However, the names of nearby items on the exporting site are not easy to predict, and prefetching could probably only be accomplished via an application overlay that indexes exporting sites.

Current approaches also fail to exploit temporal locality as much as they might. None of Chord, CAN [7], Pastry [5], or Tapestry [4] currently use caching. Repeated accesses by one site to the same data item require repeated traversals through the overlay network. However, caching could easily be added to these systems, and is used in some applications built on top of these systems (e.g., Past [10], CFS). Further, some systems (most notably Pastry) attempt to exploit locality in charting a route through the overlay network's nodes. The result is that the total IP hop count may be only a small constant higher than a native IP message.

Note that there is an implicit assumption of spatial locality in all of the virtualization work. Virtualization distributes load well only assuming that the only form of locality present is spatial, not temporal. Stated another way, virtualization can not improve load balance in the presence of single-item hot-spots; it can only distribute multiple (possibly related) data items across the network.

By contrast, current systems use replication both to provide high availability and to distribute load when single items become hot spots.

3   Searching

Most distributed object stores require some search capability in order to be used effectively. We distinguish two types of searching: indexing and keyword/attribute searching. "Indexing" refers to indexing entire documents, e.g. Google's index of the web. Indexing of documents served by a distributed overlay system requires local indexes to be created, combined, and then served, presumably through the same overlay system, although this could also be done at a higher level. The difficulty of indexing is not affected by whether the namespace is virtualized; it is a hard problem for any distributed system.

Section 2 considered spatial locality in terms of browsing and prefetching. A similar argument can be made regarding attribute searching (see Figure 1).

Assume that we wish to search for "red cars", where 'red' and 'car' are encoded as attributes of certain documents. Virtualized namespaces do not encode attributes in the overlay structure, so this search could only be accomplished by: (i) visiting every node and thus flooding the network or (ii) resorting to some higher-level protocol that would capture additional object relationships. Such relationships (topological constraints like hierarchies for instance) are orthogonal to the overlay layer where the lookup routing is performed. Therefore virtualized namespace approaches cannot benefit from such extra information in optimizing the routing procedure.

We discuss how embedding attributes in the overlay structure of a non-virtualized system allows efficient attribute searching in Section 5.

4   Adding Information Back In

There are two approaches to adding application-specific information and support for locality back into a virtualized system: use of higher-level application layers, and eliminating virtualization entirely. We discuss the former here, and one approach to the latter in the next section.

Support for locality in the query stream can be added back at higher levels. As an example, both Past and CFS cache data along paths to highly popular items. The advantage of doing so at the file system layer rather than the routing layer is that both location information, and the file's data itself, are cached together. If address caching were performed at the routing level, file caching would be less effective.

File system prefetching can also be accomplished at this level. For example, one could prefetch the rest of a directory after one file is accessed by (1) deriving the file's directory name from the target file's name, (2) routing a message to the directory object, (3) reading the directory object to get the set of other files in the directory, and then (4) sending prefetches to each of those files.

However, not only is this inefficient, but it only makes use of information about accesses to a single file. Consider how a system with a virtualized namespace would support a policy that only prefetches entire directories if two or more files of the directory are accessed within a short time.

5   Eschewing Virtualization

Consider the types of applications that are being built on top of the overlay networks discussed so far: file systems, event notification systems, distributed auctions, and cooperative web proxies. All of these applications organize their data hierarchically, and any locality in these applications is local in the framework of this hierarchy. Browsing the hierarchy is, therefore, the only way of extracting and exploiting locality. Yet, this information is discarded by virtualized namespaces.

Another approach is to encode this hierarchy directly into the overlay layer. While this paper is not about TerraDir, we discuss it as an example approach that addresses some of the shortcomings discussed above. A TerraDir [8] is a non-virtualized overlay in the form of a rooted hierarchy that explicitly codifies the application hierarchy. By default, routing is performed via tree traversal, taking $O(logN)$ hops in the worst case[1]. Availability, load balance, and latency are all addressed further by caching and replication. The degree to which a node is replicated is dependent on the node's level in the tree. This approach helps load balance and only adds a constant amount of overhead per node, regardless of the size of the system.

TerraDir provides comparable performance to the other systems (probably somewhat better latency because of the caching, probably a bit worse load distribution), but leaves application locality and hierarchical information intact.

Locality is retained because a given data item is mapped to the node that exports it, rather than to another randomized host. Not only does this save a level of indirection, but co-located items are mapped to the same locations, meaning locality can be recognized and exploited without network communication. Note that replication is per-node. All items exported by a node are replicated together, so any replica can perform prefetching.

Caching addresses both spatial and temporal locality. Repeated accesses to the same remote object are serviced by a local cache if caching of data is turned on. Otherwise, the cache provides the network address of the exporting node, limiting routing to a single hop through the overlay network.

Accesses to items "near" each other in the application hierarchy are handled efficiently because they are also near each other, or co-located, in the overlay network. Hence, the number of hops in the overlay network are again minimized.

Data item keywords are explicitly coded into the overlay hierarchy, so searching for keywords is handled efficiently. For example, consider searching for red cars in the hierarchy shown in Figure 2. The query would be of the form "/vehicles/cars/red/*", and would be routed to the smallest subtree on the left side of the figure. The wildcard will then cause the query to be split and to flood that subtree. However, the rest of the hierarchy, aside from the path from the query initiator to the "cars/red" subtree, is untouched. By contrast, searching for "red cars" can only be accomplished efficiently via some higher-level service in a virtualized system. To be fair, note that the query is handled efficiently only because

---

[1] Assuming a relatively well-balanced tree.

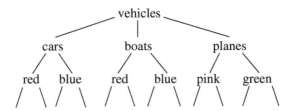

Fig. 2. An example TerraDir. Searching for "red cars" is more efficient than searching for "anything red".

the query structure matches the hierarchy's structure. Searching for "anything red" would cause all leaves to be visited. This problem is addressed by allowing "views" to be dynamically materialized. A client that expects to make multiple queries with a different structure (e.g. "all red things", then "all blue things", etc.) inserts a *view query* into the system. The view queries specifies an ordering on the set of attributes[2], which is used to build a new overlay hierarchy. Building the new hierarchy requires the entire tree to be visited once; subsequent queries will be handled efficiently.

The TerraDir approach has at least two other important advantages. First, maintenance overhead is significantly less. The virtualized approaches generally require $log(n)$ operations to allow a node to leave or join the overlay, whereas these operations require only a constant number of operations under TerraDir.

Finally, TerraDir nodes "maps" the key of a node back to that same node, meaning that the data item and its mapping are not distributed across administrative boundaries.

6   Summary

Distributed lookup services using virtualized namespaces can be important building blocks for building sophisticated P2P applications. Namespace virtualization provides load balance and tight bounds on latency at low cost.

In doing so, however, it discards potentially useful information (application hierarchies) and object relationships (proximity within the hierarchy). This is not always a problem: certain types of functionality are more efficiently provided at higher layers (this is merely the end-to-end argument [11]). However, many applications can benefit from increased functionality in the lookup layer.

We advocate encoding application hierarchies, where needed, directly into the structure of the overlay network. This approach allows systems to exploit locality between objects and to provide searching without centralized indexing or flooding.

---

[2] View queries can also name *tag functions*, which can be seen as dynamic attributes synthesized from the static attributes.

References

[1] Plaxton, C.G., Rajaraman, R., Richa, A.W.: Accessing nearby copies of replicated objects in a distributed environment. In: Proc. of the ACM Symposium on Parallel Algorithms and Architectures, Newport, RI (1997) 311–320

[2] Petersen, K., Spreitzer, M., Terry, D.B., Theimer, M., Demers, A.J.: Flexible update propagation for weakly consistent replication. In: Symposium on Operating Systems Principles. (1997) 288–301

[3] Kubiatowicz, J., Bindel, D., Chen, Y., Czerwinski, S., Eaton, P., Geels, D., Gummadi, R., Rhea, S., Weatherspoon, H., Weimer, W., Wells, C., Zhao., B.: Oceanstore: An Architecture for Global-Scale Persistent Storage. In: Proceedings of the Ninth International Conference on Architectural Support for Programming Languages and Operating Systems. (2000)

[4] Zhao, B., Kubiatowicz, K., Joseph, A.: Tapestry: An infrastructure for fault-resilient wide-area location and routing. Technical Report UCB//CSD-01-1141, University of California at Berkeley Technical Report (2001)

[5] Rowstron, A., Druschel, P.: Pastry: Scalable, distributed object location and routing for large-scale peer-to-peer systems. In: Proceedings of the 18th IFIP/ACM International Conference on Distributed Systems Platforms. (2001)

[6] Stoica, I., Morris, R., Karger, D., Kaashoek, M.F., Balakrishnan, H.: Chord: A scalable peer-to-peer lookup service for internet applications. In: Proceedings of the ACM SIGCOMM '01 Conference, San Diego, California (2001)

[7] Ratnasamy, S., Francis, P., Handley, M., Karp, R., Shenker, S.: A scalable content addressable network. In: In Proceedings of the ACM SIGCOMM 2001 Technical Conference. (2001)

[8] Bhattacharjee, B., Keleher, P., Silaghi, B.: The design of TerraDir. Technical Report CS-TR-4299, University of Maryland, College Park, MD (2001)

[9] Dabek, F., Kaashoek, M.F., Karger, D., Morris, R., Stoica, I.: Wide-area cooperative storage with CFS. In: Proceedings of the 18th ACM Symposium on Operating Systems Principles, Chateau Lake Louise, Banff, Canada (2001)

[10] Rowstron, A., Druschel, P.: Storage management and caching in PAST, a large-scale, persistent peer-to-peer storage utility. In: Proceedings of the 18th ACM Symposium on Operating Systems Principles. (2001)

[11] Saltzer, J.H., Reed, D.P., Clark, D.D.: End-to-end arguments in system design. Computer Systems **2** (1984) 277–288

Locating Data in (Small-World?) Peer-to-Peer Scientific Collaborations

Adriana Iamnitchi[1], Matei Ripeanu[1], and Ian Foster[1,2]

[1] Department of Computer Science, The University of Chicago
1100 E. 58th Street, Chicago, IL 60637, USA
{anda, matei, foster}@cs.uchicago.edu
[2] Mathematics and Computer Science Division, Argonne National Laboratory
Argonne, IL 60439, USA

Abstract. Data-sharing scientific collaborations have particular characteristics, potentially different from the current peer-to-peer environments. In this paper we advocate the benefits of exploiting emergent patterns in self-configuring networks specialized for scientific data-sharing collaborations. We speculate that a peer-to-peer scientific collaboration network will exhibit small-world topology, as do a large number of social networks for which the same pattern has been documented. We propose a solution for locating data in decentralized, scientific, data-sharing environments that exploits the small-worlds topology. The research challenge we raise is: what protocols should be used to allow a self-configuring peer-to-peer network to form small worlds similar to the way in which the humans that use the network do in their social interactions?

1   Introduction

Locating files based on their names is an essential mechanism for large-scale data sharing collaborations. A peer-to-peer (P2P) approach is preferable in many cases due to its ability to operate robustly in dynamic environments.

Existing P2P location mechanisms focus on specific data sharing environments and, therefore, on specific requirements: in Gnutella [1], the emphasis is on easy sharing and fast file retrieval, with no guarantees that files will always be located. In Freenet [2], the emphasis is on ensuring anonymity. In contrast, systems such as CAN [3], Chord [4] and Tapestry [5] guarantee that files are always located, while accepting increased overhead for file insertion and removal.

Data usage in scientific communities is different than in, for example, music sharing environments: data usage often leads to creation of new files, inserting a new dimension of dynamism into an already dynamic system. Anonymity is not typically a requirement, being generally undesirable for security and monitoring reasons.

Among the scientific domains that have expressed interest in building data-sharing communities are physics (e.g., GriPhyN project [6]), astronomy (Sloan Digital Sky Survey project [7]) and genomics [8]. The Large Hadron Collider (LHC) experiment at CERN is a proof of the physicists' interest and pressing

P. Druschel, F. Kaashoek, and A. Rowstron (Eds.): IPTPS 2002, LNCS 2429, pp. 232–241, 2002.
© Springer-Verlag Berlin Heidelberg 2002

need for large-scale data-sharing solutions. Starting 2005, the LHC will produce Petabytes of raw data a year that needs to be pre-processed, stored, and analyzed by teams comprising 1000s of physicists around the world. In this process, even more derived data will be produced. 100s of millions of files will need to be managed, and storage at 100s of institutions will be involved.

In this paper we advocate the benefits of exploiting emergent patterns in self-configuring networks specialized for scientific data-sharing collaborations. We speculate that a P2P scientific collaboration network will exhibit small-world topology, as do a large number of social networks for which the same pattern has been documented.

We sustain our intuition by observing the characteristics of scientific data-sharing collaborations and studying the sharing patterns of a high-energy physics community (Section 2). In Section 3 we propose a solution for locating data in decentralized, scientific, data-sharing environments that exploits the small-worlds topology. The research challenge we raise is: what protocols should be used to allow a self-configuring P2P network to form small worlds similar to the way in which the humans that use the network do in their social interactions? While we do not have a complete solution, we discuss this problem in Section 5.

2   Small Worlds in Scientific Communities

In many network-based applications, topology determines performance. This observation captivated researchers who started to study large real networks and found fascinating results: recurring patterns emerge in real networks [9]. For example, social networks, in which nodes are people and edges are relationships; the world wide web, in which nodes are pages and edges are hyperlinks; and neural networks, in which nodes are neurons and edges are synapses or gap junctions, are all small-world networks [10]. Two characteristics distinguish small-world networks: first, a small average path length, typical of random graphs (here 'path' means shortest node-to-node path); second, a large clustering coefficient that is independent of network size. The clustering coefficient captures how many of a node's neighbors are connected to each other. One can picture a small world as a graph constructed by loosely connecting a set of almost complete subgraphs.

The small world example of most interest to us is the scientific collaboration graph, where the nodes are scientists and two scientists are connected if they have written an article together. Multiple studies have shown that such graphs have a small-world character in scientific collaborations spanning a variety of different domains, including physics, biomedical research, neuroscience, mathematics, and computer science.

Typical uses of shared data in scientific collaborations have particular characteristics:

- *Group locality.* Users tend to work in groups: a group of users, although not always located in geographical proximity, tends to use the same set of resources (files). For example, members of a science group access newly produced data to perform analyses or simulations. This work may result

into new data that will be of interest to all scientists in the group, e.g., for comparison. File location mechanisms such as those proposed in CAN, Chord, or Tapestry [5] do not attempt to exploit this behavior: each member of the group will hence pay the cost of locating a file of common interest.
- *Time locality.* The same user may request the same file multiple times within short time intervals. This situation is different, for example, from Gnutella usage patterns, where a user seldom downloads a file again if it downloaded it in the past. (We mention that this characteristic is influenced by the perceived costs of storing vs. downloading, which may change in time.)

It is the intuition provided by the small-world phenomenon in real networks and the typical use of scientific data presented above that lead us to the following questions. Let us consider the following network: a node is formed of data and its provider (the scientist who produced the data), and two nodes are connected if the humans in those nodes are interested in each other's data. The first question is: is this a small-world network? Based on the analysis of data sharing patterns in a physics collaboration (presented in Section 2.1) we speculate that this network will be a small world. Second, how can such small-world topology be exploited for performance in the data-sharing environments of interest to us? Finally, how do we translate the dynamics of scientific collaborations into self-configuring network protocols (such as joining the network, finding the right group of interests, adapting to changes in user's interests, etc.)?

We believe this last question is relevant and challenging in the context of self-configuring P2P networks. We support this idea by answering the second question: in Section 3 we sketch a file location strategy that exploits the small-world topology in the context of scientific data-sharing collaborations. Once we show that a small-world topology can be effectively exploited, designing self-configuring topology protocols to induce specific topology patterns becomes more interesting.

2.1 Data Sharing in a Physics Collaboration

The D0 collaboration [11] involves hundreds of physicists from 18 countries that share large amounts of data. Data is accessed from remote locations through a software layer (SAM [12]) that provides file-based data management. We analyzed data access traces logged by this system during January 2002.

We considered the graph whose nodes are users and whose links connect users that shared at least one file during a specified interval. We found that the graphs generated for various interval lengths exhibit small-world characteristics: short average path lengths and large clustering coefficients. Although these graphs are relatively small compared to our envisioned target (e.g., 155 users accessed files through SAM in January), we expect similar usage patterns for larger graphs.

Table 1 presents the characteristics of the graphs of users who shared data within various time intervals ranging from 1 day to 30 days. The small-world pattern is evident when comparing the clustering coefficient and average path length with those of a random graph of the same size (same number of nodes

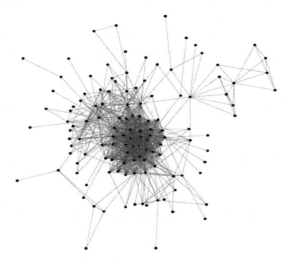

Fig. 1. The file-sharing graph of January 2002.

and edges): the clustering coefficient of a small-world graph is significantly larger than that of a similar random graph, while the average path length is about the same.

Table 1. File-sharing graph characteristics for intervals from 1 to 30 days.

Interval	Whole Graph		Largest Connected Component				Random Graph	
	# Nodes	# Links	# Nodes	# Links	Clustering	Path Lenght	Clustering	Path Lenght
1 day	20	38	12	34	0.827	1.61	0.236	2.39
2 days	20	77	15	75	0.859	1.29	0.333	1.68
7 days	63	331	58	327	0.816	2.21	0.097	2.35
14 days	87	561	81	546	0.777	2.56	0.083	2.30
30 days	128	1046	126	1045	0.794	2.45	0.067	2.29

3   Locating Files in Small-World Networks

We consider an environment with potentially hundreds of thousands of geographically distributed nodes that provide location information as <logical filename, physical location> pairs.

Locating files in this environment is challenging because of scale and dynamism: the number of nodes, logical files, requests, and concurrent users (seen as file location requesters) may all be large. The system has multiple sources of variation over time: files are created and removed frequently; nodes join and leave the system without a predictable pattern. In such a system with a large number of components (nodes and files), even a low variation rate at the individual level may aggregate into frequent group level changes.

We exploit the two environmental characteristics introduced in Section 2—group and time locality—to advance our performance objective of minimizing file location latency. We also build on our assumption that small-world structures eventually emerge in P2P scientific collaborations.

Consider a small world of C clusters, each comprising, on average, G nodes. A cluster is defined as a community with overlapping data interests, independent of geographical or administrative proximity. Clusters are linked together in a connected network. In this structure, we combine information dissemination techniques with request-forwarding search mechanisms: location information is propagated aggressively within clusters, while inter-cluster search uses request forwarding techniques.

We chose gossip [13] as the information dissemination mechanism: nodes gossip location information to other nodes within the cluster. Eventually, with high probability, all nodes will learn about all other nodes in the cluster. They will also know, with high probability, all location information provided by all nodes within the cluster. Hence, a request addressed to any node in the cluster can be satisfied at that node, if the answer exists within the cluster.

A request that cannot be answered by the local node is forwarded to other cluster(s), by unicast, multicast, or flooding. Ideally, clusters can organize themselves dynamically in search-optimized structures, thus allowing a low cost inter-cluster file retrieval. Since any node in a cluster has all information provided in that cluster, the search space reduces from $C \times G$ to C.

In this context, nodes need to store the total amount of information provided by the cluster to which they belong. In order to reduce storage costs, we use a compact, probabilistic representation of information based on Bloom Filters (Section 4.2). Nodes can trade off the amount of memory used for the accuracy in representing information.

Each node needs to have sufficient topology knowledge to forward requests outside the cluster. Not every node needs to be connected to nodes from remote clusters, but, probabilistically, every node needs to know a local node that has external connections. The question of how to form and maintain inter-cluster connections pertains to the open question we raise in this paper and discuss in Section 5: what topology protocols can induce the small-world phenomenon?

4   Gossiping Bloom Filters for Information Dissemination

In this section we briefly explain how we use the mechanisms mentioned above: gossip for information dissemination and Bloom filters for reducing the amount of communication. We also provide an intuitive quantitative estimation of the system we consider.

4.1 Gossip Mechanism

Gossip protocols have been employed as scalable and reliable information dissemination mechanisms for group communication. Each node in the group knows a partial, possibly inaccurate set of group members. When a node has information to share, it sends it to a number of f nodes (fanout) in its set. A node that receives new information will process it (for example, combine it with or update its own information) and gossip it further to f nodes chosen from its set.

We use gossip protocols for two purposes: (1) to maintain accurate membership information in a potentially dynamic cluster and (2) to disseminate file location information to nodes in the local cluster. We rely on soft-state mechanisms to remove stale information: a node not heard about for some time is considered departed; a logical file not advertised for some time is considered removed.

4.2 Bloom Filters

Bloom filters [14] are compact data structures used for probabilistic representation of a set in order to support membership queries ("Is element x in set X?"). The cost of this compact representation is a small rate of false positives: the structure sometimes incorrectly recognizes an element as member of the set.

Bloom filters describe membership of a set A by using a bit vector of length m and k hash functions, h_1, h_2, ..., h_k with $h_i : X \rightarrow 1..m$. For a fixed size (n) of the set to be represented, the tradeoff between accuracy and space (m bits) is controlled by the number of hash functions used (k). The probability of a false positive is:

$$p_{err} \approx \left(1 - e^{-kn/m}\right)^k$$

Here p_{err} is minimized for $m/n \ln 2$ hash functions. In practice, however, a smaller number of hash functions is used: the computational overhead of each additional hash function is constant while the incremental benefit of adding a new hash function decreases after a certain threshold. Experience shows that Bloom filters can be successfully used to compress a set to 2 bytes per entry with false positive rates of less than 0.1% and lookup time of about $100\mu s$.

A nice feature of Bloom filters is that they can be built incrementally: as new elements are added to a set, the corresponding positions are computed through the hash functions and bits are set in the filter. Moreover, the filter expressing the reunion of multiple sets is simply computed as the bit-wise OR applied over the corresponding filters.

Bloom filters can be compressed when transferred across the network and, in this case, filter parameters can be chosen to maximize compression rate, as shown in [15].

4.3 Advantages of Building the System around Shared Data Interests

We model this system built on group and time locality assumptions as follows:

1. *Zipf distribution for request popularity.* In Zipf distributions, the number of requests for the k-th most popular item is proportional to $k^{-\alpha}$, where α is a constant. Zipf distributions are widely present in the Internet world. For example, the popularity of documents requested from an Internet proxy cache (with $0.65 < \alpha < 0.85$), Web server document popularity ($0.75 < \alpha < 0.85$), and Gnutella query popularity ($0.63 < \alpha < 1.24$) all exhibit Zipf distributions. For our problem we assume that file popularity in each cluster (group) follows a Zipf distribution.
2. *Locality of interests.* As discussed above, clusters are formed based on shared interest. We therefore assume that information on the most popular files is available within the cluster and only requests for not-so-popular files are forwarded.

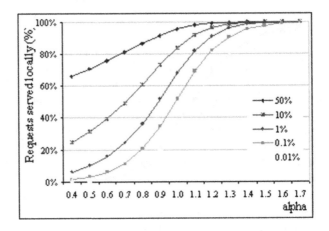

Fig. 2. Fraction of requests served locally (by one member of the group) assuming various values of α.

With these assumptions, we can estimate the fraction of file requests served by the group as a function of the distribution parameter α and the fraction of files about which the group maintains information. For example, as Figure 2 shows, 68% of all requests are served by the group when information about only top 1% most popular files is available at group level, for $\alpha = 1$. Figure 2 strongly emphasizes the need for efficient, interest-based cluster creation.

We estimate 100s of clusters with 1,000s of nodes in a cluster, sharing information on about 10 million files per cluster. Using Bloom filters, for 0.1% false positives rate, each node needs 2 bytes per file or 20MB of memory to store information about all files available in the cluster. Assuming a 10-day average lifetime for a file at a node, and a self-imposed threshold of 0.1% false positives, then the generated traffic needed to maintain this accuracy level within the cluster can be estimated at about 24 KBps at each node.

False negatives may have two sources: the probabilistic information dissemination mechanism and inaccuracy in the inter-cluster search algorithm. By appropriately tuning the gossip periodicity and fanout, the system can control the rate of false negatives by increasing communication costs.

5   Creating a Small World

The question raised and not answered in this paper is: what protocols should be used for allowing a self-configuring network to reflect the small-world properties that exist at the social (as in a scientific collaboration) level? There are at least two ways to attempt to answer this question. The first approach is to look at existing small worlds and to identify the characteristics that foster the small-world phenomenon. The second approach is to start from theoretical models that generate small worlds [10] and mirror them into protocol design.

The Gnutella network is an interesting case study as it is a P2P self-configuring technological network that exhibits (moderate) small-world characteristics [16]. How are the small-world characteristics generated? One possible answer is that the social network formed by the Gnutella users reflects its small-world patterns onto the technological network. While this is not impossible, we observe that a user has a very limited contribution to the Gnutella network topology. Hence, we believe the social influence on the Gnutella topology is insignificant.

More significant for the small-world phenomenon may be Gnutella's network exploration protocol based on **ping** and **pong** messages: a **ping** is sent to all neighbors and each neighbor forwards it further to its own neighbors, and so on. The **pong** messages return on the same path, allowing a node to learn of its neighbor's neighbors, and hence to improve clustering. However, the influence of this mechanism is limited by the (comparatively) small number of connections per node. This fact explains why, despite an aggressive exploration of the network, the clustering coefficient in Gnutella is not large (e.g., it is an order of magnitude lower than the clustering coefficients in coauthorship networks).

The theoretical model for building small-world graphs [10] starts from a highly clustered graph (e.g., a lattice) and randomly adds or rewires edges to connect different clusters. This methodology would be relevant to us if we had the clusters already formed and connected. Allowing clusters to form dynamically based on shared interests, allowing them to learn about each others, to adapt to users' changing interests (e.g., divide or merge with other clusters) are parts of the problem we formulate and do not answer. However, let us assume that clusters form independently based on out of band information (the way the Gnutella network forms) and let us assume further that they do eventually learn about each other. Possible approaches for transforming a loosely connected graph of clusters into a small world (hence, with small average path length) are:

1. The hands-off approach: random graphs have small average path length. It is thus intuitive that "randomly" connected clusters will form a small world.
2. The centralized approach at the cluster level: in each cluster, one or multiple nodes are assigned the task of creating external connections.
3. The agent-based approach: allow an agent to explore the network and rewire it where necessary. This approach is usually rejected due to associated security issues.

6   Summary

We studied the file location problem in decentralized, self-configuring P2P networks associated with scientific data sharing collaborations. A qualitative analysis of the characteristics of these collaborations, quantitative analysis of file sharing information from one such collaboration, and previous analyses of various social networks lead us to speculate that a P2P scientific collaboration may benefit from a small-world topology. We sketch a mechanism for low-latency file retrieval that benefits from the particularities of the scientific collaboration environments and a small-world topology. While we do not provide a solution for building topology protocols flexible enough to resemble the dynamics and patterns of social interactions, we stress the relevance of this problem and we discuss some possible directions for research.

Acknowledgements

We are grateful to John Weigand, Gabriele Garzoglio, and their colleagues at Fermi National Accelerator Laboratory for their generous help. This work was supported by the National Science Foundation under contract ITR-0086044.

References

[1] Clip2. The gnutella protocol specifications v0.4, http://www.clip2.com.
[2] Ian Clarke, Oskar Sandberg, Brandon Wiley, and Theodore W. Hong. Freenet: A distributed anonymous information storage and retrieval system. In *ICSI Workshop on Design Issues in Anonymity and Unobservability*, Berkeley, California, 2000.
[3] Sylvia Ratnasamy, Paul Francis, Mark Handley, Richard Karp, and Scott Shenker. A scalable content-addressable network. In *SIGCOMM*, San Diego USA, 2001.
[4] Ion Stoica, Robert Morris, David Karger, M. Frans Kaashoek, and Hari Balakrishnan. Chord: A scalable peer-to-peer lookup service for internet applications. In *SIGCOMM 2001*, San Diego, USA, 2001.
[5] Ben Y. Zhao, John D. Kubiatowicz, and Anthony D. Joseph. Tapestry: An infrastructure for fault-tolerant wide-area location and routing. Technical Report CSD-01-1141, Berkeley, 2001.

[6] The GriPyN Project, http://www.griphyn.org.

[7] Sloan Digital Sky Survey, http://www.sdss.org/sdss.html.

[8] The Human Genome Project, http://www.nhgri.nih.gov.

[9] Reka Albert and Albert-Laszlo Barabasi. Statistical mechanics of complex networks. *Reviews of Modern Physics*, 74:47–97, January 2002.

[10] Duncan J. Watts. *Small Worlds: The Dynamics of Networks between Order and Randomness*. Princeton University Press, 1999.

[11] The D0 Experiment, http://www-d0.fnal.gov.

[12] Lauri Loebel-Carpenter, Lee Lueking, Carmenita Moore, Ruth Pordes, Julie Trumbo, Sinisa Veseli, Igor Terekhov, Matthew Vranicar, Stephen White, and Victoria White. SAM and the particle physics data grid. In *Proceedings of Computing in High-Energy and Nuclear Physics*, Beijing, China, 2001.

[13] Anne-Marie Kermarrec, Laurent Massoulie, and Ayalvadi Ganesh. Reliable probabilistic communication in large-scale information dissemination systems. Technical Report MSR-TR-2000-105, Microsoft Research Cambridge, Oct. 2000 2000.

[14] Burton Bloom. Space/time trade-offs in hash coding with allowable errors. *Communications of the ACM*, 13(7):422–426, 1970.

[15] Michael Mitzenmacher. Compressed bloom filters. In *Twentieth ACM Symposium on Principles of Distributed Computing (PODC 2001)*, Newport, Rhode Island, 2001.

[16] Mihajlo A. Jovanovic, Fred S. Annexstein, and Kenneth A. Berman. Scalability issues in large peer-to-peer networks - a case study of gnutella. Technical report, University of Cincinnati, 2001.

Complex Queries in DHT-based Peer-to-Peer Networks

Matthew Harren[1], Joseph M. Hellerstein[1], Ryan Huebsch[1], Boon Thau Loo[1], Scott Shenker[2], and Ion Stoica[1]

[1] UC Berkeley, Berkeley CA 94720, USA,
{matth, jmh, huebsch, boonloo, istoica}@cs.berkeley.edu
[2] International Computer Science Institute, Berkeley CA 94704,
shenker@icsi.berkeley.edu

Abstract. Recently a new generation of P2P systems, offering *distributed hash table* (DHT) functionality, have been proposed. These systems greatly improve the scalability and exact-match accuracy of P2P systems, but offer only the exact-match query facility. This paper outlines a research agenda for building complex query facilities on top of these DHT-based P2P systems. We describe the issues involved and outline our research plan and current status.

1 Introduction

Peer-to-peer (P2P) networks are among the most quickly-growing technologies in computing. However, the current technologies and applications of today's P2P networks have (at least) two serious limitations.

Poor Scaling: From the centralized design of Napster, to the notoriously inefficient search process of Gnutella, to the hierarchical designs of FastTrack [4], the scalability of P2P designs has always been problematic. While there has been significant progress in this regard, scaling is still an issue in the currently deployed P2P systems.

Impoverished query languages: P2P networks are largely used for filesharing, and hence support the kind of simplistic query facility often used in filesystem "search" tools: *Find all files whose names contain a given string*. Note that "search" is a limited form of querying, intended for identifying ("finding") individual items. Rich query languages should do more than "find" things: they should also allow for combinations and correlations among the things found. As an example, it is possible to search in Gnutella for music by J. S. Bach, but it is not possible to ask specifically for all of Bach's chorales, since they do not typically contain the word "chorale" in their name.

The first of these problems has been the subject of intense research in the last few years. To overcome the scaling problems with unstructured P2P systems such as Gnutella where data-placement and overlay network construction are essentially random, a number of groups have proposed *structured* P2P designs.

P. Druschel, F. Kaashoek, and A. Rowstron (Eds.): IPTPS 2002, LNCS 2429, pp. 242–250, 2002.

These proposals support a *Distributed Hash Table* (DHT) functionality [11, 14, 12, 3, 18]. While there are significant implementation differences between these DHT systems (as we will call them), these systems all support (either directly or indirectly) a hash-table interface of `put(key,value)` and `get(key)`. Moreover, these systems are extremely scalable; lookups can be resolved in $\log n$ (or n^α for small α) overlay routing hops for an overlay network of size n hosts. Thus, DHTs largely solve the first problem. However, DHTs support only "exact match" lookups, since they are hash tables. This is fine for fetching files or resolving domain names, but presents an even more impoverished query language than the original, unscalable P2P systems. Hence in solving the first problem above, DHTs have aggravated the second problem.

We are engaged in a research project to address the second problem above by studying the design and implementation of complex query facilities over DHTs; see [7] for a description of a related effort. Our goal is not only bring the traditional functionality of P2P systems – filesharing – to a DHT implementation but also to push DHT query functionality well beyond current filesharing search, while still maintaining the scalability of the DHT infrastructures. This note offers a description of our approach and a brief discussion of our current status.

2   Background

Before describing our approach, we first discuss some general issues in text retrieval and hash indexes and then explain why we are not proposing a P2P database.

2.1   Text Retrieval and Hash Indexes

As noted above, DHTs only support exact-match lookups. Somewhat surprisingly, it has been shown that one can use the exact-match facility of hash indexes as a substrate for textual similarity searches, including both strict substring searches and more fuzzy matches as well [17]. The basic indexing scheme is to split each string to be indexed into "n-grams": distinct n-length substrings. For example, a file with ID I and filename "Beethovens 9th" could be split into twelve trigrams: Bee, eet, eth, tho, hov, ove, ven, ens, ns%, s%9, %9t, 9th (where '%' represents the space character). For each such n-gram g_i, the pair (g_i, I) is inserted into the hash index, keyed by g_i. One can build an index over n-grams for various values of n; it is typical to use a mixture of bigrams and trigrams.

Given such an index, a substring lookup like "thoven" is also split into n-grams (*e.g.*, tho, hov, ove, ven), and a lookup is done in the index for each n-gram from the query. The resulting lists of matches are concatenated and grouped by file ID; the count of copies of each file ID in the concatenated list is computed as well. For strict substring search, the only files returned are those for which the count of copies is as much as the number of n-grams in the query (four, in our example). This still represents a small superset of the correct answer, since the n-grams may not occur consecutively and in the correct order in the results. To account for this, the resulting smallish list can be postprocessed directly to test for substrings.

While the text-search literature tends not to think in relational terms, note that the query above can be represented nearly directly in SQL:

```
SELECT H.fileID, H.fileName
  FROM hashtable H
  WHERE H.text IN (<list-of-n-grams-in-search>)
GROUP BY H.fileID
  HAVING COUNT(*) >= <#-of-n-grams-in-search>
    AND H.fileName LIKE <substring expression>
```

In relational algebra implementation terms, this requires an index access operator, a grouping operator, and selection operators[1].

The point of our discussion is not to dwell on the details of this query. The use of SQL as a query language is not important, it merely highlights the universality of these operators: they apply not only to database queries, but also to text search, and work naturally over hash indexes. If we can process relational algebra operators in a P2P network over DHTs, we can certainly execute traditional substring searches as a special case.

2.2 Why Not Peer-to-Peer Databases?

We have noted that relational query processing is more powerful than the search lookups provided by P2P filesharing tools. Of course, traditional database systems provide a great deal of additional functionality that is missing in P2P filesharing – the most notable features being reliable, transactional storage, and the strict relational data model. This combined functionality has been the cornerstone of traditional database systems, but it has arguably cornered database systems and database research into traditional, high-end, transactional applications. These environments are quite different from the P2P world we wish to study. Like the users of P2P systems, we are not focused on perfect storage semantics and carefully administered data. Instead, we are interested in ease of use, massive scalability, robustness to volatility and failures, and best-effort functionality.

---

[1] We have kept the substring match example strict, for clarity of exposition. In many cases, n-gram search can also support a useful fuzzy semantics, in which files are simply ranked by descending count of n-gram matches (or some more subtle ranking metric), down to some cutoff. This allows the searching to be more robust to misspellings, acronyms, and so on, at the expense of false positives in the answer set. Such a ranking scheme can also be represented in SQL via an ORDER BY clause containing an externally-defined ranking function.

Also note that our index can be augmented so that each entry holds the offset of the n-gram in the string – this allows tests for ordering and consecutiveness to be done without observing the actual string. This optimization usually only helps when the strings being indexed are long – e.g. for full-text documents rather than file names. It is also clumsy to express this in SQL, though the relational algebra implementation is relatively straightforward.

The explosive growth of the P2P networks show that there is a viable niche for such systems. Transactional storage semantics are important for many applications, but are not familiar to most end-users today, who typically use file systems for storage. Most users do not want to deploy or manage a "database" of any kind. We believe this is a "street reality" worth facing, in order to maintain the grassroots approach natural in P2P. As part of that, we do not see transactions as an integral part of the research thrust we describe here.

On the other hand, relational data modeling is in some sense universal, and its details can be abstracted away from the user. All data is relational at some level, inasmuch as one can think of storing it in a table of one column labeled "bits". In many cases, there is somewhat more natural structure to exploit: as noted above, sets of P2P files can be thought of as relations, with "attributes" name, ID, host, etc. The user need not go through a complex data modeling exercise to enable a system to index and query these attributes in sophisticated ways.

Hence we do not see a pressing need for users of P2P system to load their data into a database; we prefer to build a query engine that can use the natural attributes exposed in users' existing data, querying those attributes intelligently while providing the storage semantics that users have learned to live with. In fact, we wish to stress the point that it may be strategically unwise to discuss peer-to-peer *databases* at all, with their attendant complexities in software and administration associated with database storage. Instead, we focus on peer-to-peer *query processing*, and separate it from the problem of storage semantics and administration. Of course we leave the option open: our ideas could be combined with P2P transactional mechanisms, *e.g.* as suggested in [10]. However, we do not wed ourselves to the success (both technical and propagandistic) of such efforts. It is worth noting that despite current commercial packaging, relational database research from the early prototypes onward [1, 15] has separated the storage layer from the query processing layer. We are simply following in those footsteps in a new storage regime.

3   P2P Query Processing

Our design is constrained by the following goals:

Broad Applicability: A main goal for our work is that it be broadly and practically usable. In the short term, this means that it should be able to interact with user's filesystems in the same way as existing P2P filesharing systems.

Minimal Extension to DHT APIs: DHTs are being proposed for use for a number of purposes in P2P networks, and we do not want to complicate the design of a DHT with excessive hooks for the specifics of query processing. From the other direction, we need our query processing technology to be portable across DHT implementations, since a clear winner has not emerged in the DHT design space. For both these reasons, we wish to keep the DHT APIs as thin and general-purpose as possible. The relational operators we seek to implement can present challenging workloads to DHT designers. We believe this encourages synergistic research with both query processing and DHT design in mind.

3.1   Architecture

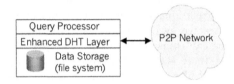

Fig. 1. The software architecture of a node implementing query processing.

Based on these design decisions, we propose a three-tier architecture as diagrammed in Figure 1. Note that all networking is handled by the DHT layer: we will use DHTs not only as an indexing mechanism, but also as a network routing mechanism. We proceed with a basic overview of the architecture and its components.

The bottom layer is a local data store, which must support the following API:

(1) An Iterator (as in Java, or STL) supporting an interface to scan through the set of objects.

(2) For each object, accessors to the attributes localID and contents. The former must be a store-wide unique identifier for the object, and the latter should be "the content" of the object, which can be a byte-array.

(3) A metadata interface to find out about additional attributes of the objects in this store.

(4) Accessors to the additional attributes.

Note that we do not specify many details of this data store, and our interface is read-only. Based on our first design goal, we expect the store to often be a filesystem, but it could easily be a wrapper over a database table or view.

The next layer is the DHT layer, which supports the put/get interface, enhanced with the following:

(1) An Iterator called ℓscan, which can be allocated by code that links to the DHT library on this machine (typically the Query Processor of Figure 1). ℓscan allows the local code to iterate through all DHT entries stored on this machine.

(2) A callback newData that notifies higher layers of the identifier for new insertions into the local portion of the DHT.

ℓscan is important because various query processing operators need to scan through all the data, as we shall see below. The addition of scanning is not unusual: other popular hashing packages support scanning through all items as well [13]. Note that we break the *location transparency* of the DHT abstraction in ℓscan, in order to allow scans to be parallelized across machines – a distributed scan interface would only have a a single caller per Iterator. The ℓscan interface allows code to run at each machine, scanning the local data in parallel with other machines. newData is desirable because we will use DHTs for temporary state during query processing, and we will want insertions into that state to be dealt with in a timely fashion.

The top layer is the query processing (QP) layer, which includes support for the parallel implementations of query operators described below, as well as support for specifying queries and iterating through query results. Our query executor will be implemented in the traditional "pull-based" *iterator* style surveyed by Graefe [6], with parallel "push-based" communication encapsulated in exchange operators [5]. We plan to support two query APIs: a graph-scripting interface for specifying explicit query plans, and a simplified SQL interface to support declarative querying. Common query types (such as keyword search) can be supported with syntatic sugar for SQL to make application programming easier.

3.2   Namespaces and Multicast

DHT systems assume a flat identifier space which is not appropriate to manage multiple data structures, as will be required for query processing. In particular, we need to be able to name tables and temporary tables, tuples within a table, and fields within a tuple. One approach is to implement an hierarchical name space on top of the flat identifier space provided by DHTs, by partitioning the identifiers in multiple fields and then have each field identify objects of the same granularity.

A hierarchical name space also requires more complex routing primitives such as multicast. Suppose we wish to store a small temporary table on a subset of nodes in the network. Then we will need to route queries to just that subset of nodes. One possibility would be to modify the routing protocol such that a node forwards a query to all neighbors that make progress in the identifier space towards *any* of the identifiers covered by the query.

3.3   Query Processing Operators

We will focus on the traditional relational database operators: selection, projection, join, grouping and aggregation, and sorting. A number of themes arise in our designs. First, we expect communication to be a key bottleneck in P2P query processing, so we will try to avoid excessive communication. Second, we wish to harness the parallelism inherent in P2P, and we will leverage traditional ideas both in intra-operator parallelism and in pipelined parallelism to achieve these goals. Third, we want answers to stream back in the style of *online query processing* [9, 8]: P2P users are impatient, they do not expect perfect answers, and they often ask broad queries even when they are only interested in a few results. Next, we we focus on the example of joins; grouping and other unary hashing operators are quite analogous to joins, with only some subtle differences [2].

Our basic join algorithm on two relations R and S is based on the pipelined or "symmetric" hash join [16], using the DHT infrastructure to route and store tuples. The algorithm begins with the query node initializing a unique temporary DHT namespace, T_{joinID}. We assume that data is iterating in from relations R and S, which each may be generated either by an ℓscan or by some more complex query subplan. The join algorithm is fully symmetric with respect to R and S, so we describe it without loss of generality from one perspective only. The join operator repeatedly gets a datum from relation R, extracts the join attribute from the datum, and uses that attribute as the insertion key for DHT T_{joinID}. When new data is inserted into T_{joinID} on some node, the newData call notifies the QP layer, which takes that datum and uses its join key to probe local data in T_{joinID} for matches. Matches are pipelined to the next iterator in the query plan (or to the client machine in the case of final results). In the case where one table (say S) is already stored hashed by the join attribute, there is no need to rehash it – the R tuples can be scanned in parallel via ℓscans, and probe the S DHT.[2]

Selection is another important operator. Relational selection can either be achieved by a table-scan followed by an explicit test of the selection predicate, or by an index lookup (which can optionally also be followed by a predicate test). Clearly explicit tests can be pushed into the network to limit the flow of data back. Index-supported selections further limit network utilization by sending requests only to those nodes that will have data. DHT indexes currently support only equality predicates. An interesting question will be to try and develop range-predicate support in a manner as efficient as current DHTs.

4   Status

We have implemented the join operation by modifying the existing CAN simulator [12] and performed exhaustive simulations. In addition to the solution presented in the previous section, we have implemented several other join variants. For example, in one of the variants, we rehash only one of the tables (say S) by the join attribute. Then each node scans locally the other table, R, and for each tuple it queries the tuples of S_t with the same join attribute value and performs local joins. The main metric we consider in our simulations is the join latency function $f(x)$, which is defined as the fraction of the total result tuples that the join initiator receives by time x. One interesting result is that this function is significantly smoother when we use a Fair Queueing like algorithm to allocate the communication and process resources. Other metrics we consider in our simulations are data placement and query processing hotspots, as well as routing hotspots. Preliminary results show that for realistic distributions of the

---

[2] This degenerate case of hash join is simply a parallel index-nested-loops join over a DHT. It also suggests an index-on-the-fly scheme, in which only one table (S) is rehashed – after S is rehashed, R probes it. Index-on-the-fly blocks the query pipeline for the duration of the rehashing, however, and is unlikely to dominate our pipelining scheme.

join attribute values, there are significant hotspots in all dimensions: storage, processing, and routing.

5 Acknowledgments

The authors have benefited from discussions with Michael J. Franklin, Petros Maniatis, Sylvia Ratnasamy, and Shelley Zhuang. We thank them for their insights and suggestions.

References

[1] ASTRAHAN, M. M., BLASGEN, M. W., CHAMBERLIN, D. D., ESWARAN, K. P., GRAY, J., GRIFFITHS, P. P., III, W. F. K., LORIE, R. A., MCJONES, P. R., MEHL, J. W., PUTZOLU, G. R., TRAIGER, I. L., WADE, B. W., AND WATSON, V. System r: Relational approach to database management. *ACM Transactions on Database Systems (TODS) 1*, 2 (1976), 97–137.

[2] BRATBERGSENGEN, K. Hashing Methods and Relational Algebra Operations. In *Proc. of the International Conferrence on Very Large Data Bases (VLDB)* (1984), pp. 323–333.

[3] DRUSCHEL, P., AND ROWSTRON, A. Past: Persistent and anonymous storage in a peer-to-peer networking environment. In *Proceedings of the 8th IEEE Workshop on Hot Topics in Operating Systems (HotOS 2001)* (Elmau/Oberbayern, Germany, May 2001), pp. 65–70.

[4] Fsttrack. `http://www.fasttrack.nu/`.

[5] GRAEFE, G. Encapsulation of Parallelism in the Volcano Query Processing System. In *Proc. ACM-SIGMOD International Conference on Management of Data* (Atlantic City, May 1990), pp. 102–111.

[6] GRAEFE, G. Query Evaluation Techniques for Large Databases. *ACM Comput. Surv. 25*, 2 (June 1993), 73–170.

[7] GRIBBLE, S., HALEVY, A., IVES, Z., RODRIG, M., AND SUCIU, D. What can p2p do for database, and vice versa? In *Proc. of WebDB Workshop* (2001).

[8] HAAS, P. J., AND HELLERSTEIN, J. M. Online Query Processing: A Tutorial. In *Proc. ACM-SIGMOD International Conference on Management of Data* (Santa Barbara, May 2001). Notes posted online at http://control.cs.berkeley.edu.

[9] HELLERSTEIN, J. M., HAAS, P. J., AND WANG, H. J. Online Aggregation. In *Proc. ACM SIGMOD International Conference on Management of Data* (1997).

[10] KUBIATOWICZ, J., BINDEL, D., CHEN, Y., CZERWINSKI, S., EATON, P., GEELS, D., GUMMADI, R., RHEA, S., WEATHERSPOON, H., WEIMER, W., WELLS, C., AND ZHAO, B. OceanStore: An architecture for global-scale persistent storage. In *Proceeedings of the Ninth international Conference on Architectural Support for Programming Languages and Operating Systems (ASPLOS 2000)* (Boston, MA, November 2000), pp. 190–201.

[11] PLAXTON, C., RAJARAMAN, R., AND RICHA, A. Accessing nearby copies of replicated objects in a distributed environment. In *Proceedings of the ACM SPAA* (Newport, Rhode Island, June 1997), pp. 311–320.

[12] RATNASAMY, S., FRANCIS, P., HANDLEY, M., KARP, R., AND SHENKER, S. A scalable content-addressable network. In *Proc. ACM SIGCOMM* (San Diego, CA, August 2001), pp. 161–172.

[13] SELTZER, M. I., AND YIGIT, O. A new hashing package for unix. In *Proc. Usenix Winter 1991 Conference* (Dallas, Jan. 1991), pp. 173–184.

[14] STOICA, I., MORRIS, R., KARGER, D., KAASHOEK, M. F., AND BALAKRISHNAN, H. Chord: A scalable peer-to-peer lookup service for internet applications. In *Proceedings of the ACM SIGCOMM '01 Conference* (San Diego, California, August 2001).

[15] STONEBRAKER, M., WONG, E., KREPS, P., AND HELD, G. The design and implementation of ingres. *ACM Transactions on Database Systems (TODS) 1*, 3 (1976), 189–222.

[16] WILSCHUT, A. N., AND APERS, P. M. G. Dataflow Query Execution in a Parallel Main-Memory Environment. In *Proc. First International Conference on Parallel and Distributed Info. Sys. (PDIS)* (1991), pp. 68–77.

[17] WITTEN, I. H., MOFFAT, A., AND BELL, T. C. *Managing Gigabytes: Compressing and Indexing Documents and Images*, second ed. Morgan Kaufmann, 1999.

[18] ZHAO, B. Y., KUBIATOWICZ, J., AND JOSEPH, A. Tapestry: An infrastructure for fault-tolerant wide-area location and routing. Tech. Rep. UCB/CSD-01-1141, University of California at Berkeley, Computer Science Department, 2001.

The Sybil Attack

John R. Douceur

Microsoft Research, One Microsoft Way, Redmond, WA, 98052-6399, USA
johndo@microsoft.com
http://www.research.microsoft.com/~johndo

Abstract. Large-scale peer-to-peer systems face security threats from faulty or hostile remote computing elements. To resist these threats, many such systems employ redundancy. However, if a single faulty entity can present multiple identities, it can control a substantial fraction of the system, thereby undermining this redundancy. One approach to preventing these "Sybil attacks" is to have a trusted agency certify identities. This paper shows that, without a logically centralized authority, Sybil attacks are always possible except under extreme and unrealistic assumptions of resource parity and coordination among entities.

1 Introduction

We[*] argue that it is practically impossible, in a distributed computing environment, for initially unknown remote computing elements to present convincingly distinct identities. With no logically central, trusted authority to vouch for a one-to-one correspondence between entity and identity, it is always possible for an unfamiliar entity to present more than one identity, except under conditions that are not practically realizable for large-scale distributed systems.

Peer-to-peer systems commonly rely on the existence of multiple, independent remote entities to mitigate the threat of hostile peers. Many systems [3, 4, 8, 10, 17, 18, 29, 34, 36] *replicate* computational or storage tasks among several remote sites to protect against integrity violations (data loss). Others [5, 6, 7, 16, 28] *fragment* tasks among several remote sites to protect against privacy violations (data leakage). In either case, exploiting the redundancy in the system requires the ability to determine whether two ostensibly different remote entities are actually different.

If the local entity has no direct physical knowledge of remote entities, it perceives them only as informational abstractions that we call *identities*. The system must ensure that distinct identities refer to distinct entities; otherwise, when the local entity selects a subset of identities to redundantly perform a remote operation, it can be duped into selecting a single remote entity multiple times, thereby defeating the redundancy. We term the forging of multiple identities a *Sybil attack* [30] on the system.

---

[*] Use of the plural pronoun is customary even in solely authored research papers; however, given the subject of the present paper, its use herein is particularly ironic.

P. Druschel, F. Kaashoek, and A. Rowstron (Eds.): IPTPS 2002, LNCS 2429, pp. 251-260, 2002.
© Springer-Verlag Berlin Heidelberg 2002

It is tempting to envision a system in which established identities vouch for other identities, so that an entity can accept new identities by trusting the collective assurance of multiple (presumably independent) signatories, analogous to the PGP web of trust [37] for human entities. However, our results show that, in the absence of a trusted identification authority (or unrealistic assumptions about the resources available to an attacker), a Sybil attack can severely compromise the initial generation of identities, thereby undermining the chain of vouchers.

Identification authorities can take various forms, not merely that of an explicit certification agency such as VeriSign [33]. For example, the CFS cooperative storage system [8] identifies each node (in part) by a hash of its IP address. The SFS network file system [23] names remote paths by appending a host identifier to a DNS name. The EMBASSY [22] platform binds machines to cryptographic keys embedded in device hardware. These approaches may thwart Sybil attacks, but they implicitly rely on the authority of a trusted agency (such as ICANN [19] or Wave Systems [35]) to establish identity.

In the following section, we define a model of a distributed computing environment that lacks a central authority. Building on this model, Section 3 proves a series of lemmas that severely limit the ability of an entity to determine identity. Section 4 surveys related work, and Section 5 concludes.

2   Formal Model

As a backdrop for our results, we construct a formal model of a generic distributed computing environment. Our model definition implicitly limits the obstructive power of corrupt entities, thereby strengthening our negative results. The universe, shown schematically in Fig. 1, includes:

- A set E of infrastructural *entities e*
- A broadcast communication *cloud*
- A *pipe* connecting each entity to the cloud

Set E is partitioned into two disjoint subsets, C and F. Each entity c in subset C is *correct*, abiding by the rules of any protocol we define. Each entity f in subset F is *faulty*, capable of performing any arbitrary behavior except as limited by explicit resource constraints. (The terms "correct" and "faulty" are standard in the domain of Byzantine fault tolerance [21], even though terms such as "honest" and "deceptive" might be more appropriate.)

Entities communicate by means of *messages*. A message is an uninterrupted, finite-length bit string whose meaning is determined either by an explicit protocol or by an implicit agreement among a set of entities. An entity can send a message through its pipe, thereby broadcasting it to all other entities. The message will be received by all entities within a bounded interval of time. Message delivery is guaranteed, but there is no assurance that all entities will hear messages in the same order.

This model has two noteworthy qualities: First, it is quite general. By leaving the internals of the cloud unspecified, this model includes virtually any interconnection topology of shared segments, dedicated links, routers, switches, or other components.

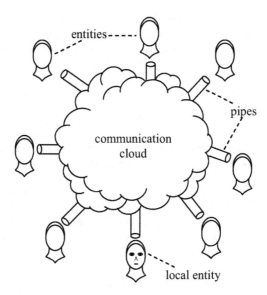

Fig. 1. Formal model of distributed environment

Second, the environment in this model is very friendly. In particular, in the absence of resource constraints, denial-of-service attacks are not possible. A message from a correctly functioning entity is guaranteed to reach all other correctly functioning entities.

We place a minimal restriction on the relative computational resources available to each entity, namely that there exists some security parameter n for which all entities can perform operations whose computational complexity is (low-order) polynomial in n but for which no entity can perform operations that are superpolynomial in n. This restriction allows entities to use public-key cryptography [24] to establish virtual point-to-point communication paths that are private and authenticated. Although these virtual paths are as secure as point-to-point physical links, they come to exist only when created by pairs of entities that have acknowledged each other. Our model excludes direct links between entities because a physical link provides a form of centrally supplied identification of a distinct remote entity. Also, in the real world, packets can be sniffed and spoofed, so the base assumption of a broadcast medium (augmented by cryptography) is not unrealistic.

An *identity* is an abstract representation that persists across multiple communication events. Each entity e attempts to *present* an identity i to other entities in the system. (Without loss of generality, we state our results with respect to a specific local entity l that is assumed to be correct.) If e successfully presents identity i to l, we say that l *accepts* identity i.

A straightforward form for an identity is a secure hash of a public key. Under standard cryptographic assumptions, such an identifier is unforgeable. Furthermore, since it can generate a symmetric key for a communication session, it is also persistent in a useful way.

Each correct entity c will attempt to present one *legitimate* identity. Each faulty entity f may attempt to present a legitimate identity and one or more *counterfeit* identities. Ideally, the system should accept all legitimate identities but no counterfeit entities.

3 Results

This section presents four simple lemmas, with nearly trivial proofs, that collectively show the impracticality of establishing distinct identities in a large-scale distributed system.

An entity has three potential sources of information about other entities: a trusted agency, itself, or other (untrusted) entities. In the absence of a trusted authority, either an entity accepts only identities that it has directly validated (by some means) or it also accepts identities vouched for by other identities it has already accepted.

For direct validation, we show:

- Even when severely resource constrained, a faulty entity can counterfeit a constant number of multiple identities.
- Each correct entity must simultaneously validate all the identities it is presented; otherwise, a faulty entity can counterfeit an unbounded number of identities.

Large-scale distributed systems are inevitably heterogeneous, leading to resource disparities that exacerbate the former result. The latter result presents a direct impediment to scalability.

For indirect validation, in which an entity accepts identities that are vouched for by already accepted identities, we show:

- A sufficiently large set of faulty entities can counterfeit an unbounded number of identities.
- All entities in the system must perform their identity validations concurrently; otherwise, a faulty entity can counterfeit a constant number of multiple identities.

Since the number of faulty entities in the system is likely to grow as the system size increases, the former result places another limit on system scale. The latter restriction becomes harder to satisfy as system size increases.

3.1 Direct Identity Validation

The only direct means by which two entities can convince a third entity that they are distinct is by performing some task that a single entity could not. If we assume that the resources of any two entities differ by at most a constant factor, a local entity can demand proof of a remote entity's resources before accepting its identity. However, this leaves us with the following limitation:

Lemma 1. If ρ is the ratio of the resources of a faulty entity f to the resources of a minimally capable entity, then f can present $g = \lfloor \rho \rfloor$ distinct identities to local entity l.

Proof. Define r_M as the resources available to a minimally capable entity. By hypothesis, g entities can present g identities to l; therefore, g r_M resources are sufficient to present g identities. Since $\rho \geq g$, f has at least g r_M resources available, so it can present g identities to l.

Lemma 1 states a lower bound on the damage achievable by a faulty entity. To show how this can be enforced as an upper bound, we present three mechanisms that can (at least theoretically) exploit limitations in three different resources: communication, storage, and computation.

If communication resources are restricted, local entity l can broadcast a request for identities and then only accept replies that occur within a given time interval.

If storage resources are restricted, entity l can challenge each identity to store a large amount of unique, uncompressible data. By keeping small excerpts of this data, entity l can verify, with arbitrarily high probability, that all identities simultaneously store the data they were sent.

If computation resources are restricted, entity l can challenge each identity to solve a unique computational puzzle. For example, the local entity can generate a large random value y and challenge the identity to find, within a limited time, a pair of values x, z such that the concatenation $x \mid y \mid z$, when run through a secure hash function, yields a value whose least significant n bits are all zero:

$$\text{given } y, \text{ find } x, z \text{ s.t. } \text{LSB}_n(\text{hash}(x \mid y \mid z)) = 0$$

The time to solve* such a puzzle is proportional to $2n-1$. The time to verify the result is constant. (The reason for allowing the challenged entity to find a prefix x and a suffix z, rather than merely one or the other, will become clear in Section 3.2.)

To be effective, these resource challenges must be issued to all identities simultaneously:

Lemma 2. If local entity l accepts entities that are not validated simultaneously, then a single faulty entity f can present an arbitrarily large number of distinct identities to entity l.

Proof. Faulty entity f presents an arbitrarily long succession of distinct identities to l. The resources required for each presentation are used and then freed for the subsequent presentation.

---

* measured in count of hash function evaluations. For a random oracle [2] hash function, the only way to find a solution is to iterate through candidate values of x and/or z; compute the hash for each $x \mid y \mid z$ triple; and test the result. Actual implementation requires a hash function that is both preimage-resistant and resistant to non-brute-force attacks such as chaining attacks [24].

Lemma 2 is insurmountable for intrinsically temporal resources, such as computation speed and communication bandwidth. However, since storage is not inherently time-based, entity l can indefinitely extend the challenge duration by periodically demanding to see newly specified excerpts of the stored data. If an accepted identity ever fails to meet a new challenge, the local entity can discard it from its acceptance list, thereby eventually catching a Sybil attack that it might have initially missed. A major practical problem with this extension is that (by assumption) the challenge consumes the majority of an entity's storage resources, so extending the challenge duration greatly impedes the ability of the entity to perform other work. (However, the challenge data itself could be valuable data, compressed and encrypted by the local entity before sending it to the remote entities, using a different key for each remote entity to maintain challenge uniqueness.)

3.2 Indirect Identity Validation

As described in the introduction, the reason for establishing the distinctness of identities is to allow the local entity to employ redundancy in operations it delegates to remote entities. One such operation it could conceivably delegate is the validation of other identities. Thus, in addition to accepting identities that it has directly validated using one of the challenge mechanisms described above, an entity might also accept identities that have been validated by a sufficient count of other identities that it has already accepted.

If an entity that has presented identity i_1 claims to have accepted another entity's identity i_2, we say that i_1 *vouches for* i_2. An obvious danger of accepting indirectly validated identities is that a group of faulty entities can vouch for counterfeit identities:

Lemma 3. If local entity l accepts any identity vouched for by q accepted identities, then a set F of faulty entities can present an arbitrarily large number of distinct identities to l if either $|F| \geq q$ or the collective resources available to F at least equal those of $q + |F|$ minimally capable entities.

Proof. Define r_F as the total resources available to set F, r_k as the resources available to each faulty entity f_k, and r_M as the resources available to a minimally capable entity. Then:

$$q + |F| \leq \frac{r_F}{r_M} = \sum_{f_k \in F} \frac{r_k}{r_M} < \sum_{f_k \in F} \left\lfloor \frac{r_k}{r_M} \right\rfloor + |F|$$

By Lemma 1, entity f_k can present $\lfloor r_k / r_M \rfloor \geq 1$ identities to l, so F can present q identities to l. Thereafter, all of F's identities vouch for an arbitrarily large number of counterfeit identities, all of which will be accepted by l.

As in the case of direct identity validation, indirect identity validation also has a concurrency requirement. In particular, all entities must perform their resource challenges concurrently:

Lemma 4. If the correct entities in set C do not coordinate time intervals during which they accept identities, and if local entity l accepts any identity vouched for by q accepted identities, then even a minimally capable faulty entity f can present $g = \lfloor |C| / q \rfloor$ distinct identities to l.

Proof. Define r_M as the resources required to present one identity. By assumption, entity f has r_M resources available. Partition set C into g disjoint subsets C_k of minimum cardinality q. Faulty entity f presents identity i_k to each entity in C_k, using r_M resources during time interval T_k. Since T_k need not overlap with $T_{k'}$ for $k \neq k'$, r_M resources are available during interval $T_{k'}$ to present identity $i_{k'} \neq i_k$ to entities in set $C_{k'}$. At least q entities in each set C_k will vouch for distinct identity i_k, so l will accept all g identities.

Lemma 4 shows the need for multiple entities to issue challenges concurrently. Whether it is possible for a correct entity to satisfy multiple concurrent challenges depends upon the resource:

In our formal model, all communication is broadcast, so an entity can simultaneously reply to communication challenges from arbitrarily many entities. (However, this is rather less practical in actual networks than in our abstract model.)

It may not be possible to satisfy multiple concurrent storage challenges, and there are information-theoretic reasons for believing that it is impossible, since every bit of data stored for one challenger consumes one bit of storage space that is thus unavailable to serve another challenger (and the data from all challengers is, of necessity, incompressible). This may prevent storage challenges from being used for indirect validation.

For computation challenges, it is possible for an entity to solve multiple puzzles simultaneously by combining them. If an entity receives m puzzles $y_1, y_2, \dots y_m$, it can find a w such that:

$$\mathrm{LSB}_n(\mathrm{hash}(0 \mid y_1 \mid y_2 \mid \dots y_m \mid w)) = 0$$

Then, the solution to each puzzle y_k is:

$$x_k = 0 \mid y_1 \mid y_2 \mid \dots y_{k-1} \text{ and } z_k = y_{k+1} \mid \dots y_m \mid w$$

An obvious danger here is that if a validating entity issues challenges to multiple identities that have been counterfeited by a single faulty entity, the faulty entity could combine the challenges and solve them together. However, the challenger can identify this attempted Sybil attack by checking whether $x_1 \mid y_1 \mid z_1 = x_2 \mid y_2 \mid z_2$ for any two solutions from putatively different identities.

Like Lemma 1, the result of Lemma 4 is that a faulty entity can amplify its influence. A system that can tolerate a fraction φ of all identities being faulty can tolerate only φ/g of all entities being faulty. In some systems, this may be acceptable.

4 Related Work

Most prior research on electronic identities has focused on persistence and unforgeability [14, 15, 27, 31], rather than on distinctness.

Computational puzzles are an old technique [25] that has become popular recently for resisting denial-of-service attacks [1, 9, 20] by forcing the attacker to perform more work than the victim.

Dingledine et al. [11] suggest using puzzles to provide a degree of accountability in peer-to-peer systems, but this still allows a resourceful attacker to launch a substantial attack, especially if the potential for damage is disproportionate to the fraction of the system that is compromised.

The issue of establishing on-line identities for humans has been studied for some time [12, 32], with solutions that generally depend on some direct interaction in the physical world [13, 37].

5 Summary and Conclusions

Peer-to-peer systems often rely on redundancy to diminish their dependence on potentially hostile peers. If distinct identities for remote entities are not established either by an explicit certification authority (as in Farsite [3]) or by an implicit one (as in CFS [8]), these systems are susceptible to Sybil attacks, in which a small number of entities counterfeit multiple identities so as to compromise a disproportionate share of the system.

Systems that rely upon implicit certification should be acutely mindful of this reliance, since apparently unrelated changes to the relied-upon mechanism can undermine the security of the system. For example, the proposed IPv6 privacy extensions [26] obviate much of the central allocation of IP addresses assumed by CFS.

In the absence of an identification authority, a local entity's ability to discriminate among distinct remote entities depends on the assumption that an attacker's resources are limited. Entities can thus issue resource-demanding challenges to validate identities, and entities can collectively pool the identities they have separately validated. This approach entails the following conditions:

- All entities operate under nearly identical resource constraints.
- All presented identities are validated simultaneously by all entities, coordinated across the system.
- When accepting identities that are not directly validated, the required number of vouchers exceeds the number of system-wide failures.

We claim that in a large-scale distributed system, these conditions are neither justifiable as assumptions nor practically realizable as system requirements.

Acknowledgements

The author thanks Miguel Castro for issuing the challenge that led to this paper, Jon Howell and Sandro Forin for reviewing drafts, Dan Simon for sanity checking the combinable computational puzzle described in Section 3.2, the anonymous IPTPS reviewers for their helpful suggestions for improving this presentation, and Brian Zill for suggesting the term "Sybil attack."

References

1. T. Aura, P. Nikander, J. Leiwo, "DoS-Resistant Authentication with Client Puzzles", *Cambridge Security Protocols Workshop*, Springer, 2000.
2. M. Bellare and P. Rogaway, "Random Oracles are Practical: A Paradigm for Designing Efficient Protocols", *1st Conference on Computer and Communications Security*, ACM, 1993, pp. 62-73.
3. W. J. Bolosky, J. R. Douceur, D. Ely, M. Theimer, "Feasibility of a Serverless Distributed File System Deployed on an Existing Set of Desktop PCs", *SIGMETRICS 2000*, 2000, pp. 34-43.
4. M. Castro, B. Liskov, "Practical Byzantine Fault Tolerance", 3rd OSDI, 1999.
5. D. Chaum, "Untraceable Electronic Mail, Return Addresses, and Digital Pseudonyms", *CACM* 4 (2), 1982.
6. B. Chor, O. Goldreich, E. Kushilevitz, M. Sudan, "Private Information Retrieval", *36th FOCS*, 1995.
7. I. Clarke, O. Sandberg, B. Wiley, T. Hong, "Freenet: A Distributed Anonymous Information Storage and Retrieval System", *Design Issues in Anonymity and Unobervability*, ICSI, 2000.
8. F. Dabek, M. F. Kaashoek, D. Karger, R. Morris, I. Stoica, "Wide-Area Cooperative Storage with CFS", *18th SOSP*, 2001, pp. 202-215.
9. D. Dean, A. Stubblefield, "Using Client Puzzles to Protect TLS", *10th USENIX Security Symp.*, 2001.
10. R. Dingledine, M. Freedman, D. Molnar "The Free Haven Project: Distributed Anonymous Storage Service", *Design Issues in Anonymity and Unobservability*, 2000.
11. R. Dingledine, M. J. Freedman, D. Molnar "Accountability", *Peer-to-Peer: Harnessing the Power of Disruptive Technologies*, O'Reilly, 2001.
12. J. S. Donath, "Identity and Deception in the Virtual Community", *Communities in Cyberspace*, Routledge, 1998.
13. C. Ellison, "Establishing Identity Without Certification Authorities", *6th USENIX Security Symposium*, 1996, pp. 67-76.
14. U. Feige, A. Fiat, A. Shamir, "Zero-Knowledge Proofs of Identity", *Journal of Cryptology* 1 (2), 1988, pp. 77-94.
15. A. Fiat, A. Shamir, "How to Prove Yourself: Practical Solutions of Identification and Signature Problems", *Crypto '86*, 1987, pp. 186-194.
16. Y. Gertner, S. Goldwasser, T. Malkin, "A Random Server Model for Private Information Retrieval", *RANDOM '98*, 1998.
17. A. Goldberg, P. Yianilos, "Towards an Archival Intermemory", *International Forum on Research and Technology Advances in Digital Libraries*, IEEE, 1998, pp. 147-156.
18. J. H. Hartman, I. Murdock, T. Spalink, "The Swarm Scalable Storage System", *19th ICDCS*, 1999, pp. 74-81.

19. ICANN, Internet Corporation for Assigned Names and Numbers, 4676 Admiralty Way, Suite 330, Marina del Rey, CA 90292-6601, www.icann.org.

20. A. Juels, J. Brainard, "Client Puzzles: A Cryptographic Defense against Connection Depletion Attacks", *NDSS '99*, ISOC, 1999, pp. 151-165.

21. L. Lamport, R. Shostak, M. Pease, "The Byzantine Generals Problem", *TPLS* 4(3), 1982.

22. K. R. Lefebvre, "The Added Value of EMBASSY in the Digital World", Wave Systems Corp. white paper, www.wave.com, 2000.

23. D. Mazières, M. Kaminsky, M. F. Kaashoek, E. Witchel, "Separating Key Management from File System Security", *17th SOSP*, 1999, pp. 124-139.

24. A. J. Menezes, P. C. van Oorschot, S. A. Vanstone. *Handbook of Applied Cryptography.* CRC Press, 1997.

25. R. C. Merkle, "Secure Communications over Insecure Channels", *CACM 21*, 1978, pp. 294-299.

26. T. Narten, R. Draves, "Privacy Extensions for Stateless Address Autoconfiguration in IPv6", *RFC 3041*, 2001.

27. K. Ohta, T. Okamoto, "A Modification to the Fiat-Shamir Scheme", *Crypto '88*, 1990, pp. 232-243.

28. M. K. Reiter, A. D. Rubin, "Crowds: Anonymous Web Transactions", *Transactions on Information System Security* 1 (1), ACM, 1998.

29. A. Rowstron, P. Druschel, "Storage Management and Caching in PAST, a Large-Scale, Persistent Peer-to-Peer Storage Utility", *18th SOSP*, 2001, pp. 188-201.

30. F. R. Schreiber, *Sybil*, Warner Books, 1973.

31. A. Shamir, "An Efficient Identification Scheme Based on Permuted Kernels", *Crypto '89*, 1990, pp. 606-609.

32. S. Turkle, *Life on the Screen: Identity in the Age of the Internet*, Simon & Schuster, 1995.

33. VeriSign, Inc. 487 East Middlefield Road, Mountain View, CA 94043, www.verisign.com.

34. M. Waldman, A. D. Rubin, L. F. Cranor, "Publius: A Robust, Tamper-Evident Censorship-Resistant Web Publishing System", *9th USENIX Security Symposium*, 2000, pp. 59-72.

35. Wave Systems Corp. 480 Pleasant Street, Lee, MA 01238, www.wave.com

36. J. J. Wylie, M. W. Bigrigg, J. D. Strunk, G. R. Ganger, H. Kilite, P. K. Khosla, "Survivable Information Storage Systems", *IEEE Computer* 33 (8), IEEE, 2000, pp. 61-68.

37. P. Zimmerman, *PGP User's Guide*, MIT, 1994.

Security Considerations for Peer-to-Peer Distributed Hash Tables

Emil Sit and Robert Morris

Laboratory for Computer Science, MIT
200 Technology Square, Cambridge, MA 02139, USA
{sit,rtm}@lcs.mit.edu

Abstract. Recent peer-to-peer research has focused on providing efficient hash lookup systems that can be used to build more complex systems. These systems have good properties when their algorithms are executed correctly but have not generally considered how to handle misbehaving nodes. This paper looks at what sorts of security problems are inherent in large peer-to-peer systems based on distributed hash lookup systems. We examine the types of problems that such systems might face, drawing examples from existing systems, and propose some design principles for detecting and preventing these problems.

1 Introduction

Peer-to-peer systems present an interesting security problem as there is no central system to protect. Instead, the nodes must work together to ensure correct and secure behavior. Unfortunately, deployment on an open network, such as the Internet, implies that there will be malicious nodes in the system. These nodes will try and disrupt the system or subvert it to their advantage. Peer-to-peer systems must be designed to operate correctly even in these situations.

A number of recent systems are built on top of peer-to-peer distributed hash lookup systems [7, 8, 11, 12]. Lookups for keys are performed by routing queries through a series of nodes; each of these nodes uses a local routing table to forward the query towards the node that is ultimately responsible for the key. These systems can be used to store data, for example, as a distributed hash table or file system [3, 9]. Other projects take advantage of other aspects of the lookup system, such as the properties of lookup routing [10]. In this paper, we will examine security concerns that are particular to distributed hash tables.

One class of attacks on distributed hash tables causes the system to return incorrect data to the application. Fortunately, the correctness and authenticity of data can be addressed using cryptographic techniques such as self-certifying path names [5]. These techniques allow the system to detect and ignore inauthentic data.

This paper focuses on the remaining attacks — those that threaten the liveness of the system by preventing participants from finding data. We begin by presenting a common framework for distributed hash tables in Section 2, followed by the basic adversary model in Section 3. The core of the paper is in

P. Druschel, F. Kaashoek, and A. Rowstron (Eds.): IPTPS 2002, LNCS 2429, pp. 261–269, 2002.
© Springer-Verlag Berlin Heidelberg 2002

Table 1. Design Principles

Define verifiable system invariants (and verify them!)
Allow the querier to observe lookup progress.
Assign keys to nodes in a verifiable way.
Server selection in routing may be abused.
Cross-check routing tables using random queries.
Avoid single points of responsibility.

Section 4, where we present a series of examples of particular weaknesses in existing distributed hash lookup systems. From these discussions, we derive a set of general design principles, shown in Table 1. We conclude in Section 5.

2   Background

Typical distributed hash tables consist of a storage API layered on top of a lookup protocol. Lookup protocols share a few basic components:

1. a key identifier space,
2. a node identifier space,
3. rules for associating keys to particular nodes,
4. per-node routing tables that refer to other nodes, and
5. rules for updating routing tables as nodes join and fail.

The lookup protocol maps a desired key identifier to the IP address of the node responsible for that key. A storage protocol layered on top of the lookup protocol then takes care of storing, replicating, caching, retrieving, and authenticating the data. CAN [7], Chord [11] and Pastry [8] all fit into this general framework.

Routing in the lookup is handled by defining a distance function on the identifier space so that distance can be measured between the current node and the desired key; the responsible node is defined to be the node closest to the key.

Lookup protocols typically have an invariant that must be maintained in order to guarantee that data can be found. For example, the Chord system arranges nodes in a one-dimensional (but circular) identifier space; the required invariant is that every node knows the node that immediately follows it in the identifier space. If an attacker could break this invariant, Chord would not be able to look up keys correctly.

Similarly, the storage layer will also maintain some invariants in order to be sure that each piece of data is available. In the case of DHash [3], a storage API layered on Chord used by CFS, there are two important invariants. First, it must ensure that the node that Chord believes is responsible for a key actually stores the data associated with that key. Since nodes can fail, it is also important that DHash maintain replicas of each piece of data, and that those replicas be at predictable nodes. An attacker could potentially target either of these invariants.

3   Adversary Model

The adversaries that we consider in this paper are participants in a distributed hash lookup system that do not follow the protocol correctly. Instead, they seek to mislead legitimate nodes by providing them with false information.

We assume that a malicious node is able to generate packets with arbitrary contents (including forged source IP addresses), but that a node is only able to examine packets addressed to itself. That is, malicious nodes are *not* able to overhear or modify communication between other nodes. The fact that a malicious node can only receive packets addressed to its own IP address means that an IP address can be used as a weak form of node identity; if a node receives a packet from an IP address, it can verify that the packet's sender owns the address by sending a request for confirmation to that address. We also consider malicious nodes that conspire together, but where each one is limited as above. This allows an adversary to gather additional data and act more deviously by providing false but "confirmable" information.

The rest of the paper will examine the ways in which malicious nodes can use these abilities to subvert the system.

4   Attacks and Defenses

This section is organized into attacks against the routing, attacks against the data storage system, and finally some general considerations.

The first line of defense for any attack is detection. Many attacks can be detected by the node being attacked because they involve violating invariants or procedure contracts. However, it is less clear what to do once an attack has been detected. A node may genuinely be malicious, or it may have failed to detect that it was being tricked. Thus, our discussion focuses on methods to detect and possibly correct inconsistent information. We will see that achieving *verifiability* underlies all of our detection techniques.

4.1   Routing Attacks

The routing portion of a lookup protocol involves maintaining routing tables and then dispatching requests to the nodes in the routing table. It is critical that routing is correct in a distributed hash table. However, there is considerable room for an adversary to play in existing systems. These attacks can be detected if the system *defines verifiable system invariants (and verifies them)*. When invariants fail, the system must have a recovery mechanism.

Incorrect Lookup Routing. An individual malicious node could forward lookups to an incorrect or non-existent node. Since the malicious node would be participating in the routing update system in a normal way, it would appear to be alive, and would not ordinarily be removed from the routing tables of other nodes. Thus re-transmissions of the misdirected lookups would also be sent to the malicious node.

Fortunately, blatantly incorrect forwarding can often be easily detected. At each hop, the querier knows that the lookup is supposed to get "closer" to the key identifier. For example, in Pastry, each hop should match the key identifier in at least one more digit than the last. The querier should check for this so that this attack can be detected. If such an attack is detected, the querier might recover by backtracking to the last good hop and asking for an alternate step.

In order for the querier to be able to perform this check, however, each step of progress must be visible to the querier. For example, CAN proposes an optimization where each node keeps track of the network RTTs to neighbor nodes and forwards to the neighbor with the best ratio of progress to RTT. This implies that queries are generally forwarded without interacting with the querier. Thus in CAN, a querier simply can not verify forward progress. One should *allow the querier to observe lookup progress*.

A malicious node might also simply declare (incorrectly) that a random node is the node responsible for a key. Since the querying node might be far away in the identifier space, it might not know that this node is, in fact, not the closest node. This could cause a key to be stored on an incorrect node or prevent the key from being found. This can be fixed with two steps.

First, the querier should ensure that the destination itself agrees that it is a correct termination point for the query. In Chord, the predecessor returns the address of the query endpoint (the "successor") instead of the endpoint itself, making this attack possible — a malicious node can cause the query to overshoot the correct successor. Since the querier Q does not know about the true successor S, a malicious predecessor P could forward to some node $S' > S$. This can cause DHash to violate its storage location invariant. However, if S' is good, then it can see that it should not be responsible for this key and can raise an error.

Second, the system should *assign keys to nodes in a verifiable way*. In particular, in some systems, keys are assigned to the node that is closest to them in the identifier space. Thus in order to assign keys to nodes verifiably, it is sufficient to derive node identifiers in a verifiable way. Contrast this to CAN, which allows any node to specify its own identity. This makes it impossible for another node to verify that a node is validly claiming responsibility for a key. Some systems, like Chord, make an effort to defend against this by basing a node's identifier on a cryptographic hash of its IP address and port.[1] Since this is needed to contact the node, it is easy to tell if one is speaking to the correct node.

Systems may want to consider deriving long-term identities based on public keys. This has performance penalties due to the cost of signatures, but would allow systems to have faith on the origin of messages and the validity of their contents. That is, public keys would facilitate the verifiability of the system. For example, a certificate with a node's public key and address can be used by new nodes to safely join the system.

---

[1] The hash actually also includes a virtual node identifier, which will be relevant in Section 4.2

Incorrect Routing Updates. Since each node in a lookup system builds its routing table by consulting other nodes, a malicious node could corrupt the routing tables of other nodes by sending them incorrect updates. The effect of these updates would be to cause innocent nodes to misdirect queries to inappropriate nodes, or to non-existent nodes. However, if the system knows that correct routing updates have certain requirements, these can be verified. For example, Pastry updates require that each table entry has a correct prefix. Blatantly incorrect updates can be easily identified and dropped. Other updates should only be incorporated into a node's routing table after it has verified itself that the remote node is reachable.

A more subtle attack would be to take advantage of systems that allow nodes to choose between multiple correct routing entries. For example, CAN's RTT optimization allows precisely this in order to minimize latency. A malicious node can abuse this flexibility and provide nodes that are undesirable. For example, it might choose an unreliable node, one with high latency, or even a fellow malicious node. While this may not affect strict correctness of the protocol, it may impact applications that may wish to use the underlying lookup system to find nodes satisfying certain criteria. For example, the Tarzan anonymizing network [4] proposes the use of Chord as a way of discovering random nodes to use in dynamic anonymizing tunnels. Any flexibility in Chord might allow an adversary to bias the nodes chosen, compromising the design goals of Tarzan. Applications should be aware that *server selection in routing may be abused*.

Partition. In order to bootstrap, a new node participating in any lookup system must contact some existing node. At this time, it is vulnerable to being partitioned into an incorrect network. Suppose a set of malicious nodes has formed a parallel network, running the same protocols as the real, legitimate network. This parallel network is entirely internally consistent and may even contain some of the data from the real network. A new node may join this network accidentally and thus fail to achieve correct results. One of the malicious nodes might also be cross-registered in the legitimate network and may be able to cause new participants to be connected to the parallel network even if they have a valid bootstrap node.

Partitions can be used by malicious nodes to deny service or to learn about the behavior of clients that it would otherwise be unable to observe. For example, if a service was made available to publish documents anonymously, an adversary could establish a malicious system that shadows the real one but allows it to track clients who are reading and storing files.

In order to prevent a new node from being diverted into an incorrect network, it must bootstrap via some sort of trusted source. This source will likely be out-of-band to the system itself. When rejoining the system, a node can either use these trusted nodes, or it can use one of the other nodes it has previously discovered in the network. However, building trust metrics for particular nodes can be dangerous in a network with highly transient nodes that lack any strong sense of identity. If a particular address is assigned via DHCP, for example, a node could be malicious one day but benign the next. Again, use of public keys may reduce this risk.

If a node believes it has successfully bootstrapped in the past, then it can detect *new* malicious partitions by cross-checking results. A node can maintain a set of other nodes that it has used successfully in the past. Then, it can *cross-check routing tables using random queries*.[2] By asking those nodes to do random queries and comparing their results with its own results, a node can verify whether its view of the network is consistent with those other nodes. Note that randomness is important so that a malicious partition can not distinguish verification probes from a legitimate query that it would like to divert. Conversely, a node that has been trapped in a malicious partition might accidentally discover the correct network in this manner, where the "correct" network here is defined as the one which serves desired data.

4.2   Storage and Retrieval Attacks

A malicious node could join and participate in the lookup protocol correctly, but deny the existence of data it was responsible for. Similarly, it might claim to actually store data when asked, but then refuse to serve it to clients. In order to handle this attack, the storage layer must implement replication. Replication must be handled in a way so that no single node is responsible for replication or facilitating access to the replicas; that node would be a single point of failure. So, for example, clients must be able to independently determine the correct nodes to contact for replicas. This would allow them to verify that data is truly unavailable with all replica sites. Similarly, all nodes holding replicas must ensure that the replication invariant (e.g. at least r copies exist at all times) is maintained. Otherwise, a single node would be able to prevent all replication from happening. In summary, *avoid single points of responsibility*.

Clients doing lookups must be prepared for the possibility of malicious nodes as well. Thus, they must consult at least two replica sites in order to be sure that either all of the replicas are bad or that the data is truly missing.

As an example, DHash does not follow this principle: only the node immediately associated with the key is responsible for replication. However, even if the storing node performed replication, DHash would still be vulnerable to the actual successor lying about the r later successors. Replication with multiple hash functions, as proposed in CAN, is one way to avoid this reliance on a single machine.

This attack can be further refined in a system that does not assign nodes verifiable identifiers. In such a system, a node can choose to become responsible for data that it wishes to hide. DHash continues to be at risk here, despite Chord having verifiable node identifiers, because the identifier is derived from a hash of the node's IP address, port number and virtual node number. Since a person in control of a single node can run a large number of virtual nodes, they can still effect some degree of choice in what data they wish to hide. IPv6 or sparsely used IPv4 networks may also allow a single host to appear to run many nodes.

---

[2] Of course, without a sense of node identity that is stronger than IP address, this is still dangerous.

4.3   Miscellaneous Attacks

Inconsistent Behavior. Any of the attacks here can be made more difficult to detect if a malicious node presents a good face to part of the network. That is, a malicious node may choose to maximize its impact by ensuring that it behaves correctly for certain nodes. One possible target would be nodes near it in the identifier space. These nodes will not see any reason to remove the node from their routing tables despite the fact that nodes that are distant see poor or invalid behavior. This may not be a serious problem if queries must be routed through close nodes before reaching the target node. However, most routing systems have ways of jumping to distant points in the identifier space in order to speed up queries.

Ideally, distant nodes would be able to convince local nodes that the "locally good" malicious node is in fact malicious. However, without public keys and digital signatures, it is not possible to distinguish a report of a "locally good" node being malicious, from a malicous report trying to tarnish a node that is actually benign. On the other hand, with public keys, this can be proven by requiring nodes to sign all of their responses. Then a report would contain the incorrect response and the incongruity could be verified. Lacking this, each node must make its own determination as to whether another node is malicious.

Overload of Targeted Nodes. Since an adversary can generate packets, it can attempt to overload targetted nodes with garbage packets. This is a standard denial of service attack and not really a subversion of the system. This will cause the node to appear to fail and the system will be able to adapt to this as if the node had failed in some normal manner. A system must use some degree of data replication to handle even the normal case of node failure. This attack may be effective if the replication is weak (i.e. the malicious nodes can target all replicas easily) or if the malicious node is one of the replicas or colluding with some of the replicas.

The impact of denial of service attacks can be partially mitigated by ensuring that the node identifier assignment algorithm assigns identifiers to nodes randomly with respect to network topology. Additionally, replicas should be located in physically disparate locations. These would prevent a localized attack from preventing access to an entire portion of the key space. If an adversary did wish to shut out an entire portion of the key space, it would have to flood packets all over the Internet.

Rapid Joins and Leaves. As nodes join and leave the system, the rules for associating keys to nodes imply that new nodes must obtain data (from replicas) that was stored by nodes that have left the system. This rebalancing is required in order for the lookup procedures to work correctly. A malicious node could trick the system into rebalancing unnecessarily causing excess data transfers and control traffic. This will reduce the efficiency and performance of the system; it may even be possible to overload network segments. This attack would work best if the attacker could avoid being involved in data movement since that would consume the bulk of the bandwidth. An adversary might try to convince the system that a particular node was unavailable or that a new node had (falsely) joined. However, our model allows the adversary no (low-bandwidth) way of accomplishing the former; the latter case presumably will involve acknowledged data transfers that the adversary can not correctly acknowledge. Any other rebalancing would involve the adversary node itself, requiring it to be involved in the data movement.

Note that any distributed hash table must provide a mechanism for dealing with this problem, regardless of whether there are malicious nodes present. Early studies have shown that in some file sharing systems, peers join and leave the system very rapidly [6]. The rate of replication and amount of data stored at each node must be kept at levels that allow for timely replication without causing network overload when even regular nodes join and leave the network.

Unsolicited Messages. A malicious node may be able to engineer a situation where it can send an unsolicited response to a query. For example, consider an iterative lookup process where querier Q is referred by node E to node A. Node E knows that Q will next contact A, presumably with a follow-up to the query just processed by E. Thus, E can attempt to forge a message from A to Q with incorrect results.

The best defense against this would be to employ standard authentication techniques such as digital signatures or message authentication codes. However, digital signatures are currently expensive and MACs require shared keys. A more reasonable defense may be to include a random nonce with each query and have the remote end echo the nonce in its reply. This would essentially ensure the origin of the response.

5   Conclusion

This paper presents a categorization of the basic attacks that peer-to-peer hash lookup systems should be aware of. It discusses details of those attacks as applied to some specific systems, and suggests defenses in many cases. It abstracts these defenses into this set of general design principles: 1) define verifiable system invariants, and verify them; 2) allow the querier to observe lookup progress; 3) assign keys to nodes in a verifiable way; 4) be wary of server selection in routing; 5) cross-check routing tables using random queries; and 6) avoid single points of responsibility.

References

[1] *Proceedings of ACM SIGCOMM* (San Diego, California, Aug. 2001).

[2] *Proceedings of the 18th ACM Symposium on Operating Systems Principles (SOSP '01)* (Banff, Canada, Oct. 2001).

[3] DABEK, F., KAASHOEK, M. F., KARGER, D., MORRIS, R., AND STOICA, I. Wide-area cooperative storage with CFS. In *Proceedings of the 18th ACM SOSP* [2], pp. 202–215.

[4] FREEDMAN, M. J., SIT, E., CATES, J., AND MORRIS, R. Tarzan: A peer-to-peer anonymizing network layer. In *Proceedings of the First International Workshop on Peer-to-Peer Systems* (Cambridge, MA, Mar. 2002).

[5] FU, K., KAASHOEK, M. F., AND MAZIÈRES, D. Fast and secure distributed read-only file system. In *Proceedings of the 4th USENIX Symposium on Operating Systems Design and Implementation (OSDI)* (Oct. 2000), pp. 181–196.

[6] KRISHNAMURTHY, B., WANG, J., AND XIE, Y. Early measurements of a cluster-based architecture for P2P systems. In *Proceedings of the First ACM SIGCOMM Internet Measurement Workshop* (San Francisco, California, Nov. 2001), pp. 105–109.

[7] RATNASAMY, S., FRANCIS, P., HANDLEY, M., KARP, R., AND SHENKER, S. A scalable content-addressable network. In *Proceedings of ACM SIGCOMM* [1], pp. 161–172.

[8] ROWSTRON, A., AND DRUSCHEL, P. Pastry: Scalable, distributed object location and routing for large-scale peer-to-peer systems. In *Proceedings of the 18th IFIP/ACM International Conference on Distributed Systems Platforms (Middleware 2001)* (Nov. 2001).

[9] ROWSTRON, A., AND DRUSCHEL, P. Storage management and caching in PAST, a large-scale, persistent peer-to-peer storage utility. In *Proceedings of the 18th ACM SOSP* [2], pp. 188–201.

[10] ROWSTRON, A., KERMARREC, A.-M., CASTRO, M., AND DRUSCHEL, P. SCRIBE: The design of a large-scale event notification infrastructure. In *Networked Group Communication: Third International COST264 Workshop* (Nov. 2001), J. Crowcroft and M. Hofmann, Eds., vol. 2233 of *Lecture Notes in Computer Science*, Springer-Verlag, pp. 30–43.

[11] STOICA, I., MORRIS, R., KARGER, D., KAASHOEK, M. F., AND BALAKRISHNAN, H. Chord: A scalable peer-to-peer lookup service for internet applications. In *Proceedings of ACM SIGCOMM* [1], pp. 149–160.

[12] ZHAO, B., KUBIATOWICZ, J., AND JOSEPH, A. Tapestry: An infrastructure for fault-tolerant wide-area location and routing. Tech. Rep. UCB/CSD-01-1141, Computer Science Division, U. C. Berkeley, Apr. 2001.

Dynamically Fault-Tolerant Content Addressable Networks

Jared Saia[1], Amos Fiat[2], Steve Gribble[1], Anna R. Karlin[1], and Stefan Saroiu[1]

[1] Department of Computer Science and Engineering, University of Washington,
Seattle, WA 98195;
{saia, gribble, karlin, tzoompy}@cs.washington.edu
[2] Department of Computer Science, Tel Aviv University, Tel Aviv, Israel
fiat@cs.tau.ac.il

Abstract. We describe a content addressable network which is robust in the face of massive adversarial attacks and in a highly dynamic environment. Our network is robust in the sense that at any time, an arbitrarily large fraction of the peers can reach an arbitrarily large fraction of the data items. The network can be created and maintained in a completely distributed fashion.

1 Introduction

Distributed denial-of-service attacks on the Internet are highly prevalent, targeting a wide-range of victims [3]. Peer-to-peer systems are particularly vulnerable to such attacks, since peers lack the technical expertise and resources needed for maintaining a high level of protection. In addition to being vulnerable to such attacks, we can expect peer-to-peer systems to be confronted with a highly dynamic peer turnover rate [8]. For example, in both Napster and Gnutella, half of the peers participating in the system will be replaced by new peers within one hour. Thus, maintaining fault-tolerance in the face of massive targeted attacks and in a highly dynamic environment is critical to the success of a peer-to-peer system.

The contributions of this paper are two-fold. First, we define the notion of *dynamically strong fault-tolerance*. Our definition captures the properties that a peer-to-peer system must have to be robust to orchestrated attacks and in a highly dynamic environment. Second, we present a content addressable network [9] which is dynamically strong fault-tolerant.

1.1 Dynamic Fault Tolerance

To better address fault-tolerance in peer-to-peer networks, we define a new notion of *dynamically strong fault-tolerance*. First, we assume an adversarial fail-stop model – at any time, the adversary has complete visibility of the entire state of the system and can choose to "delete" any peer it wishes. A "deleted" peer stops functioning immediately, but is not assumed to be Byzantine. Second, we

P. Druschel, F. Kaashoek, and A. Rowstron (Eds.): IPTPS 2002, LNCS 2429, pp. 270–279, 2002.

require our network to remain "robust" at all times provided that in any time interval during which the adversary deletes some number of peers, some larger number of new peers join the network.

More formally, we say that an *adversary is limited* if for some constants $\gamma > 0$ and $\delta > \gamma$, during any period of time in which the adversary deletes γn peers from the network, at least δn new peers join the network (where n is the number of peers initially in the network). Each new peer that is inserted knows only one other random peer currently in the network.

For such a limited adversary, we seek to maintain a robust network for indexing up to n data items. Although the number of indexed data items remains fixed, the number of peers in the network will fluctuate as nodes are inserted and deleted by the adversary.

We say that a content addressable network (CAN) is ϵ-*robust* at some particular time if all but an ϵ fraction of the peers in the CAN can access all but an ϵ fraction of the data items.

Finally, we say that a CAN (initially containing n peers) is ϵ-*dynamically strong fault-tolerant* (or simply ϵ-*dynamically fault-tolerant*) if, with high probability, the CAN is always ϵ-robust during a period when a limited adversary deletes a number of peers polynomial in n.

In section 2, we present an ϵ-dynamically fault-tolerant CAN for any arbitrary $\epsilon > 0$, and any constants γ and δ such that $\gamma < 1$ and $\delta > \gamma + \epsilon$. Our CAN stores n data items[1], and has the following characteristics:

1. With high probability, at any time, an arbitrarily large fraction of the nodes can find an arbitrarily large fraction of the data items.
2. Search takes time $O(\log n)$ and requires $O(\log^3 n)$ messages in total.
3. Every peer maintains pointers to $O(\log^3 n)$ other peers.
4. Every peer stores $O(\log n)$ data items.
5. Peer insertion takes time $O(\log n)$.

The constants in these resource bounds are functions of ϵ, γ and δ. The technical statement of this result is presented in Theorem 1.

We note that, as we have defined it, an ϵ-*dynamically fault-tolerant* CAN is ϵ-robust for only a polynomial number of peer deletions by the limited adversary. To address this issue, we imagine that very infrequently, there is an all-to-all broadcast among all live peers to reconstruct the CAN(details of how to do this are in [1]). Even with these infrequent reconstructions, the amortized cost per insertion will be small. Our main theorem is provided below.

Theorem 1. *For all $\epsilon > 0$ and value P which is polynomial in n, there exist constants $k_1(\epsilon)$, $k_2(\epsilon)$ and $k_3(\epsilon)$ and $k_4(\epsilon)$ such that the following holds with high probability for the CAN for deletion of up to P peers by the limited adversary:*

---

[1] For simplicity, we've assumed that the number of items and the number of initial nodes is equal. However, for any n nodes and $m \geq n$ data items, our scheme will work, where the search time remains $O(\log n)$, the number of messages remains $O(\log^3 n)$, and the storage requirements are $O(\log^3 n \times m/n)$ per node.

- *At any time, the CAN is ϵ-robust*
- *Search takes time no more than $k_1(\epsilon) \log n$.*
- *Peer insertion takes time no more than $k_2(\epsilon) \log n$.*
- *Search requires no more than $k_3(\epsilon) \log^3 n$ messages total.*
- *Every node stores no more than $k_4(\epsilon) \log^3 n$ pointers to other nodes and $k_3(\epsilon) \log n$ data items.*

1.2   Related Work

Fiat and Saia [1] present a content addressable network for which even after adversarial removal of a linear number of nodes in the network, an arbitrarily large fraction of the remaining nodes can access an arbitrarily large fraction of the original data items. While the Fiat-Saia network is an important first step towards the goal of a strongly fault-tolerant CAN, this scheme is inherently static. Thus, even if many new peers join the network, the CAN ceases to be ϵ-robust when all the original peers die.

Weaker forms of static fault-tolerance are known to exist for other peer-to-peer systems. Experimental measurements of a connected component of the real Gnutella network have been studied [8], and it has been found to still contain a large connected component even with a $1/3$ fraction of random peer deletions.

Several content addressable networks are robust under random node deletions [4, 9, 2]. For example, Chord correctly routes queries in $O(\log(n))$ expected time even after each node fails with probability $1/2$. However, it is unclear whether it is possible to extend any of these systems to remain robust under orchestrated attacks. In addition, many known network topologies are known to be vulnerable to adversarial deletions. For example, with a linear number of node deletions, the hypercube can be fragmented into components all of which have size no more than $O(n/\sqrt{\log n})$ ([5]).

2   A Dynamically Fault-Tolerant Content Addressable Network

Our scheme is most easily described by imagining a "virtual CAN". The specification of this CAN consists of describing the network connections between virtual nodes, the mapping of data items to virtual nodes, and some additional auxiliary information. In Section 2.1, we describe the virtual CAN. In Section 2.2, we go on to describe how the virtual CAN is implemented by the peers.

2.1   The Virtual CAN

The virtual CAN, consisting of n virtual nodes, is closely based on the [1] scheme. We make use of a butterfly network of depth $\log n - \log \log n$, we call the nodes of the butterfly network *supernodes* (see Figure 1). Every supernode is associated with a set of virtual nodes. We call a supernode at the topmost level of the butterfly a top supernode, one at the bottommost level of the network a

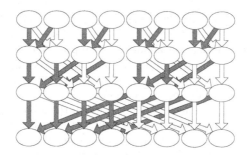

Fig. 1. The butterfly network of supernodes.

bottom supernode and one at neither the topmost or bottommost level a middle supernode.

We use a set of hash functions for mapping virtual nodes to supernodes of the butterfly and for mapping data items to supernodes of the butterfly. We assume these hash functions are approximately random. [2]

The virtual network is constructed as follows:

- We choose an error parameter $\epsilon > 0$, and as a function of ϵ we determine constants C, D, α and β. (See [1] for detailed information on how this is done).
- Every virtual node v is hashed to C random top supernodes (we denote by $T(v)$ the set of C top supernodes v hashes to), C random bottom supernodes (denoted $B(v)$) and $C \log n$ random middle supernodes (denoted $M(v)$) to which the virtual node will belong.
- All the virtual nodes associated with any given supernode are connected in a clique. (We do this only if the set of virtual nodes in the supernode is of size at least $\alpha C \ln n$ and no more than $\beta C \ln n$.)
- Between two sets of virtual nodes associated with two supernodes connected in the butterfly network, we have a complete bipartite graph. (We do this only if both sets of virtual nodes are of size at least $\alpha C \ln n$ and no more than $\beta C \ln n$.)
- We map the n data items to the $n/\log n$ bottom supernodes in the butterfly: each data item, say d, is hashed to D random bottom supernodes; we denote by $S(d)$ the set of bottom supernodes that data item d is mapped to. (Typically, we would not hash the entire data item but only it's title, e.g., "Singing in the Rain").
- The data item d is then stored in all the component virtual nodes of $S(d)$ (if any bottom supernode has more than $\beta B \ln n$ data items hashed to it, it drops out of the network.)

---

[2] We use the random oracle model ([6]) for these hash function, it would have sufficed to have a weaker assumption such as that the hash functions are expansive.

- Finally, we map the meta-data associated with each of the n virtual nodes in the network to the $n/\log n$ bottom supernodes in the butterfly. For each virtual node v, information about v is mapped to D bottom supernodes. We denote by $I(v)$ the set of bottom supernodes storing information about virtual node v. (if any bottom supernode has more than $\beta B \ln n$ virtual nodes hashed to it, it drops out of the network.)
- For each virtual node v in the network, we do the following:
    1. We store the id of v on all component virtual nodes of $I(v)$.
    2. A complete bipartite graph is maintained between the virtual nodes associated with supernodes $I(v)$ and the virtual nodes in supernodes $T(v)$, $M(v)$ and $B(v)$.

2.2   Implementation of Virtual CAN by Peers

Each peer that is currently live will map to exactly one node in the virtual network and each node in the virtual network will be associated with at most one live peer. At all times we will maintain the following two invariants:

1. If peers $p1$ and $p2$ map to virtual nodes x and y and x links to y in the virtual network, then $p1$ links to $p2$ in the physical overlay network.
2. If peer p maps to virtual node x, then p stores the same data items that x stores in the virtual network.

Recall that each virtual node in the network participates in C top, $C \log n$ middle and C bottom supernodes. When a virtual node v participates in a supernode s in this way, we say that v is a *member* of s. For a supernode s, we define $V(s)$ to be the set of virtual nodes which are members of s. Further we define $P(s)$ to be the set of live peers which map to virtual nodes in $V(s)$.

2.3   Search for a Data Item

We will now describe the protocol for searching for a data item from some peer p in the network. We will let v be the virtual node p maps to and let d be the desired data item.

1. Let $b_1, b_2, \ldots, b_D$ be the bottom supernodes in the set $S(d)$.
2. Let $t_1, t_2, \ldots, t_C$ be the top supernodes in the set $T(v)$.
3. Repeat in parallel for all values of k between 1 and C:
    (a) Let $\ell = 1$.
    (b) Repeat until successful or until $\ell > B$:
        i. Let $s_1, s_2, \ldots s_m$ be the supernodes in the path in the butterfly network from t_k to the bottom supernode b_ℓ.
            - Transmit the query to all peers in the set $P(s_1)$.
            - For all values of j from 2 to m do:
                - The peers in $P(s_{j-1})$ transmit the query to all the peers in $P(s_j)$.

  – When peers in the bottom supernode are reached, fetch the content from whatever peer has been reached.
  – The content, if found, is transmitted back along the same path as the query was transmitted downwards.
  ii. Increment ℓ.

2.4   Content and Peer Insertion

An algorithm for inserting new content into the network is presented in [1]. In this section, we describe the new algorithm for peer insertion. We assume that the new peer knows one other random live peer in the network. We call the new peer p and the random, known peer p'.

1. p first chooses a random bottom supernode, which we will call b. p then searches for b in the manner specified in the previous section. The search starts from the top supernodes in $T(p')$ and ends when we reach the node b(or fail).
2. If b is successfully found, we let W be the set of all virtual nodes, v , such that meta-data for v is stored on the peers in $P(b)$. We let W' be the set of all virtual nodes in W which are not currently mapped to some live peer.
3. If b can not be found, or if W' is empty, p does not map to any virtual node. Instead it just performs any desired searches for data items from the top supernodes, $T(p')$.
4. If there is some virtual node v in W', p takes over the role of v as follows:
   (a) Let $S = T(v) \cup M(v) \cup B(v)$. Let F be the set of all supernodes, s in S such that $P(s)$ is not empty. Let $E = S - F$.
   (b) For each supernode s in F:
       i. Let R be the set of supernodes that neighbor s in the butterfly.
       ii. p copies the links to all peers in $P(r)$ for each supernode r in R. These links can all be copied at once from one of the peers in $P(s)$. Note that each peer in $P(b)$ contains a pointer to some peer in $P(s)$.
       iii. p notifies all peers to which it will be linking to also link to it. For each supernode r in R, p sends a message to one peer in $P(r)$ notifying it of p's arrival. The peer receiving the message then relays the message to all peers in $P(r)$. These peers then all point to p.
       iv. If s is a bottom supernode, p copies all the data items that map to s. It copies these data items from some peer in $P(s)$.
   (c) If E is non-empty, we will do one broadcast to all peers that are reachable from p. We will first broadcast from the peers in all top supernodes in $T(p)$ to the peers in all reachable bottom supernodes. We will then broadcast from the peers in these bottom supernodes back up the butterfly network to the peers in all reachable top supernodes. [3]:

---

[3] This broadcast takes $O(\log n)$ time but requires a large number of messages. However, we anticipate that this type of broadcast will occur infrequently. In particular, under the assumption of random failures, this broadcast will never occur with high probability.

i. p broadcasts the id of v along with the ids of all the supernodes in E. All peers that receive this message, which are in supernodes neighboring some supernode in E will connect to p.

ii. In addition to forging these links, we seek to retrieve data items for each bottom supernode which is in the set E. Hence, we also broadcast the ids for these data items. We can retrieve these data items if they are still stored on other peers.[4]

3   Conclusion

In this paper, we have introduced the notion of a dynamically strong fault-tolerance and have described a content addressable network that has this property. Future directions include reducing the number of messages sent for search and node insertion and reducing the number of pointers stored at each peer.

References

[1] Amos Fiat and Jared Saia. Censorship Resistant Peer-to-Peer Content Addressable Networks. In *Symposium on Discrete Algorithms*, 2002.

[2] B.Y. Zhao, K.D. Kubiatowicz and A.D. Joseph. Tapestry: An Infrastructure for Fault-Resilient Wide-Area Location and Routing. Technical Report UCB//CSD-01-1141, University of California at Berkeley Technical Report, April 2001.

[3] David Moore, Geoffrey Voelker and Stefan Savage. Inferring internet denial-of-service activity. In *Proceedings of the 2001 USENIX Security Symposium*, 2001.

[4] Ion Stoica, Robert Morris, David Karger, Frans Kaashoek and Hari Balakrishnan. Chord: A Scalable Peer-to-peer Lookup Service for Internet Applications. In *Proceedings of the ACM SIGCOMM 2001 Technical Conference*, San Diego, CA, USA, August 2001.

[5] Johan Hastad, Thomson Leighton and Mark Newman. Fast computation using faulty hypercubes. In *Proceedings of the 21st Annual ACM Symposium on Theory of Computing*, 1989.

[6] Mihir Bellare and Phillip Rogaway. Random oracles are practical: a paradigm for designing efficient protocols. In *The First ACM Conference on Computer and Communications Security*, pages 62–73, 1993.

[7] Noga Alon, Haim Kaplan, Michael Krivelevich, Dahlia Malkhi and Julien Stern. Scalable secure storage when half the system is faulty. In *Proceedings of the 27th International Colloquium on Automata, Languages and Programming*, 2000.

[8] Stefan Saroiu, P. Krishna Gummadi and Steven D. Gribble. A Measurement Study of Peer-to-Peer File Sharing Systems. In *Proceedings of Multimedia Computing and Networking*, 2002.

[9] Sylvia Ratnasamy, Paul Francis, Mark Handley, Richard Karp and Scott Shenker. A Scalable Content-Addressable Network. In *Proceedings of the ACM SIGCOMM 2001 Technical Conference*, San Diego, CA, USA, August 2001.

---

[4] We note that, using the scheme in [7], we can retrieve the desired data items, even in the case where we are connected to no more than $n/2$ live peers. To use this scheme, we need to store, for each data item of size s, some extra data of size $O(s/n)$ on each node in the network. Details on how to do this are ommitted.

A   Appendix

In this appendix, we provide proofs for statements made in the paper.

A.1   Dynamic Fault-Tolerance

We will be using the following two lemmas which follow from results in [1]. We first define a peer as ϵ-*good* if it is connected to all but $1 - \epsilon$ of the bottom supernodes.

Lemma 1. *Assume at any time, at least κn of the virtual nodes map to live peers for some $\kappa < 1$. Then for any ϵ, we can choose appropriate constants C and D for the virtual network such that at all times, all but an ϵ fraction of the top supernodes are connected to all but an ϵ fraction of the bottom nodes.*

Proof. This lemma follows directly from Theorem 4.1 in [1] by plugging in appropriate values.

Lemma 2. *Assume at any time, at least κn of the virtual nodes map to live peers for some $\kappa < 1$. Then for any $\epsilon < 1/2$, we can choose appropriate constants C and D for the virtual network such that at all times, all ϵ-good nodes are connected in one component with diameter $O(\log n)$.*

Proof. By Lemma 1, we can choose C and D such that all ϵ-good peers can reach more than a $1/2$ fraction of the bottom supernodes. Then for any two ϵ-good peers, there must be some bottom supernode such that both peers are connected to that same supernode. Hence, any two ϵ-good peers must be connected. In addition, the path between these two ϵ-good peers must be of length $O(\log n)$ since the path to any bottom supernode is of length $O(\log n)$

We now give the proof of Theorem 1 which is restated here.
Theorem 1: For all $\epsilon > 0$ and value P which is polynomial in n, there exist constants $k_1(\epsilon)$, $k_2(\epsilon)$ and $k_3(\epsilon)$ and $k_4(\epsilon)$ such that the following holds with high probability for the CAN for deletion of up to P peers by the limited adversary:

- *At any time, the CAN is ϵ-robust*
- *Search takes time no more than $k_1(\epsilon) \log n$.*
- *Peer insertion takes time no more than $k_2(\epsilon) \log n$.*
- *Search requires no more than $k_3(\epsilon) \log^3 n$ messages total.*
- *Every node stores no more than $k_4(\epsilon) \log^3 n$ pointers to other nodes and $k_3(\epsilon) \log n$ data items.*

Proof. We briefly sketch the argument that our CAN is dynamically fault-tolerant. The proofs for the time and space bounds are given in the next two subsections.

For concreteness, we will prove dynamic fault-tolerance with the assumption that $2n/10$ peers are added whenever $(1/10 - \epsilon)n$ peers are deleted by the adversary. The argument for the general case is similar. Consider the state of the system when exactly $2n/10$ virtual nodes map to no live peers. We will focus on what happens for the time period during which the adversary kills off $(1/10-\epsilon)n$ more peers. By assumption, during this time, $2n/10$ new peers join the network. In this proof sketch, we will show that with high probability, the number of virtual nodes which are not live at the end of this period is no more than $2n/10$. The general theorem follows directly.

We know that Lemma 1 applies during the time period under consideration since there are always at least $n/2$ live virtual nodes. Let R be the set of virtual nodes that at some point during this time period are not ϵ-good. By Lemma 2, peers in virtual nodes that are not in the set R have been connected in the large component of ϵ-good nodes throughout the considered time interval. Thus these peers have received information broadcasted during successful peer insertions. However, the peers mapping to virtual nodes in R may at some point have not been connected to all the other ϵ-good nodes and so may not have have received information broadcasted by inserted peers. We note that $|R|$ is no more than ϵn by Lemma 1 (since even with no insertions in the network, no more than ϵn virtual nodes would be not be ϵ-good at any point in the time period under consideration). Hence we will just assume that those peers with stale information, i.e. the peers in R, are dead. To do this, we will assume that the number of adversarial node deletions is $n/10$. (We further note that all peers which are not ϵ-good will actually be considered dead by all peers which are ϵ-good. This is true since no bottom supernode reachable from an ϵ-good node will have a link to a peer which is not ϵ-good. Hence, such a virtual node will be fair game for a new peer to map to.)

We claim that during the time interval, at least $n/10$ of the inserted peers will map to virtual nodes. Assume not. Then there is some subset, S, of the $2n/10$ peers that were inserted such that $|S| = n/10$ and all peers in S did not reach any bottom supernodes with information on virtual nodes that had no live peers. Let S' be the set of peers in S that both 1) had an initial connection to an ϵ-good peer and 2) reached the bottom supernode which they searched for after connecting. We note that with high probability, $|S'| = \theta(n)$ since each new peer connects to a random peer (of which most are ϵ-good) and since most bottom supernodes are reachable from an ϵ-good peer.

Now let B' be the set of bottom supernodes that are visited by peers in S'. With high probability $|B'| = \theta(n/\log n)$. Finally let V' be the set of virtual nodes that supernodes in B' have information on. For D (the constant defined in the virtual network section) chosen sufficiently large, $|V'|$ must be greater than $9n/10$ (by expansion properties between the bottom supernodes and the virtual nodes they have information on). But by assumption, there must be some subset V of virtual node ids which are empty after the insertions where $|V| \geq n/10$. But this is a contradiction since we know that the set of virtual nodes that the new peers in S' tried to map to was of size greater than $9n/10$

Hence during the time that $n/10$ peers were deleted from the network, at least $n/10$ virtual nodes were newly mapped to live peers. This implies that the number of virtual peers not mapped to live nodes can only have decreased. Thus the number of virtual peers not mapped to live nodes will not increase above $2n/10$ after any interval with high probability.

Time That the algorithm for searching for data items takes $O(\log n)$ time and $O(\log^2 n)$ messages is proven in [1].

The common and fast case for peer insertion is when all supernodes to which the new peer's virtual node belongs already have some peer in them. In this case, we spend constant time processing each one of these supernodes so the total time spent is $O(\log n)$.

In the degenerate case where there are supernodes which have no live nodes in them, a broadcast to all nodes in the network is required. Insertion time will still be $O(\log n)$ since the connected component of ϵ-good nodes has diameter $O(\log n)$. However we will need to send $O(n)$ messages for the insertion. Unfortunately, the adversary can force this degenerate case to occur for a small (less than ϵ) fraction of the node insertions. However if the node deletions are random instead of adversarial, this case will never occur in the interval in which some polynomial number of nodes are deleted.

Space Each node participates in C top supernodes. The number of links that need to be stored to play a role in a particular top supernode is $O(\log n)$. This includes links to other nodes in the supernode and links to the nodes that point to the given top supernode.

Each node participates in $C \log n$ middle supernodes. To play a role in a particular middle supernode takes $O(\log n)$ links to point to all the other nodes in the supernode and $O(\log n)$ links to point to nodes in all the neighboring supernodes. In addition, each middle supernode has $O(\log n)$ roles associated with it and each of these roles is stored in D bottom supernodes. Hence each node in the supernode needs $O(log^2 n)$ links back to all the nodes in the bottom supernodes which store roles associated with this middle supernode.

Each node participates in C bottom supernodes. To play a role in a bottom supernode requires storing $O(\log n)$ data items. It also requires storing $O(\log n)$ links to other nodes in the supernode along with nodes in neighboring supernodes. In addition, it requires storing $O(\log n)$ links for each of the $O(\log n)$ supernodes for each of the $O(\log n)$ roles that are stored at the node. Hence the total number of links required is $O(\log^3 n)$.

Scalable Management and Data Mining Using Astrolabe*

Robbert van Renesse, Kenneth Birman, Dan Dumitriu, and Werner Vogels

Department of Computer Science
Cornell University, Ithaca, NY 14853
{rvr,ken,dumitriu,vogels@cs.cornell.edu}

Abstract. Astrolabe is a new kind of peer-to-peer system implementing a hierarchical distributed database abstraction. Although deigned for scalable management and data mining, the system can also support wide-area multicast and offers powerful aggregation mechanisms that permit applications to build customized virtual databases by extracting and summarizing data located throughout a large network. In contrast to other peer-to-peer systems, the Astrolabe hierarchy is purely an abstraction constructed by running our protocol on the participating hosts – there are no servers, and the system doesn't superimpose a specialized routing infrastructure or employ a DHT. This paper focuses on wide-area implementation challenges.

1   Introduction

To a growing degree, applications are expected to be self-configuring and self-managing, and as the range of permissible configurations grows, this is becoming an enormously complex undertaking. Indeed, the management subsystem for a distributed system is often more complex than the application itself. Yet the technology options for building management mechanisms have lagged. Current solutions, such as cluster management systems, directory services, and event notification services, either do not scale adequately or are designed for relatively static settings.

In this paper, we describe a new information management service called Astrolabe. Astrolabe monitors the dynamically changing state of a collection of distributed resources, reporting summaries of this information to its users. Like DNS, Astrolabe organizes the resources into a hierarchy of domains, and associates attributes with each domain. Unlike DNS, no servers are associated with domains, the attributes may be highly dynamic, and updates propagate quickly; typically, in tens of seconds.

Astrolabe continuously computes summaries of the data in the system using on-the-fly aggregation. The aggregation mechanism is controlled by SQL queries, and can be understood as a type of data mining capability. For example, Astrolabe aggregation can be used to monitor the status of a set of servers scattered within the network, to locate a desired resource on the basis of its attribute values, or to compute a summary description

* This research was funded in part by DARPA/AFRL-IFGA grant F30602-99-1-0532, in part by a grant under NASA's REE program administered by JPL, in part by NSF-CISE grant 9703470, and in part by the AFRL/Cornell Information Assurance Institute.

P. Druschel, F. Kaashoek, and A. Rowstron (Eds.): IPTPS 2002, LNCS 2429, pp. 280–294, 2002.
© Springer-Verlag Berlin Heidelberg 2002

of loads on critical network components. As this information changes, Astrolabe will automatically and rapidly recompute the associated aggregates and report the changes to applications that have registered their interest.

The Astrolabe system looks to a user much like a database, although it is a virtual database that does not reside on a centralized server. This database presentation extends to several aspects. Most importantly, each domain can be viewed as a relational table containing the attributes of its child domains, which in turn can be queried using SQL. Also, using database integration mechanisms like ODBC and JDBC standard database programming tools can access and manipulate the data available through Astrolabe.

The design of Astrolabe reflects four principles:

1. *Scalability through hierarchy:* Astrolabe achieves scalability through its domain hierarchy. Given bounds on the size and amount of information in a domain, the computational, storage and communication costs of Astrolabe are also bounded.
2. *Flexibility through mobile code:* A restricted form of mobile code, in the form of SQL aggregation queries, allows users to customize Astrolabe on the fly.
3. *Robustness through a randomized peer-to-peer protocol:* Systems based on centralized servers are vulnerable to failures, attacks, and mismanagement. Astrolabe agents run on each host, communicating through an epidemic protocol that is highly tolerant of failures, easy to deploy, and efficient.
4. *Security through certificates:* Astrolabe uses digital signatures to identify and reject potentially corrupted data and to control access to potentially costly operations.

This paper is organized as follows. Section 2 describes the Astrolabe system itself, while its use is illustrated in Section 3. To avoid overlap with work published elsewhere, we focus on wide-area communication challenges here. Accordingly, the basic low-level communication and addressing mechanisms used by Astrolabe are the subject of Section 4. Section 5 describes Astrolabe's self-configuration strategy. How Astrolabe operates in the presence of firewalls is the topic of Section 6. In Section 7 we describe various related work in the peer-to-peer area. Finally, Section 8 concludes.

2 Astrolabe

The goal of Astrolabe is to maintain a dynamically updated data structure reflecting the status and other information contributed by hosts in a potentially large system. For reasons of scalability, the hosts are organized into a domain hierarchy, in which each participating machine is a leaf domains (see Figure 1). The leafs may be visualized as tuples in a database: they correspond to a single host and have a set of attributes, which can be base values (integers, floating point numbers, etc) or an XML object. In contrast to these leaf attributes, which are directly updated by hosts, the attributes of an internal domain (those corresponding to "higher levels" in the hierarchy) are generated by *aggregating* (summarizing) attributes of its child domains.

The implementation of Astrolabe is entirely peer-to-peer: the system has no servers, nor are any of its agents configured to be responsible for any particular domain. Instead, each host runs an agent process that communicates with other agents through an epidemic protocol or *gossip* [DGH+87]. The data structures and protocols are designed such that the service scales well:

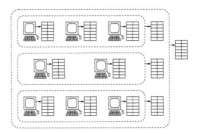

Fig. 1. An example of a three-level Astrolabe tree. The top-level *root domain* has three child domains. Each domain, including the leaf domains (the hosts), has an attribute list. Each host runs a Astrolabe agent.

– The memory used by each host grows logarithmically with the membership size;
– The size of gossip messages grows logarithmically with the size of the membership;
– If configured well (more on this in Section 5), the gossip load on network links grows logarithmically with the size of the membership, and is independent of the update rate;
– The *latency* grows logarithmically with the size of the membership. Latency is defined as the time it takes to take a snapshot of the entire membership and aggregate all its attributes;
– Astrolabe is tolerant of severe message loss and host failures, and deals with network partitioning and recovery.

In practice, even if the gossip load is low (Astrolabe agents are typically configured to gossip gossip only once every few seconds), updates propagate very quickly, and is typically within a minute even for very large deployments [vRB02].

In the description of the implementation of Astrolabe below, we omit all details except those that are necessary in order to understand the function of Astrolabe, and the issues that relate to peer-to-peer communication in the Internet. A much more detailed description and evaluation of Astrolabe appears in [vRB02].

As there is exactly one Astrolabe agent for each leaf domain, we name Astrolabe agents by their corresponding domain names. Each Astrolabe agent maintains, for each domain that it is a member of, a relational table called the *domain table*. For example, the agent "/nl/amsterdam/vu" has domain tables for "/", "/nl", and "/nl/amsterdam" (see Figure 2). A domain table of a domain contains a row for each of its child domains, and a column for each attribute name. One of the rows in the table is the agent's *own* row, which corresponds to that child domain that the agent is a member of as well. Using a SQL aggregation function, a domain table may be aggregated by computing some form of summary of the contents to form a single row. The rows from the children of a domain are concatenated to form that domain's table in the agent, and this repeats to the root of the Astrolabe hierarchy.

Since multiple agents may be in the same domain, the corresponding domain table is replicated on all these agents. For example, both the agents "/nl/amsterdam/vu" and

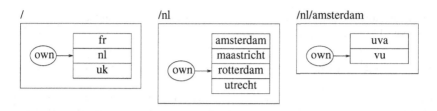

Fig. 2. A simplified representation of the data structure maintained by the agent corresponding to /nl/amsterdam/vu.

"/nl/utrecht/uu" maintain the "/" and "/nl" tables. A replicated table is kept approximately consistent using an epidemic protocol. Each agent in a domain calculates, using an aggregation function, a small set of representative agents for its domain. Typically, Astrolabe is configured to use up to three representative agents for each domain. The representative agents of the child domains of a parent domain run the epidemic protocol for the parent's domain table. On a regular basis, say once a second, each agent X that is a representative for some child domain chooses another child domain at random, and then a representative agent Y within the chosen child domain, also at random. X sends the parent's table to Y. Y merges this table with its own table, and sends the result back to X, so that X and Y now agree on the contents of their tables.

The rule by which such tables are merged is central to the behavior of the system as a whole. The basic idea is as follows. Y adopts into the merged table rows from X for child domains that were not in Y's original table, as well as rows for child domains that are more current than in Y's original table. To determine currency, agents timestamp rows each time they are updated by writing (in case of leaf domains) or by generation (in case of internal domains). Unfortunately, this requires all clocks to be synchronized which is, at least in today's Internet, far from being the case.

To solve this problem, each row is tagged with the *generator*: the domain name of the agent that wrote or generated the row (in addition to the timestamp). Agents also maintain, for each row in each table, the set of generators from which they received updates for that row, along with the timestamp on the last received update. The merge operation is now executed as follows. When receiving a row in a gossip message, the agent adopts it, as is, if it is created by a previously unknown generator. If it is a known generator, the row is adopted if and only if the row's timestamp is more recent than the last received timestamp from that generator. This way, only timestamps from the same agent are compared with one another, and thus no clock synchronization is necessary.

When no update has been received from a particular generator for some time period T, that generator is considered faulty, and eventually forgotten (after time $2T$ for reasons that go beyond the scope of this paper, but which are described in [vRMH98]). T should be chosen so that the probability of any *old* gossips from this generator still going around is very low [vRB02, vRMH98]. Since gossips disseminate in time $O(\log n)$, this typically is not very long, and can be determined by techniques of epidemic analysis or simulation. If there are no more known generators for a domain, the domain itself is removed from the agent's domain table.

3   Using Astrolabe

Applications invoke Astrolabe interfaces through calls to a library (see Table 1). The library allows applications to peruse all the information in the Astrolabe tree, setting up new connections as necessary. The creation and termination of connections is transparent to application processes, so the programmer can think of Astrolabe as a ubiquitous service, somewhat analogous to the DNS.

Table 1. Application Programmer Interface.

Method	Description
find_contacts(time, scope)	search for Astrolabe agents in the given *time* and *scope*
set_contacts(addresses)	specify addresses of initial agents to connect to
get_attributes(domain, event_queue)	report updates to attributes of *domain*
get_children(domain, event_queue)	report updates to domain membership
set_attribute(domain, attribute, value)	update the given attribute

In addition to its native interface, the library has an SQL interface that allows applications to view each node in the domain tree as a relational database table, with a row for each child domain and a column for each attribute. The programmer can then simply invoke SQL operators to retrieve data from the tables. Using selection, join, and union operations, the programmer can create new views of the Astrolabe data that are independent of the physical hierarchy of the Astrolabe tree. An ODBC driver is available for this SQL interface, so that many existing database tools can use Astrolabe directly, and many databases can import data from Astrolabe.

New aggregation functions can be installed dynamically, and their dissemination piggybacks on the gossip protocol. This way an Astrolabe hierarchy can be customized for the applications that use it. The code of these functions is embedded in so-called *aggregation function certificates* (AFCs), which are signed certificates installed as domain attributes.

We also use AFCs for purposes other than aggregation. An *Information Request AFC* specifies what information the application wants to retrieve at each participating host, *and* how to aggregate this information in the domain hierarchy. (both are specified using SQL queries). A *Configuration AFC* specifies run-time parameters that applications may use for dynamic on-line configuration.

Example: Peer-to-Peer Multicast

In [vRB02] we present a number of possible uses for Astrolabe, such as for locating "nearby" resources by using ODBC to query the local domain, monitoring the dynamically evolving state of a subset of hosts in a large network using an aggregation function, or tracking down desired resources in very large settings. To avoid repeating that material here, we now present a different example of how Mariner might be used. Many distributed games and other applications require a form of multicast that scales well, is fairly reliable, and does not put a TCP-unfriendly load on the Internet. In the face of slow participants, the multicast protocol's flow control mechanism should not force the entire system to grind to a halt. This section describes such a multicast facility. It uses Astrolabe to track the set of multicast recipients, but then but sets up a separate tree of TCP connections for actually transporting multicast messages.

Each multicast group has a name, say "game". A participant expresses its interest in receiving messages for this group by installing its TCP/IP address (or addresses) in the attribute "game" of its leaf domain's MIB. This attribute is aggregated using the query

```
SELECT FIRST(3, game) AS game
```

That is, each domain selects three of its participants' TCP/IP addresses. (FIRST is an often-used Astrolabe extension to SQL.)

Participants exchange messages of the form (domain, data). A participant that wants to initiate a multicast lists the child domains of the root domain, and, for each child that has a non-empty "game" attribute, sends the message (child-domain, data) to a selected participant for that child domain (more on this selection later). Each time a participant receives a message (domain, data), it finds the child domains of the given domain that have non-empty "game" attributes and recursively continues the dissemination process.

The TCP connections that are created are cached. This effectively constructs a tree of TCP connections that spans the set of participants. This tree is automatically updated as Astrolabe reports domain membership updates.

To make sure that the dissemination latency does not suffer from slow participants in the tree, some measures must be taken. First, a participant could post (in Astrolabe) the rate of messages that it is able to process. The aggregation query can then be updated as follows to select only the highest performing participants for "internal routers."

```
SELECT FIRST(3, game) AS game ORDER BY rate
```

Senders can also monitor their outgoing TCP pipes. If one fills up, they may want to try another participant for the corresponding domain. It is even possible to use more than one participant to construct a "fat tree" for dissemination, but then care should be taken to reconstruct the order of messages. These mechanisms together effectively route messages around slow parts of the Internet, much like Resilient Overlay Networks [ABKM01] accomplishes for point-to-point traffic. Notice that this makes our multicast "TCP-friendly", in the sense that if a router becomes overloaded and starts dropping messages, the multicast protocol will reduce the load imposed on that router. This property is rarely seen in Internet multicast protocols.

Our solution can also be used to implement the Publish/Subscribe paradigm. In these systems [OPSS93], receivers subscribe to topics of interest, and publishers post messages to topics. The multicast protocol described above can easily implement Publish/Subscribe, as well as a generalized concept that we call *selective multicast* or *selective Publish/Subscribe*.

The idea is to tag messages with a SQL condition, chosen by the publishers. For example, a publisher that wants to send an update to all hosts that have a version of some object that is less than 3.1, this condition could be "MIN(version) < 3.1". The participants in the multicast protocol above use this condition to decide to which other participants should receive the message. In the example used above, when receiving a message (domain, data, condition), a participant executes the following SQL query to find out which participants to forward the message to:

```
SELECT game
FROM domain
WHERE condition
```

In this idea, the publisher specifies the set of receivers. Basic Publish/Subscribe can then be expressed as the publisher specifying that a message should be delivered to all subscribers to a particular topic.

The simplest way to accomplish this type of routing is to create a new attribute by the name of the topic. Subscribers set this attribute to 1, and the attribute is aggregated by taking the sum. The condition is then "attribute > 0". However, such an approach would scale poorly if a system has large number of possible topics, since it requires one bit each.

A solution that scales much better is to use a Bloom filter [Blo70].[1] This solution uses a single attribute that contains a fixed-size bit map. The attribute is aggregated using bitwise OR. Topic names are hashed to a bit in this bit map. The condition tagged to the message is "BITSET(HASH(topic))". In the case of hash collisions, a message may reach some non-subscribing destinations, but would be filtered and ignored at the last hop.

4   Low-Level Communication

The preceding example showed how a multicast protocol could be layered over Astrolabe, running on TCP channels. However, for its own communication, Astrolabe uses UDP/IP, HTTP (on top of TCP/IP or SSL), or both. To support HTTP, Astrolabe agents act both as HTTP servers and clients. To enable communication through firewalls, Astrolabe makes use of Application Level Gateways, but for scalability concerns this is only done as a last resort (see Section 6). This section discusses some of the challenges that arose in running Astrolabe in wide-area settings. Before we discuss the actual communication in more detail, we will first describe the concept of *realms*, and how addressing is done in Astrolabe.

A *realm* is a set of hosts and a communication protocol. For example, the tuple ("Cornell Computer Science Department", UDP) forms a realm, as does ("Core Internet", HTTP). ("Core Internet" is the set of those hosts on the main Internet that do not reside behind firewalls.) Each realm has a unique identifier of the form name:protocol,

---

[1] This idea is also used in various other distributed systems, including the directory service of the Ninja system [GWVB+01].

for example "cornellcs:udp" and "internet:http". The hosts in a realm form an equivalence class, in that they can all be accessed using the realm's protocol in the same way. A host can be in more than one realm, and can have more than one address in the same realm. No two hosts in the same realm can have the same address, but the same address may be used in different realms.

UDP addresses are of the form "IP-address:port" (e.g., "10.0.0.4:6422") or "DNS-name:port" (e.g., "rome.cs.cornell.edu:6422"). HTTP addresses are of the form "agent-name@TCP-address", where "agent-name" is the Astrolabe domain name of the agent, and "TCP-address," as in UDP addresses, consists of a port and either an IP address or a DNS name. For example, "/usa/ny/ithaca/cornell/cs/rome@10.0.0.4:2246".

We define an *extended address* to be the triple (realm identifier, address, preference). For example, ("cornellcs:udp", 10.0.0.4:6422, 5). A host has a set of these addresses, and can indicate its preference for certain addresses using the *preference* field. We call the set of extended addresses of a host the *contact* for that host. The contact is dynamic as addresses may appear and disappear over time as the administrator of the host connects to, or disconnects from, ISPs and VPNs.

Unlike agent's contacts, agent's names are constant. In order to simplify configuration, we observed that realms often coincide with Astrolabe domains, and thus named realms using their corresponding domain name. Thus, rather than "cornellcs:udp", we would use "/usa/ny/Ithaca/cornell/cs:udp". The core Internet coincides with the root domain, and is thus called "/:http".

Each domain in Astrolabe has an attribute called *contacts*, which contains the contacts of those agents in the domain that have been elected as representatives. (This uses the FIRST function that was described in Section 3.) The *contacts* attribute of a leaf domain contains the singleton set with the contact of the agent of that leaf domain.

We will now briefly revisit Astrolabe's gossip protocol to show how this works in practice. When an agent wants to gossip the table of some domain, it has to come up with an address. First, the agent picks one of the table's rows at random and retrieves the *contacts* attribute from that row. The agent then picks one of the contacts at random. The resulting contact is a set of extended addresses. The agent removes the addresses of realms that it cannot reach (more on this below). If there is more than one remaining address, the agent has to make one more choice.

In order to make intelligent choices, each Astrolabe agent maintains statistics about addresses. This is simple to do, as each gossip message is followed by a response. Currently, an agent maintains for each extended address the following three values:

1. outstanding: the number of gossips sent since the last response was received;
2. last_sent: time of last send of a gossip;
3. last_received: time of last reception of a response.

If there is more than one extended address to choose from, the agent *scores* each address:

1. If there is no outstanding gossip, the score is the preference;
2. If it has been more than a minute since the last gossip was sent, the score is the preference;
3. If there is just one outstanding gossip, the score is the preference minus one;
4. In all other cases, the score is zero.

This results in the following behavior. In the normal case, when gossips are followed by responses, the address of the highest preference is always used. If a single response got lost, the score becomes only slightly smaller. The intention is that if there is more than one address of the same preference, the ones that are only somewhat flaky become less preferential. If there are more losses, the score becomes such that the address is only used as a last resort. Once a minute, the score is, for a single send operation, reset to the original preference. This allows addresses to be occasionally re-tested.

5   Configuration

In order for Astrolabe to scale well, the domain hierarchy has to be set up with care. Each domain in Astrolabe runs an instance of the gossip protocol among the representatives of its child domains. The first concern to think about is the size of domains, that is, the number of child domains in a domain. If very large, the size of gossip messages, as well as the generated gossip load, will be large as well (they both grow linearly with the domain size). If chosen to be very small, the hierarchy becomes very deep, and latency will suffer. In practice, we find that a size of 25-100 child domains in a domain works well. Smaller sizes are possible too, at the cost of some additional latency, but larger sizes make the load unacceptably large.

Locality is a second important consideration. Domains should be constructed (if possible) so that the number of network hops between its child domains' representatives is minimized, and so that independent domains (one domain is not an ancestor of the other) do not share any network links. If the Internet were a tree topology, the Astrolabe hierarchy should be preferably identical to this tree. In reality the edges of the Internet often resemble a tree topology, but the internal Internet is a complicated mesh of links that defies any resemblance to a tree.

In practice, this means that there is considerable freedom in designing the higher levels of the Astrolabe hierarchy, within the limits of the branching factor, but the lower levels should be mapped closely to the topology of the Internet edge. If we ignore the branching factor, this is not much different from the DNS hierarchy design. In DNS, too, the low levels often correspond closely to the network topology, while the high levels of the hierarchy have little correspondence to the actual network topology. Thus the main difference between DNS and Astrolabe configuration is the constrained branching factor of the Astrolabe hierarchy.

Astrolabe supports two forms of configuration: manual and automatic. The manual configuration supports various notions of security, including an integrated PKI infrastructure for the Astrolabe service. The automatic configuration is not secure. In order to foil all but the simplest forms of compromise, the communication is scrambled and signed using secret keys. Below, we will focus on the insecure automatic configuration. More on Astrolabe security can be found in [vRB02].

In an insecure version of Astrolabe, all an agent needs to know is

- Its domain name;
- The set of realms that it can send messages to;
- How to find peer agents to gossip with.

In the remainder of this section, we will look at the automatic configuration of domain names and realms.

Currently, we generate the Astrolabe domain name of an agent from the DNS domain name of the host, and the process identifier of the agent. We will first explain how this is done, and then provide the rational for this design. Say that the DNS domain name is $C_0.C_1.....C_k$, and the process id of the agent is P. We use a one-way hash function on $C_1...C_k$ (all but the first component of the domain name) to construct three 6-bit integers, A_1, A_2, and A_3. Finally, the Astrolabe domain name is constructed to be $/C_k/A_1/A_2/A_3/C_{k-1}/.../C_0/P$.

For example, say an agent runs as process 4365 on host "rome.cs.cornell.edu". By hashing "cs.cornell.edu" onto three 6-bit integers, we have effectively split the ".edu" domain up into 2^{18} pieces, as the ".edu" domain itself is much too large for a single Astrolabe domain. Using this construction, the Astrolabe "/edu" domain itself has at most 64 child domains. Say the three generated integers in our example are 25, 43, and 4 respectively. Then the domain name of the agent is "/edu/25/43/4/cornell/cs/rome/4365". (The three generated domains can be hidden from view if so desired.)

The hope is that each of the domains following "/edu/25/43/4" are of relatively limited size that can be supported by the Astrolabe protocol, and that these domains reflect the network topology to a close enough approximation. If in the future this turns out to be insufficient, we can update the downloadable executable to use more levels, or perhaps come up with an adaptive scheme. The addition of the process identifier makes it possible to run multiple agents on the same host.

Next we have to determine the set of realms that the agent can reach. We assume that any agent can communicate to the "/:http" realm, that is, any agent can use HTTP to reach another agent on the core Internet. (Agents may use WPAD to determine the existence and location of an HTTP proxy automatically.) Furthermore, we assume that $/C_k/A_1/A_2/A_3/C_{k-1}/.../C_1$:udp ("/edu/25/43/4/cornell/cs:udp" in our example) is a realm, and that all agents within this realm can communicate with one another.

Currently, these are all the assumptions we make about realms. In practice this is sometimes conservative, and in that case agents that can communicate directly using UDP will use an Application Level Gateway (ALG) instead. In the next section, we describe how ALGs are configured and used.

6   Communication through an ALG

An Application Level Gateway (ALG) may be the only possibility for two agents to communicate (see Figure 3). Significant care should be taken in deploying and configuring ALGs. The number of ALGs is likely to be small compared to the number of agents using them, and thus they should be used judiciously in order not to overload them or the network links that connect them. Also, care should be taken that the peer-to-peer network remains tolerant of failures, and does not get partitioned when a single ALG server crashes or otherwise becomes unavailable, and that network security is not compromised. Finally, in order for the system to scale, it should be possible to add new ALG servers dynamically to the system as the number of Astrolabe agents grows. These new servers should automatically be discovered and used by the existing agents as well as the new ones.

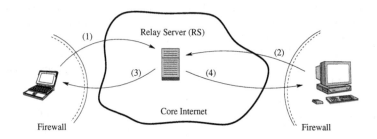

Fig. 3. Application Level Gateway. (1) Receiver sends a RECEIVE request using an HTTP POST request; (2) Sender sends the message using a SEND request using an HTTP POST request; (3) ALG forwards the message to the receiver using an HTTP 200 response; (4) ALG sends an empty HTTP 200 response back to the sender.

Ideally, an ALG is located on the network path between a sender and a receiver, so that the number of hops that a message has to travel is not severely affected by the presence of an ALG. Since many senders may send messages to the same receiver, it follows that the ALG should be located as close to the receiver as possible. Thus, ideally, each firewall or NAT box has a companion ALG that serves receivers behind the firewall. In practice, we suspect that far fewer ALGs will be deployed, but it is still important for receivers to connect to the nearest-by ALG or ALGs.

Each receiver that wishes to receive messages through an ALG has to use HTTP requests to the ALG. In practice, this happens over a persistent TCP connection. In order to reduce the number of such connections to an ALG, not every host behind a firewall has to connect to the ALG. In Astrolabe, only representatives for the realm corresponding to the firewalled site gossip beyond the firewall boundaries, and only these agents (typically, two or three), need to receive through an ALG. The other agents learn indirectly of updates outside the realm through gossip with the representatives (see Figure 4).

In order for agents to locate ALGs, the ALGs themselves are situated in the Astrolabe hierarchy itself. Each ALG has a companion Astrolabe agent with a configured domain name. The *relays* attribute of the corresponding leaf domain is set to the singleton set containing the TCP/IP address of the ALG. This attribute is aggregated into internal domains in the same way as the *contacts* attribute (*i.e.*, using the FIRST aggregation operator).

An agent determines whether it is a representative for a firewalled site by monitoring the *contacts* attribute of the corresponding realm domain and noticing whether its contact is in there. When this becomes the case, the agent finds ALGs by traveling up the Astrolabe hierarchy starting in its realm and finding the *relays* attributes, stopping when it has located k ALGs or when it reaches the root domain. To ensure fault

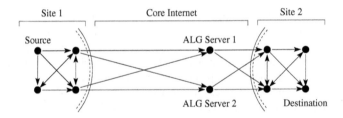

Fig. 4. The many ways gossip can travel from a source host in Site 1 to a destination host in Site 2. Each site has four hosts, two of which are representatives, behind a firewall. The representatives of Site 2 connect to two different ALG servers to receive messages from outside their firewall.

tolerance, k is typically chosen to be a small integer such as 2 (as in Figure 4). If no ALGs are found, agents resort to using a set of built-in addresses of ALG servers that we deployed for this purpose.

For each ALG in the set, the agent generates a new extended address of the form ("domain-name@ALG", "/:http", preference), and adds this address to its contact set. The preference is chosen to be relatively low compared to its other addresses, so as to discourage its use. Finally, the agent sends an HTTP request to the ALG to receive the first message on this address.

7   Related Work

The most popular peer-to-peer systems such as Chord [SMKK95], Pastry [RD01], and Tapestry [ZKJ01] implement distributed hash tables (DHT), then use these tables to locate desired objects. Just as a conventional hash table maps a key to a value, a DHT maps a key to a location in the network. The host associated with that location stores a copy of the value associated with the key. The hosts that implement the DHT maintain routing tables of $O(\log N)$ size that allow messages to be routed in $O(\log N)$ steps to one of these locations.

Astrolabe also implements a distributed data structure, although it is neither a DHT nor a distributed file system. Instead, Astrolabe most closely resembles a spreadsheet, particularly because updates of an attribute causes other attributes to change. However, the Astrolabe interface is more like that of a database. Athough Astrolabe uses a peer-to-peer protocol with scalability properties similar to those of DHTs, it differs in important ways from the best known DHT solutions.

In particular, Astrolabe reflects the physical organization of the hosts in its domain hierarchy, while DHT implementations hide the hosts and present a uniform key-value mapping. This difference is also apparent in the underlying protocols. Astrolabe exploits the domain organization so that most message exchanges are between nearby hosts, while the DHT implementations require special mechanisms to avoid doing large numbers of long distance exchanges. Pastry and Tapestry come closest to exploiting proximity in their message routing, but the protocols that maintain the routing tables themselves still require mostly long distance messages. Also, these protocols treat each network link as equal, and are expected to suffer from significant problems with slow modem links.

The designers of Tapestry are investigating a two-level architecture, called Brocade [JK02], that exploits the network topology. Basically, each site (*e.g.*, an organization or campus) deploys a number of so-called supernodes. The supernodes are organized in an ordinary Tapestry network. Each site then deploys another Tapestry network locally, including its supernodes. Messages are now routed in three superhops: first to a local supernode, then to the remote supernode, and lastly to its final destination. Simulation shows dramatic improvements in performance.

Astrolabe is perhaps most similar to a multi-level Brocade architecture. Astrolabe's aggregation facilities are used to elect, for each domain, the most appropriate supernodes. Election may be done based on quality of connectivity, resource capacity, host security, etc., and these policies may in fact be changed on the fly. Another difference between Astrolabe and other peer-to-peer protocols is therefore that Astrolabe exploits the available heterogeneity in the network, rather than hiding it.

Much of Astrolabe is concerned with the details of peer-to-peer communication in the actual Internet, an environment rife with Network Address Translation and other inconveniences. Well-known technologies in this field are Groove Networks (groove.net) and JXTA (jxta.org). Groove Networks provide a peer-to-peer communications technology for distributed collaboration. Although in theory peers in Groove can communicate directly with one another (assuming they are not separated by firewalls or NAT), they heavily rely on their proprietary ALG, called the *Groove Relay Server* [Ora01]. Unless peers are explicitly configured to communicate directly with one another, they will use the Relay Server for communication. They also use the Relay Server for other functions. This includes message queuing for off-line peers and resource discovery. This makes most Groove applications heavily dependent on the Relay Server, and direct peer-to-peer interactions are rarely used.

JXTA [Gon01] is an open platform for peer-to-peer interactions in the network, intended for pervasive use from servers to workstations to PDAs and cell phones. JXTA offers a variety of services such as Peer Discovery and Peer Membership. In order to allow peer discovery and peer-to-peer communication, the notion of an ALG (*Rendez-Vous Server* in JXTA terminology) has been proposed, but this is still an ongoing research effort.

8   Conclusions

By combining peer-to-peer and gossip protocols and using the resulting mechanism to implement a scalable database abstraction, Astrolabe plugs a gap in the existing Internet infrastructure. Today, far too many applications are forced to operate in the dark: they lack a good way to sense the status of component systems and of the network, and yet need to adapt their behavior on the basis of such status. More broadly, there is an important need for better scalable computing tools addressing the communications and configuration requirements of large-scale applications. Astrolabe offers such tools, packaged in an easily used database abstraction. Because Astrolabe itself is stable under stress and extremely scalable, it promotes the development of new kinds of applications sharing these properties.

Acknowledgements

We would like to thank the following people for various contributions to the Astrolabe design and this paper: Tim Clark, Al Demers, Terrin Eager, Johannes Gehrke, Barry Gleeson, Indranil Gupta, Kate Jenkins, Anh Look, Yaron Minsky, Andrew Myers, Venu Ramasubramanian, Richard Shoenhair, Emin Gun Sirer, Lidong Zhou, and the anonymous reviewers.

References

[ABKM01]   D.G. Andersen, H Balakrishnan, M.F. Kaashoek, and R. Morris. Resilient overlay networks. In *Proc. of the Eighteenth ACM Symp. on Operating Systems Principles*, pages 131–145, Banff, Canada, October 2001.

[Blo70]   B. Bloom. Space/time tradeoffs in hash coding with allowable errors. *CACM*, 13(7):422–426, July 1970.

[DGH⁺87]   A. Demers, D. Greene, C. Hauser, W. Irish, J. Larson, S. Shenker, H. Sturgis, D. Swinehart, and D. Terry. Epidemic algorithms for replicated database maintenance. In *Proc. of the Sixth ACM Symp. on Principles of Distributed Computing*, pages 1–12, Vancouver, BC, August 1987.

[Gon01]   L. Gong. JXTA: A network programming environment. *IEEE Internet Computing*, 5(3):88–95, May/June 2001.

[GWVB⁺01]   S.D. Gribble, M. Welsh, R. Von Behren, E.A. Brewer, D. Culler, N. Borisov, S. Czerwinski, R. Gummadi, J. Hill, A. Joseph, R.H. Katz, Z.M. Mao, S. Ross, and B. Zhao. The Ninja architecture for robust internet-scale systems and services. *To appear in a Special Issue of Computer Networks on Pervasive Computing*, 2001.

[JK02]   A.D. Joseph and J.D. Kubiatowicz. Brocade: Landmark routing on overlay networks. In *Proc. of the First International Workshop on Peer-to-Peer Systems*, Cambridge, MA, March 2002.

[OPSS93]   B. M. Oki, M. Pfluegl, A. Siegel, and D. Skeen. The Information Bus—an architecture for extensible distributed systems. In *Proc. of the Fourteenth ACM Symp. on Operating Systems Principles*, pages 58–68, Asheville, NC, December 1993.

[Ora01]   A. Oram, editor. *Peer-To-Peer: Harnessing the Power of Disruptive Technologies*. O'Reilly, 2001.

[RD01]   A. Rowstron and P. Druschel. Pastry: Scalable, distributed object location and routing for large-scale peer-to-peer systems. In *Proc. of the Middleware 2001*, November 2001.

[SMKK95]   I. Stoica, R. Morris, D. Karger, and M.F. Kaashoek. Chord: A scalable peer-to-peer lookup service for Internet applications. In *Proc. of the '95 Symp. on Communications Architectures & Protocols*, Cambridge, MA, August 1995. ACM SIGCOMM.

[vRB02]   R. van Renesse and K.P. Birman. Astrolabe: A robust and scalable technology for distributed system monitoring, management, and data mining. *ACM Transactions on Computer Systems*, 2002. Submitted for review.

[vRMH98]   R. van Renesse, Y. Minsky, and M. Hayden. A gossip-style failure detection service. In *Proc. of Middleware'98*, pages 55–70. IFIP, September 1998.

[ZKJ01]   B.Y. Zhao, J. Kubiatowicz, and A. Joseph. Tapestry: An infrastructure for fault-tolerant wide-area location and routing. Technical Report UCB/CSD-01-1141, University of California, Berkeley, Computer Science Department, 2001.

Atomic Data Access in Distributed Hash Tables

Nancy Lynch[1], Dahlia Malkhi[2], and David Ratajczak[3]

[1] MIT
[2] Hebrew University
[3] UC Berkeley

Abstract. While recent proposals for *distributed hashtables* address the crucial issues of communication efficiency and load balancing in dynamic networks, they do not guarantee strong semantics on concurrent data accesses. While it is well known that guaranteeing availability and consistency in an asynchronous and failure prone network is impossible, we believe that guaranteeing atomic semantics is crucial for establishing DHTs as a robust middleware service. In this paper, we describe a simple DHT algorithm that maintains the atomicity property regardless of timing, failures, or concurrency in the system. The liveness of the algorithm, while not dependent on the order of operations in the system, requires that node failures do not occur and that the network eventually delivers all messages to intended recipients. We outline how state machine replication techniques can be used to approximate these requirements even in failure-prone networks, and examine the merits of placing the responsibility for fault-tolerance and reliable delivery below the level of the DHT algorithm.

1 Introduction

Several groups have proposed *distributed hashtables* as a building block for large-scale distributed systems, sometimes under the alias of *content addressable networks* [RFH+01], *distributed data structures* [GBH+00, LNS96], *resource lookup services* [SMK+01], or *peer-to-peer routing services* [ZKJ01]. DHTs are composed of *nodes* that are allowed to join and leave the system and that share the burden of implementing a distributed hash table of *data objects*. For large networks, only limited portions of the data set and/or membership set might be known to any particular node; thus it is possible that accesses to the data structure are forwarded between nodes until an appropriate handler of that data object is found. DHT proposals are generally distinguished by the way in which the data set is partitioned and sparse routing information is maintained.

The design of efficient DHTs is confounded by opposing design goals. First, the set of nodes is assumed to be large, dynamic, and vulnerable to failure, so it is imperative not only to manage joins and leaves to the network efficiently while maintaining short lookup path lengths and eliminating bottlenecks, but also to replicate data and routing information to increase availability. Most DHT proposals focus primarily on these objectives. However, another design goal, and

P. Druschel, F. Kaashoek, and A. Rowstron (Eds.): IPTPS 2002, LNCS 2429, pp. 295–305, 2002.

one which is essential for maintaining the illusion of a single-system image to clients, is to ensure the *atomicity* of operations on the data objects in the system. Stated simply, submissions and responses to and from objects (values) in the DHT should be consistent with an execution in which there is only one copy of the object accessed serially [Lyn96]. Because of the complexity of dynamic systems, and because many DHTs assume an environment in which leaves and failures are equivalent, most proposals focus on the first design goal and are designed to make a "best effort" with respect to atomicity. They violate the atomicity guarantee by allowing stale copies of data to be returned, or skirt around the problem by allowing only write-once semantics.

It is well-known that simultaneously guaranteeing availability (liveness) and atomicity in failure-prone networks is impossible. Therefore, we assume a system in which the network is asynchronous but reliable (messages are eventually delivered) and where servers do not fail. They can, however, initiate a join or leave routine at any time, thus admitting possible concurrent modifications along with concurrent data accesses. By tackling the problem of concurrency in the absence of failures, we aim to produce a simple and elegant algorithm that will provide correct semantics and achieve competitive scaling performance to current DHTs. We will later discuss how existing fault-tolerance techniques can be used to mask node failures and network unreliability with little impact to the simplicity of the high-level algorithm.

Given our strong system assumptions, we seek an algorithm that yields atomic access to data when there is only one copy of each piece of data in the system. The challenge will be to ensure that as data migrates (when nodes join and leave), requests do not access a residual copy nor do they arrive at the destination before the data is transferred and mistakenly think the data does not exist. Another challenge is to ensure that once a request has been initiated by a node, a result is eventually returned. Because we have assumed an asynchronous network, we must ensure that requests are not forwarded to machines that have left the system (and thus will never respond). Furthermore, we must ensure that routing information is maintained so that requests eventually reach their targets as long as there is some active node.

2   Guarantees

The goal of a DHT is to support three operations: join, leave and data update. Joins and leaves are initiated by nodes when they enter and leave the service. An update is a *targeted request* initiated at a node, and is forwarded toward its target by a series of *update-step* requests between nodes.[1]

As far as liveness is concerned, we are primarily interested in the behavior of the algorithm when the system is quiescent: when only a small number of concurrent joins and leaves are occuring during a sufficiently long period and

---

[1] The *update* operation includes updates as simple as reads and writes, as well as much more powerful data types such as compare-and-swap and consensus objects. See [Lyn96] for a thorough treatment of atomic data objects.

join(m): This operation is initiated by a node wishing to join the network, and includes as an argument the physical machine address of a currently active node.

leave(): This operation is initiated by an active node wishing to leave the network.

update(op,x): This operation is initiated by an active node wishing to perform a data operation, *op*, on a value in the DHT that is stored under the logical identifier *x*.

not all nodes have tried to leave the system. Otherwise, if too many nodes join and leave all the time, then updates may not make sufficient progress toward their target as the set of newly joined nodes lengthens the paths between nodes. The language used in the following descriptions accounts for this detail.

Stated informally, we require that the system guarantee the following properties:

Atomicity: Updates to and the corresponding values read from data objects must be consistent with a sequential execution at one copy of the object.

Termination: If after some point no new join or leave operation is initiated and not all nodes have initiated a leave, then all pending join and leave operations must eventually terminate and all updates eventually terminate (including those initiated after that point).

Stabilization: If after some point no new join or leave operation is initiated, then the data and link information at each node should eventually be the same as that prescribed by the chosen hashing and routing schemes, with the expected lookup/update performance.

3   Algorithm

In this section we describe an algorithm that implements the guarantees described above. Here we focus on a particular implementation that stems from Chord [SMK+01]. In this implementation, objects and nodes are assigned logical identifiers from the unit ring, $[0, 1)$, and objects are assigned to nodes based on the *successor* relationship which compares object and node identifiers. Moreover, nodes maintain "edge" information that enables communication (e.g., IP addresses) to some of the other nodes. Specifically, nodes maintain edges to their successor and predecessor along the ring (they can also keep track of other *long-range contacts* using the algorithm presented). We augment the basic ring construction of Chord to include support for atomic operations on the data objects, even in the face of concurrent joins and leaves.

Each node, n, keeps track of its own status (such as *joining, active, leaving,* etc.), and its identifier. Every node maintains a table of physical and logical identifiers corresponding to its "in-links" and "out-links." In-links are nodes from which requests are allowed to be entered into the input queue. Out-links are nodes to which a connection has been established and requests can be forwarded. The data objects controlled by the node are kept in a local data structure, *data*. Each node accumulates incoming requests and messages in a FIFO queue, *InQ*. Requests in the *InQ* include all action-enabling requests, including self-generated leave and join requests, other nodes' requests involved with their joining or leaving, and data update requests. Figure 1 illustrates a single node with its local data structures.

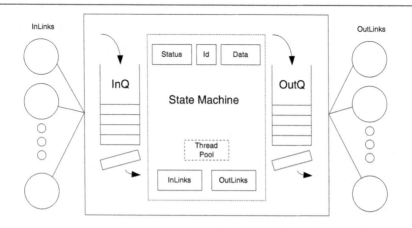

Fig. 1. In our node model, requests are performed in a serial manner at each node, with its execution determined entirely by the order in which external events are received.

Each node has a simple dispatch loop, which takes a message off the *InQ* and runs the appropriate procedure for that message, awakens any suspended procedure waiting for that message, and checks if any suspended procedures can be run due to a change of status. Thus each procedure will be initiated from some message arriving on the *InQ*, may produce outgoing messages between waiting points — when control is returned to the dispatch loop — and will eventually terminate. Only a single procedure has control at any time.

We now describe the algorithm at a high level. (The appendix provides a psuedocode description of the actions performed by each machine.) We assume that the system is initialized with one or more nodes with an initial set of edges and data objects. We will describe the LEAVE, JOIN, and UPDATE-STEP operations in order below.

When an active node wishes to leave, it places all of its data in a nice big message to its successor and changes its status to *transferring*. At this point, all requests that would be meant for the current node will be forwarded along to the successor.[2] After getting an acknowledgement that the data was received, the node changes its status to *leaving* and sends a "leaving" message to its in-links (predecessor and any others) informing that it is going away. These nodes will route "connecting" requests on the network to add an edge to their new closest active machines as a replacement for the leaving edge. When the "connecting" request finally reaches the closest active successor, it is processed, a "connecting" acknowledgment is returned, and once processed, the new edge is added. When the new edge is added, the leaving node receives an acknowledgment that the edge to it is removed, so that no more requests will be forwarded along this edge. When the leaving node has collected "leaving" acknowledgments from all in-links, and it has no more pending requests requiring a response, then it can drop out of the system. This operation is illustrated in Figure 2.

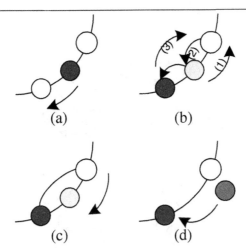

Fig. 2. For a leave operation (a) the leaving node transfers its state to its successor (b) it tells its predecessor to find its new active successor [1], which the successor does by submitting a targeted request [2] that is forwarded until a response [3] from an active node is returned, (c) an edge is added, and the leaving node is informed that its in-edges are flushed, and (d) it drops out.

A joining node will attempt to send a join request to a node for which it has a priori knowledge.[3] The request, if the node has not left, will be acknowledged

---

[2] If the successor is also leaving, messages will be forwarded even further, though this is only visible to the first node in that acknowledgements may come from a node other than the successor.

[3] Because this information may be stale, a node might never succeed in joining. However, if the joining node has knowledge of some active node, joins will complete, and in any case they will not disrupt the safety properties of the system.

and atomically put on the queue with the rest of the requests. The join message, similar to other *targeted requests*, will be routed around the system until the closest active target processes the message. At this time, the target node will separate its data, modify its bucket, and send a big message to the joining node that it is processing. It will also not be allowed to leave or handle other joins until the entire join procedure has completed. It creates a *surrogate* pointer to the joining node so that all requests for the new joining node are forwarded along this link during this period. It then contacts each of its in-neighbors telling them to update their pointers to the new node. Each of them sends a "connecting" message to the new node, updates its out-neighbors table, then sends an acknowledgment to the host node after removing the host from its out-neighbors table. When the host node collects acknowledgments from all of the neighbors in question, it can remove its surrogate pointer and start entertaining more join requests. This is illustrated in Figure 3.

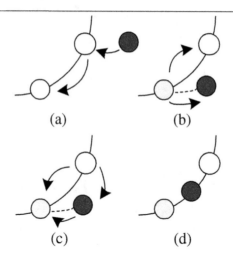

(a)   (b)

(c)   (d)

Fig. 3. For a join operation (a) the joining node initiates a targeted request to the successor, where (b) a response is returned to the joining node including the relevant state and the predecessor is notified of a new node. At this point the joining node has a surrogate edge pointing to it. After this, (c) the predecessor will contact the joining node to add an edge, it will remove an edge to the old successor, and the surrogate edge will then be removed (d).

When an UPDATE-STEP is invoked on a node, it is either forwarded or processed locally depending on its target identifier. When a response is generated, it is sent back to the return address specified in the request. All other targeted requests are either forwarded or processed locally depending on their target identifier.

4   Discussion

Certain aspects of our algorithm deserve further mention. First, the particular choice of a ring structure as the underlying routing/hashing scheme was somewhat arbitrary. The presented algorithm is readily adaptable to other routing schemes, such as the d-dimensional torii described in [RFH+01] (of which the connected ring is a special case). However, most other schemes admit the possibility that a small subset of nodes leaving at the same time could fail to make progress because they all wish to transfer state to each other.[4] In the case of the ring, the only time this occurs is when all nodes in the system leave, a scenario already excluded from consideration. For a d-dimensional torus (see [RFH+01]) this could be as few as $O(\sqrt[d]{n})$ nodes.

For routing schemes that are based on a connected ring structure with additional long-range edges [MNR02, SMK+01], the presented algorithm and pseudocode can be modified to forward targeted requests based on the routing protocol rather than merely forwarding to the successor, and to allow a joined node to send *connect* requests to connect to its appropriate long-range edges before (or just after) becoming active. The remaining code for managing incoming and outgoing links can be left unaltered.

The use of the *connect* request to add links in the network is also useful if the update messages are large, and the cost of forwarding a message through the network is prohibitive. In this case, the initiator of the update can send a connect request to the target of the update, add an edge to that node, and then forward the update to that node directly. This does not violate the correctness of the algorithm since this node could still leave and the targeted update will be forwarded appropriately. In a relatively static network this will require the update to be forwarded along only one edge, and can drastically reduce the number of hops in even the most dynamic setting. However, it also incurs additional overhead for maintaining and dismantling the resulting edge.

5   Fault-Tolerance

It is a deliberate aspect of our algorithm that we have modeled each node as a state machine dependent only on the order of its inputs. This means we can employ existing state machine replication (SMR) algorithms to produce a fault-tolerant version of our algorithm, where the abstract nodes of the algorithm are implemented by a replicated set of machines as illustrated in Figure 4. SMR algorithms ensure that a set of replicas receive inputs in the same order (thus they have the same execution) and provide mechanisms for replicas to join and leave a group, as well as to coordinate a response and a replica change when a failure is detected.[5] There are different variants that tolerate different types of failures and that are tuned for different environments.

---

[4] Note that this is an instance of the classical dining philosophers problem, and is solvable by a number of techniques beyond the scope of our algorithm or this paper.

[5] See [Sch90] for a thorough treatment of SMR techniques.

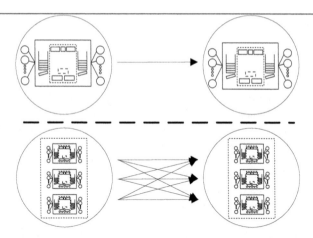

Fig. 4. Normal communication between nodes at the algorithm level (top) can be made fault-tolerant using state machine replication techniques at the node level (bottom).

The simplest way to incorporate SMR into the system is to have physical nodes form replica groups elsewhere, and join into the network as a virtual node with a virtual address encoding the set of replicas in the group.[6] The replicas in a virtual node will execute the algorithm we have already presented, though there will be occasional view change operations invoked by the SMR service when replica failures are detected. When the set of remaining active replicas in a virtual node gets below a certain threshold, the SMR service will invoke the leave operation on the remaining replicas, and when it is complete, the active replicas may go elsewhere to become a new virtual node when more participants are found. In this way, failures of individual replicas are masked and eventually turned into correct leaves at the algorithm level.

There are several benefits to this approach. First, many mature SMR implementations already exist and their behavior in various environments has been well established. Second, different replication factors and different SMR implementations may be appropriate for different deployments; a DHT on a tightly-controlled cluster of machines will have a different failure model than one that is deployed over the Internet, and will require different fault-tolerance guarantees. Third, because this scheme does not specify how replica groups are formed, they can be formed to optimize a number of different factors, such as geographic proximity, failure independence, trust, etc.

---

[6] Communication between virtual nodes will involve k^2 actual messages in a trivial implementation and will require some filtering on the part of the replicas in the receiving virtual node. This can be optimized in numerous ways which we will not discuss here.

An alternative approach is to enforce that replica groups are sets of consecutive nodes on the ring. This would work by keeping high and low thresholds for the size of replica groups about the ring. A replica group manages all data that would be managed by any of the replicas individually in the original algorithm, and thus a new physical node will join into the replication group wherever it is "hashed" onto the ring.[7] When the size of a group exceeds the high threshold, it splits into two adjacent replication groups each with roughly half the replicas and data. When the size of a group drops below the low threshold, it merges with its successor group.

When the thresholds are set at roughtly $O(\log n)$, this scheme has some interesting theoretical advantages. First, it ensures that the replica groups are composed of $O(\log n)$ independently chosen physical nodes, and thus if we assume that failures are independent of identifiers, it ensures that with high probability there is no virtual node failure in the system. Second, the size of the ring regions covered by the replica groups are balanced to within a *constant* factor with high probability. This is an improvement over the logarithmic factor normally guaranteed and requires fewer edges than the "virtual node" scheme employed by Chord [SMK+01].

The details of this scheme remain to be fully worked out due to some subtleties arising from having two replica groups communicating view changes to each other. We are in the process of finalizing these details and building the system to examine its behavior in practice.

References

[GBH+00]   S. D. Gribble, E. A. Brewer, J. M. Hellerstein, and D. Culler. "Scalable, distributed data structures for Internet service construction. In the *Fourth Symposium on Operating System Design and Implementation (OSDI 2000)*, October 2000.

[Lam79]   L. Lamport. How to make a multiprocessor computer that correctly executes multiprocessor programs. *IEEE Transactions on Computers*, C-28(9):690–691, 1979.

[LNS96]   W. Litwin, M.A. Neimat, D. A. Schneider. "LH*-A scalable, distributed data structure". *ACM Transactions on Database Systems*, Vol. 21, No. 4, pp 480-525, 1996.

[Lyn96]   Lynch, N. *Distributed Algorithms*, Morgan Kaufmann, San Francisco, CA 1996.

[MNR02]   D. Malkhi, M. Naor and D. Ratajczak. "Viceroy: A Scalable and Dynamic Lookup Scheme". Submitted for publication.

[RFH+01]   S. Ratnasamy, P. Francis, M. Handley, R. Karp and S. Shenker. "A scalable content-addressable network". In *Proceedings of the ACM SIGCOMM 2001 Technical Conference*. August 2001.

[Sch90]   F. Schneider. Implementing Fault-Tolerant Services Using the State Machine Approach. *ACM Computing Surveys* 22:4 (Dec. 1990), 299-319.

---

[7] Note that this is not the same join operation that is described for the high-level algorithm.

[SMK+01]   I. Stoica, R. Morris, D. Karger, M. F. Kaashoek, and H. Balakrishnan. "Chord: A scalable peer-to-peer lookup service for Internet applications". In *Proceedings of the SIGCOMM 2001*, August 2001.

[ZKJ01]   B. Y. Zhao, J. D. Kubiatowicz and A. D. Joseph. "Tapestry: An infrastructure for fault-tolerant wide-area location and routing". U. C. Berkeley Technical Report UCB/CSD-01-1141, April, 2001.

A   Pseudocode

local data:
  $ID = \{id, addr\}$, $id \in \mathcal{R}$ randomly chosen, $addr$ is a physical address
  $InLinks$ and $OutLinks$, set of $\{id, addr\}$ logical/physical address pairs,
    initially empty
  $data$, set of named data objects, initially empty
  $myrange = (low, high) \in \mathcal{R} \times \mathcal{R}$, initially $(ID.id, ID.id)$
  InQ and $OutQ$, FIFO queues containing requests/msgs. InQ initially contains
    JOIN($someAddr$)
  $status \in \{inactive, joining, active, transferring, leaving\}$, initially $inactive$

definitions:
  $successor = closest(OutLinks)$
  "(msg,ID)" a message of type msg from a machine with logical/physical address
    of ID

main program:
  do forever
    if there is any waiting procedure that may resume, dispatch the oldest one
    else
      remove head request from InQ
      if $status$ is $leaving$ and request is targeted, then forward request to $successor$
      else
        dispatch the oldest procedure waiting for that message (if any)
        else dispatch a new procedure to handle request

LEAVE():   ; handle self leaving
  wait until $status$ is $active$; yield
  $status \leftarrow transferring$
  send ((data-trans,$data$),ID) to $successor$
  $myrange \leftarrow (ID, ID)$
  wait for (data-trans-ack,$successor$)   ; yield
  $status \leftarrow leaving$
  send (leaving,ID) to all machines in $InLinks$
  wait for (leaving-ack,m) from all machines in $InLinks$; yield
  forward all UPDATE-STEP requests in InQ to $successor$
  $status \leftarrow inactive$

JOIN($someAddr$):   ; handle join
  wait until $status$ is $inactive$; yield
  $status \leftarrow joining$

send (joining,ID) to the machine denoted by *someAddr*
wait for ((join-ack-and-data-trans,*datainfo*),*surrogate*) yield
send (data-trans-ack,ID) to *surrogate*
include *surrogate* in *OutLinks*
set *data* and *myrange* based on *datainfo*
wait for (join-complete,*surrogate*)
status ← *active*

UPDATE-STEP(x, op)$_{retaddress}$:
    if x is in *myrange* (contained within $[low, high)$ on the unit ring)
    then perform *op* on x and send result to *retaddress*
    else forward to *successor*

receive-msg T:
    if T is (data-trans,m)
       merge *data* and *myrange* with incoming data and range information
       send (data-trans-ack,ID) to m

    else if T is (joining,m)
      if m is in *myrange*
        oldstatus ← *status*
        status ← *transferring*
        group data and modify range between m and ID into *datainfo* msg
        send ((join-ack-and-data-trans,*datainfo*),ID) to m
        wait for (data-trans-ack,m)   ; yield
        status ← *oldstatus*
        include m in *OutLinks*   ; surrogate pointer
        send ((notify-of-new,m),ID) to all machines in *InLinks*
        wait for (new-ack,m') from all machines in *InLinks*yield
        remove those machines from *InLinks*
        remove m from *OutLinks*   ; remove surrogate pointer
      else forward request to *successor*

    else if T is (leaving,m)
      send ((connect,m),ID) to *successor*
      wait for (connecting-ack,*substitute*)   ; may differ from *successor*
      replace m in *OutLinks* with *substitute'*
      send (leaving-ack,ID) to m

    else if T is ((connect,x),m)
      if x is in *myrange*
        add m to *InLinks*
        send (connecting-ack,ID) to m
      else forward to *successor*

    else if T is ((notify-of-new,x),m)
      send ((connect,x),ID) to closest link in *OutLinks*
      wait for (connecting-ack,*new*)
      replace m with *new* in *OutLinks*
      send (new-ack,ID) to m

Dynamic Replica Placement for Scalable Content Delivery

Yan Chen, Randy H. Katz, and John D. Kubiatowicz

Computer Science Division,
University of California, Berkeley
{yanchen, randy, kubitron}@cs.berkeley.edu

Abstract. In this paper, we propose the *dissemination tree*, a dynamic content distribution system built on top of a peer-to-peer location service. We present a replica placement protocol that builds the tree while meeting QoS and server capacity constraints. The number of replicas as well as the delay and bandwidth consumption for update propagation are significantly reduced. Simulation results show that the dissemination tree has close to the optimal number of replicas, good load distribution, small delay and bandwidth penalties for update multicast compared with the ideal case: static replica placement on IP multicast.

1 Introduction

The efficient distribution of Web content and streaming media is of growing importance. The challenge is to provide content distribution to clients with good *Quality of Service (QoS)* while retaining *efficient* and *balanced* resource consumption of the underlying infrastructure. Central to these goals is the careful placement of data replicas and the dissemination of updates.

Previous work on replica placement involves *static* placement of replicas – assuming that clients' distribution and access patterns are known in advance[13, 8]. These techniques ignore server capacity constraints and assume explicit knowledge of the global IP network topology.

Actual Web content distribution requires *dynamic* or *online* replica placement. Most current Content Distribution Networks (CDNs) use DNS-based redirection to route clients' requests [1, 4, 10, 17]. Due to the nature of centralized location services, the CDN name server cannot afford to keep records for the locations of each replica. Thus the CDN often places many more replicas than necessary and consumes unnecessary storage resources and update bandwidth.

For update dissemination, IP multicast has fundamental problems for Internet distribution [5]. Further, there is no widely available inter-domain IP multicast. As an alternative, Application Level Multicast (ALM) tries to build an efficient network of unicast connections and to construct data distribution trees on top of this *overlay* structure [5, 2, 3, 9, 21]. Most ALM systems have scalability problems, since they utilize a central node to maintain state for all existing children [3, 9, 11, 2], or to handle all "join" requests [21]. Replicating the

P. Druschel, F. Kaashoek, and A. Rowstron (Eds.): IPTPS 2002, LNCS 2429, pp. 306–318, 2002.

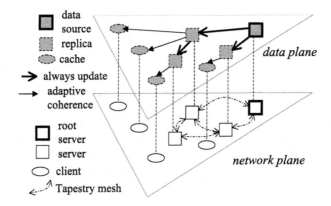

Fig. 1. Architecture of a dissemination tree.

root is the common solution [9, 21], but this suffers from consistency problems and communication overhead.

There are two crucial design issues that we try to address in this paper:

1. How to dynamically choose the number and placement of replicas while satisfying QoS requirements and server capacity constraints.
2. How to disseminate updates to these replicas with small delay and bandwidth consumption.

Both must be addressed without explicit knowledge of the global network topology. Further, we would like to scale to millions of objects, clients, and servers.

To tackle these challenges, we propose a new Web content distribution system: *dissemination tree* (in short, *d-tree*). Figure 1 illustrates a d-tree system. There are three kinds of data in the system: *sources, replicas,* and *caches.* The d-tree targets dynamic Web content distribution; hence there is a single source on the Web server. A replica is a copy of source data that is stored on the overlay server and is always kept up-to-date, while a cache is stored on clients and may be stale. These components self-organize into a d-tree and use application-level multicast to disseminate updates from source to replicas. Coherence of caches is maintained dynamically through approaches such as [15]. We assume that d-tree servers are placed in Internet Data Centers (IDC) of major ISPs with good connectivity to the backbone. These servers form a peer-to-peer overlay network called *Tapestry* [20], to find nearby replicas for the clients. Note that Tapestry is shared across objects, while each object for dissemination has a hierarchical d-tree.

We make the following contributions in the paper:

- We propose novel algorithms to dynamically place close to minimum number of replicas while meeting the clients' QoS and servers' capacity constraints.

- We self-organize these replicas into an application-level multicast tree with small delay and bandwidth consumption for update dissemination.
- We leverage Tapestry to improve scalability. Tapestry permits clients to locate nearby replica servers without contacting a root; as a result, each node in a d-tree maintains state only for its parent and direct children.

Note that all these are achieved with limited local network topology knowledge only.

The rest of the paper is organized as follows: We formulate the replica placement problem in Sec. 2 and introduce Tapestry in Sec. 3. Sec. 4 describes the protocols for building and maintaining a d-tree. Evaluation and results are given in Sec. 5, and finally conclusions and future work in Sec. 6.

2 Problem Formulation

There is a big design space for modeling Web replica placement as an optimization problem and we describe it as follows. Consider a popular Web site or a CDN hosting server, which aims to improve its performance by pushing its content to some hosting server nodes. The problem is to dynamically decide where content is to be replicated so that some objective function is optimized under a dynamic traffic pattern and set of clients' QoS and/or resource constraints. The objective function can either minimize clients' QoS metrics, such as latency, loss rate, throughput, etc., or minimize the replication cost of CDN service providers, e.g., network bandwidth consumption, or an overall cost function if each link is associated with a cost. For Web content delivery, the major resource consumption in replication cost is the network access bandwidth at each Internet Data Center (IDC) to the backbone network. Thus when given a Web object, the cost is linearly proportional to the number of replicas.

As Qiu *et al.* tried to minimize the total response latency of all the clients' requests with the number of replicas as constraint [13], we tackle the replica placement problem from another angle: minimize the number of replicas when meeting clients' latency constraints and servers' capacity constraints. Here we assume that clients give reasonable latency constraints as it can be negotiated through a service-level agreement (SLA) between clients and CDN vendors. Thus we formulate the Web content placement problem as follows. Given a network G with C clients and S server nodes, each client c_i has its *latency constraint* d_i, and each server s_j has its load/bandwidth/storage *capacity constraint* l_j. The problem is to find a smallest set of servers S' such that the distance between any client c_i and its "parent" server $s_{c_i} \in S'$ is bounded by d_i. More formally, find the minimum K, such that there is a set $S' \subset S$ with $|S'| = K$ and $\forall c \in C$, $\exists s_c \in S'$ such that distance$(c, s_c) \leq d_c$. Meanwhile, these clients C and servers S' self-organize into an application-level multicast tree with C as leaves and $\forall$ $s_i \in S'$, its fan-out degree (i.e., number of direct children) satisfies $f(s_i) \leq l_i$.

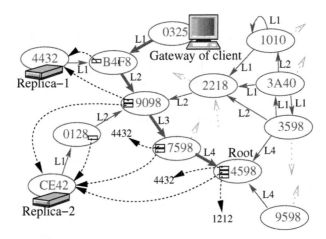

Fig. 2. The Tapestry Infrastructure: *Nodes route to nodes one digit at a time: e.g. 0325 → B4F8 → 9098 → 7598 → 4598. Objects are associated with a particular "root" node (e.g. 4598). Servers publish replicas by sending messages toward root, leaving back-pointers (dotted arrows). Clients route directly to replicas by sending messages toward root until encountering a pointer (e.g. 0325 → B4F8 → 4432).*

3   Peer-to-Peer Location Services: The Tapestry Infrastructure

Networking researchers have begun to explore decentralized peer-to-peer location services [20, 14, 18, 16]. Such services offer a distributed infrastructure for locating objects quickly, with guaranteed success and locality. Rather than depending on a single server to locate an object, a query in this model is passed around the network until it reaches a node that knows the location of the requested object. Our dissemination tree is built on top of Tapestry [20] and takes advantage of two features: *distributed location services* and *search with locality*.

Tapestry is an IP overlay network that uses a distributed, fault-tolerant architecture to track the location of objects in the network. In our architecture (Figure 1), the d-tree servers (i.e., CDN edge servers) and multicast root server (i.e., Web source server) are Tapestry nodes. Each client talks to its nearby Tapestry node (*the gateway*) to send object requests. In practice, the gateway node can be located through certain bootstrap mechanisms.

3.1   Tapestry Routing Mesh

Figure 2 shows a portion of Tapestry. Each node joins Tapestry in a distributed fashion through nearby surrogate servers and set up *neighboring* links for connection to other Tapestry nodes [20]. The neighboring links are shown as solid arrows. Such neighboring links provide a route from every node to every other node; the routing process resolves the destination address one digit at a time (e.g., ***8 $\Longrightarrow$ **98 $\Longrightarrow$ *598 $\Longrightarrow$ 4598, where *'s represent wildcards). This routing scheme is based on the hashed-suffix routing structure originally presented by Plaxton, Rajaraman, and Richa [12].

3.2 Tapestry Distributed Location Service

Tapestry employs this infrastructure for data location. Each object is associated with a *Tapestry location root* through a deterministic mapping function. This root is for location purposes only and has nothing to do with the multicast root server (such as the Web content server in Figure 1). To advertise an object o, the server s storing the object sends a publish message toward the Tapestry location root for that object, depositing *location pointers* in the form of <Object-ID(o), Server-ID(s)> at each hop. These mappings are simply pointers to the server s where o is being stored, and not a copy of the object itself. A node s that keeps location mappings for multiple replicas keeps them sorted in the order of distance from s.

Figure 2 shows two replicas and the Tapestry root for an object. Location pointers are shown as dotted arrows that point back to replica servers. To locate an object, a client sends a message toward the object's root. When the message encounters a pointer, it routes directly to the object. It is shown in [12] that the average distance traveled in locating an object is *proportional* to the distance from that object in terms of the number of hops traversed. In addition, it is proved that for any node c that requests object o, Tapestry can route the request to the asymptotically optimal node s (in terms of the shortest path network distance) that contains a replica of o [12].

4   Dissemination Tree Protocols

4.1   Replica Placement and Tree Construction

In this section, we present an algorithm that dynamically places replicas and organizes them into an application-level multicast tree with only limited knowledge of the network topology. This algorithm attempts to satisfies both client latency and server capacity constraints. Our goal is to minimize the number of replicas deployed and to self-organize the servers with replicas into a load-balanced tree. We contrast static solutions that assume global knowledge of clients and topology.

Dynamic Replica Placement We consider two algorithms: *naive placement* and *smart placement*, for comparison. We describe these as procedures for a new client c to join the tree of object o, possibly generating new replicas in the process. Following the notations in Sec. 2, the latency constraint of c is d_c and the capacity constraint of s is l_s. We define the following notations: current load

---

procedure DynamicReplicaPlacement_Naive(c, o)

1 c sends a "join" request to s with o through Tapestry, piggybacks the IP addresses, $dist_{overlay}(c, s')$ and $rc_{s'}$, for each server s' on the path

2 **if** $rc_s > 0$ **then**

3   **if** $dist_{overlay}(c, s) \leq d_c$ **then** s becomes c's parent, **exit.**
        else

4   s pings c to get $dist_{IP}(s, c)$
5   **if** $dist_{IP}(s, c) \leq d_c$ **then** s becomes c's parent, **exit.**
        end
    end
6 From the closest one to c, **foreach** *server s' on the path* **do**
        search for t that satisfies $rc_t > 0$ and $dist_{overlay}(t, c) \leq d_c$
    end
7 s puts a replica on t and becomes its parent, t becomes c's parent
8 t publishes o in Tapestry, **exit.**
9 **foreach** *path server s_i whose $rc_{s_i} > 0$* **do** s_i pings c to get $dist_{IP}(s_i, c)$
10 c chooses t which has the smallest $dist_{IP}(t, c) \leq d_c$
11 Same as steps 7 and 8.

Algorithm 1: Dynamic Replica Placement (Naive)

of s: lc_s; remaining capacity of s: $rc_s = l_s - lc_s$; overlay distance on Tapestry: $dist_{overlay}$ and IP distance: $dist_{IP}$. As periodically there are "refresh" messages going from a child server to its parent for soft state management, we assume that each parent server knows the current remaining capacity of each child server.

Naive placement: Client c sends the request for object o through Tapestry and is routed to server s. For the naive approach, s only considers itself to be c's parent server, i.e., whether $rc_s > 0$ and $dist_{IP}(s, c) \leq d_c$ are satisfied. If unsatisfied, it will try to place a replica on the overlay path server that is as *close* to c as possible (see Algorithm 1). Note that given the limited search, the naive approach may not always find the suitable parent server for every client, even when such a parent exists.

Smart placement: Essentially, the smart approach (Algorithm 2) attempts to optimize the "best" parent selection for c in a larger set: including s, its *parent*, *siblings* and its *other server children*. Among qualified candidates, c chooses the one with the lightest load as parent. If none of them meet the client's latency and server's load constraints, s will try to place a replica on the overlay path server that is as *far* from c as possible. We call it *lazy placement*. All these steps aim to distribute the load with the greedy algorithm to reduce the number of replicas needed while satisfying the constraints.

Note that we try to use the overlay latency to estimate the IP latency in order to save "ping" messages. Here the client can start a daemon program provided by its CDN service provider when launching the browser so that it can actively

---

procedure DynamicReplicaPlacement_Smart(c, o)
1 c sends a "join" request to s with o through Tapestry
2 s sends c's IP address to its parent p and other server children sc **if** $rc_{sc} > 0$
3 p forwards the request to s's siblings ss **if** $rc_{ss} > 0$
4 s, p, ss and sc send c its rc **if** its $rc > 0$
5 **if** c *gets any reply* **then**

6   c chooses the parent t which has the biggest rc and $dist_{IP}(t, c) \le d_c$, **exit**.
    else

7   c sends a message to s through Tapestry again and the message piggybacks
       the IP addresses, $dist_{overlay}(c, s')$ and $rc_{s'}$ for each server s' on the path
8   From the closest one to s, **foreach** *server s' on the path* **do**
           search for t that satisfies $rc_t > 0$ **and** $dist_{overlay}(t, c) \le d_c$
       end
9   Same as steps 7, 8 and 9 in **procedure** DynamicReplicaPlacement_Naive.
10   c chooses t which has the biggest $dist_{IP}(t, c) \le d_c$
11   Same as step 11 in **procedure** DynamicReplicaPlacement_Naive.
    end

Algorithm 2: Dynamic Replica Placement (Smart)

participate in the protocols. The locality property of Tapestry naturally leads to the locality of d-tree, i.e., the parent and children tend to be close to each other in terms of the number of IP hops between them. This provides good delay and multicast bandwidth consumption when disseminating updates, as measured in Sec. 5. The tradeoff between the smart and naive approaches is that the smart one consumes more "join" traffic to construct a tree with fewer replicas, covering more clients, with less delay and multicast bandwidth consumption. We evaluate this tradeoff in Sec. 5.

Static Replica Placement The replica placement methods given above are unlikely to be optimal in terms of the number of replicas deployed, since clients are added sequentially and with limited knowledge of the network topology. In the static approach, the root server has complete knowledge of the network and places replicas *after* getting all the requests from the clients. In this scheme, updates are disseminated through IP multicast. Static placement is not very realistic, but may provide better performance since it exploits knowledge of the client distribution and global network topology.

The problem formulated in Sec. 2 can be converted to a special case of the capacitated facility location problem [7] defined as follows. Given a set of locations i at which facilities may be built, building a facility at location i incurs a cost of f_i. Each client j must be assigned to one facility, incurring a cost of $d_j c_{ij}$ where d_j denotes the demand of the node j, and c_{ij} denotes the distance between i and j. Each facility can serve at most l_i clients. The objective is to find the number of facilities and their locations yielding the minimum total cost.

To map the facility location problem to ours, we set f_i always 1, and set c_{ij} 0 if location i can cover client j or ∞ otherwise. The best approximation algorithm known today uses the primal-dual schema and Lagrangian relaxation to achieve a guaranteed factor of 4 [7]. However, this algorithm is too complicated for practical use. Instead, we designed a greedy algorithm that has a logarithmic approximation ratio.

Besides the previous notations, we define the following variables: set of covered clients by s: C_s, $C_s \subseteq C$ and $\forall\, c \in C_s$, $dist_{IP}(c, s) \leq d_c$; set of possible server parents for client c: S_c, $S_c \subseteq S$ and $\forall\, s \in S_c$, $dist_{IP}(c, s) \leq d_c$.

---

procedure ReplicaPlacement_Greedy_DistLoadBalancing(C, S)
input   : Set of clients to be covered: C, total set of servers: S
output : Set of servers chosen for replica placement: S'
while C *is not empty* **do**
    Choose $s \in S$ which has the largest value of min(cardinality $|C_s|$, remaining capacity rc_s)
    $S' = S' \bigcup \{s\}$
    $S = S - \{s\}$
    if $|C_s| \leq rc_s$ **then** $C = C - C_s$
    else
        Sort each element $c \in C_s$ in increasing order of $|S_c|$
        Choose the first rc_s clients in C_s as $C_{sChosen}$
        $C = C - C_{sChosen}$
    end
    recompute S_c for $\forall\, c \in C$
end
return S'.

Algorithm 3: Static Replica Placement with Distributed Load Balancing

We consider two types of static replica placement: with only overlay path topology vs. with global IP topology. For the former, to each client c, the root only knows the servers on the Tapestry path from c to root which can cover that client (in IP distance). On the other hand, the latter assumes the knowledge of global IP topology and gives close-to-optimal bound on the number of replicas.

4.2   Soft State Tree Maintenance

The liveness of the tree is maintained using a soft-state mechanism. Periodically, we send "heartbeat" messages from the root down to each member. We assume that all the nodes are loosely synchronized through the Network Time Protocol (NTP) [6]. Thus if any member (except the root) gets the message within a certain threshold, it will know that it is still alive on the tree. Otherwise it will time out and start rejoining the tree. Meanwhile, each member will periodically send out a "refresh" message to its parent. If the parent does not get the "refresh" message within a certain threshold, it will kick out the child's entry.

5   Evaluation

In this section, we evaluate the performance of our d-tree algorithms. We use the GT-ITM transit-stub model to generate five 5000-node topologies [19]. The results are averaged over the experiments on the five topologies. A packet-level, priority-queue based event manager is implemented to simulate the network latency.

We utilize two strategies for placing d-tree servers. One selects all d-tree servers at random (labeled *random d-tree*). The other preferentially chooses transit and gateway nodes (labeled *backbone d-tree*). This approach mimics the strategy of placing d-tree servers strategically in the network.

We couple the server placement with four different replica placement techniques: overlay dynamic naive placement (*od_naive*), overlay dynamic smart placement (*od_smart*), overlay static placement (*overlay_s*), and static placement on IP network (*IP_s*). 500 nodes are chosen to be d-tree servers with either "random" or "backbone" approach. The rest of nodes are clients and join the d-tree in a random order. We randomly choose one non-transit d-tree server to be the multicast source and set as 50KB the size of data to be replicated. Further, we assume the latency constraint is 50ms and the load capacity is 200 clients/server.

In the following, we consider three metrics:

- **Quality of Replica Placement**: Includes number of deployed replicas and degree of load distribution, measured by the ratio of the standard deviation vs. the mean of the number of client children for each replica server. A smaller ratio implies better load distribution.
- **Multicast performance**: We measure the relative delay penalty (RDP) and the bandwidth consumption which is computed by summing the number of bytes multiplied by the transmission time over every link in the network.
- **Tree construction traffic:** We count both the number of application-level messages sent and the bandwidth consumption for constructing the d-tree.

Figure 3 shows the number of replicas placed and the load distribution on these servers. *Od_smart* approach uses only about 30% to 60% of the servers used by *od_naive*, is even better than *overlay_s*, and is very close to the optimal case: *IP_s*. Also note that *od_smart* has better load distribution than *od_naive* and *overlay_s*, close to *IP_s* for both *random* and *backbone d-tree*.

In Figure 4, *od_smart* has better RDP than *od_naive*, and 85% of *od_smart* RDPs between any member server and the root pairs are within 4. Figure 5 contrasts the bandwidth consumption of various d-tree construction techniques with optimal IP placement. The results are very encouraging: the bandwidth consumption of *od_smart* is quite close to the optimal *IP_s* and is much less than that of *od_naive*.

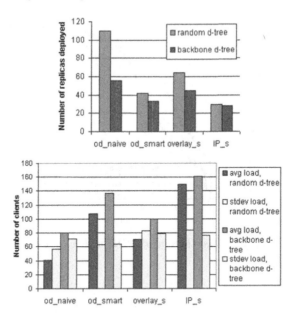

Fig. 3. *Number of replicas deployed (top) and load distribution on selected servers (bottom) (500 d-tree servers).*

The performance above is achieved at the cost of d-tree construction (Figure 6). However, for both *random* and *backbone d-tree*, *od_smart* approach produces less than three times of the messages of *od_naive* and less than six times of that for optimal case: *IP_s*. Meanwhile, *od_naive* uses almost the same amount of bandwidth as *IP_s* while *od_smart* uses about three to five times that of *IP_s*.

In short, the smart dynamic replica placement has a close-to-optimal number of replicas, better load distribution, and less delay and multicast bandwidth consumption than the naive approach, at the price of three to five times as much tree construction traffic. Usually, tree reconstruction is a much less frequent event than Web data access and update. Further, its performance is quite close to the ideal case: static placement on IP multicast. Hence, the "smart approach" is more advantageous.

Due to the limited number and/or distribution of servers, there may exist some clients who cannot be covered when facing the QoS and capacity requirements. In this case, our algorithm can provide hints as where to place more servers. And the experiments show that the naive scheme has many more uncovered clients than the smart one, due to the nature of its unbalanced load.

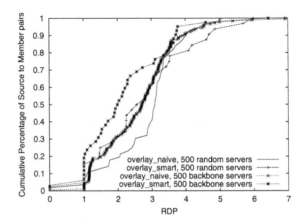

Fig. 4. *Cumulative distribution of RDP with various approaches (500 d-tree servers).*

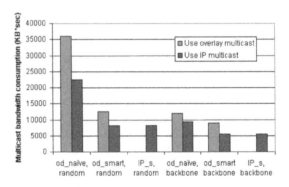

Fig. 5. *Bandwidth consumption when multicast 1MB update data (500 d-tree servers).*

6   Conclusions and Future Work

In this paper, we explore techniques for building the dissemination tree, a dynamic content distribution network. First, we propose and compare several replica placement algorithms which reduce the number of replicas deployed and self-organize them into a balanced dissemination tree. Second, we use Tapestry, a peer-to-peer location service, for better scalability and locality. In the future, we would like to continue evaluation with more diverse topologies and workloads, add dynamic replica deletion to d-tree, and investigate how to build a better CDN with other peer-to-peer techniques.

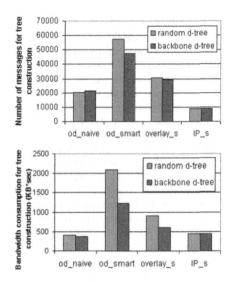

Fig. 6. *Number of application-level messages (top) and total bandwidth consumed (bottom) for d-tree construction (500 d-tree servers).*

7 Acknowledgments

We graciously acknowledge sponsorship and grants from DARPA (grant N66061-99-2-8913), California Micro Grant #01-042, Ericsson, Nokia, Siemens, Sprint, NTTDoCoMo and HRL laboratories. We thank Hao Chen, Matthew Caesar and Chen-nee Chuah for reviewing the draft of the paper and thank the anonymous reviewers for their valuable suggestions.

References

[1] Akamai Technologies Inc. http://www.akamai.com.

[2] Y. Chawathe, S. McCanne, and E. Brewer. RMX: Reliable multicast for heterogeneous networks. In *Proceedings of IEEE INFOCOM*, 2000.

[3] Y. Chu, S. Rao, and H. Zhang. A case for end system multicast. In *Proceedings of ACM SIGMETRICS*, June 2000.

[4] Digital Island Inc. http://www.digitalisland.com.

[5] P. Francis. Yoid: Your own Internet distribution. Technical report, ACIRI, http://www.aciri.org/yoid, April, 2000.

[6] J. D. Guyton and M. F. Schwartz. Experiences with a survey tool for discovering network time protocol servers. In *Proc. of USENIX*, 1994.

[7] K. Jain and V. Varirani. Approximation algorithms for metric facility location and *k*-median problems using the primal-dual schema and lagrangian relaxation. In *Proc. of FOCS*, 1999.

[8] S. Jamin, C. Jin, A. Kurc, D. Raz, and Y. Shavitt. Constrained mirror placement on the Internet. In *Proceedings of IEEE Infocom*, 2001.

[9] J. Jannotti et al. Overcast: Reliable multicasting with an overlay network. In *Proceedings of OSDI*, 2000.

[10] Mirror Image Internet Inc. `http://www.mirror-image.com`.

[11] D. Pendarakis, S. Shi, D. Verma, and M. Waldvogel. ALMI: An application level multicast infrastructure. In *Proceedings of 3rd USITS*, 2001.

[12] C. G. Plaxton, R. Rajaraman, and A. W. Richa. Accessing nearby copies of replicated objects in a distributed environment. In *Proc. of the SCP SPAA*, 1997.

[13] L. Qiu, V. N. Padmanabhan, and G. Voelker. On the placement of Web server replicas. In *Proceedings of IEEE Infocom*, 2001.

[14] S. Ratnasamy, P. Francis, M. Handley, R. Karp, and S. Shenker. A scalable content-addressable network. In *Proceedings of ACM SIGCOMM*, 2001.

[15] P. Rodriguez and S. Sibal. SPREAD: Scaleable platform for reliable and efficient automated distribution. In *Proceedings of WWW*, 2000.

[16] A. Rowstron and P. Druschel. Pastry: Scalable, distributed object location and routing for large-scale peer-to-peer systems. In *Proc. of Middleware 2001*.

[17] Speedera Inc. http://www.speedera.com.

[18] I. Stoica et al. Chord: A scalable peer-to-peer lookup service for Internet applications. In *Proceedings of ACM SIGCOMM*, 2001.

[19] E. Zegura, K. Calvert, and S. Bhattacharjee. How to model an Internetwork. In *Proceedings of IEEE INFOCOM*, 1996.

[20] B. Y. Zhao, J. Kubiatowicz, and A. Joseph. Tapestry: An infrastructure for fault-tolerant wide-area location and routing. UCB Tech. Report UCB/CSD-01-1141.

[21] S. Q. Zhuang et al. Bayeux: An architecture for scalable and fault-tolerant wide-area data dissemination. In *Proceedings of ACM NOSSDAV*, 2001.

Peer-to-Peer Resource Trading in a Reliable Distributed System

Brian F. Cooper and Hector Garcia-Molina

Department of Computer Science
Stanford University
{cooperb,hector}@db.stanford.edu

Abstract. Peer-to-peer architectures can be used to build a robust, fault tolerant infrastructure for important services. One example is a peer-to-peer data replication system, in which digital collections are protected from failure by being replicated at multiple peers. We argue that such community-based redundancy, in which multiple sites contribute resources to build a fault-tolerant system, is an important application of peer-to-peer networking. In such a system, there must be flexible, effective techniques for managing resource allocation. We propose data trading, a mechanism where a site acquires remote resources in the community by trading away its own local resources. We discuss the application of data trading to the data replication problem, and examine other applications of trading. A general trading infrastructure is a valuable part of a peer-to-peer, community-based redundancy system.

1 Introduction

Peer-to-peer systems form a useful architecture for a wide range of important applications. Although the term "peer-to-peer" is often associated in the public imagination with Napster and related file-sharing systems, other important services that can be built on a peer-to-peer framework. For example, a group of digital libraries may cooperate with each other to provide preservation by storing copies of each other's digital materials. In this system, each library acts as an autonomous peer in a distributed, heterogeneous collection replication mechanism. Such a community does not require a central controller to manage the replication of data; instead, each peer can communicate with other peers to replicate its own collections. The result of individual libraries seeking locally to preserve their own information by working with other peers is a global community in which every library's collections are protected.

Such a replication network is an example of a *community-based redundancy system*: a group of peers collaborate to provide resource redundancy and thus reliability and fault tolerance. There are several benefits to community-based redundancy. First, each peer is able to take advantage of a system with large aggregate resources simply by contributing its own, relatively small set of resources. Second, the distribution of resources in the system means that the value of the aggregate resources is larger than the sum of the individual contributions. For

P. Druschel, F. Kaashoek, and A. Rowstron (Eds.): IPTPS 2002, LNCS 2429, pp. 319–327, 2002.

example, it is more valuable to the Stanford library to have one copy of its collections locally and one copy at say MIT than it is for Stanford to have two copies locally. If Stanford experiences a failure (such as a hardware fault, a malicious attack such as a virus or trojan, or a natural disaster), then it can recover from the failure by using the unaffected copy at MIT. Third, the heterogeneity inherent in a community of autonomous sites is valuable; if all sites are homogeneous then a software bug or security vulnerability that afflicts one site would afflict all sites. Because of these advantages, several systems have been built on the model of community-based redundancy, including Archival Intermemory [6], OceanStore [17], LOCKSS [3], and SAV [9].

A central question in community-based redundancy systems revolves around the contribution and allocation of resources. Peers must determine how many resources they can reasonably expect from the community, and how many resources they themselves must contribute. Moreover, because there is no central allocation mechanism, participants must make careful decisions when determining how resources are used, in order to avoid a situation where the needs of some participants are not met by the community despite the fact that there are nominally enough resources available to meet everyone's needs.

In order to deal with these allocation issues, we are investigating a mechanism that we call *data trading*. One application of data trading is digital archiving, where sites protect their collections from failures by making multiple copies at remote sites. When a site has a digital collection it wishes to replicate, the site contacts a remote site and proposes a trade. For example, Stanford's library may have a collection of technical reports that it wants to preserve, and thus Stanford contacts MIT and proposes a trade. MIT might respond that it is willing to store Stanford's collection, if in turn Stanford is willing to store a copy of a collection of scientific measurement data owned by MIT. A series of such binary trades creates a peer-to-peer trading network. Each peer tries to maximize its own local reliability, but the effect is that the whole network is a reliable infrastructure for archiving data.

A trading-based peer-to-peer system has several advantages. First, it preserves the autonomy of individual peers. Each site makes local decisions about who to trade with, how many resources to contribute to the community, how many trades to try to make, and so on. Sites are more willing to participate in a peer-to-peer scheme if they can retain this local decision making. Second, the symmetric nature of trading ensures fairness and discourages free-loading. In order to acquire resources from the community, each peer must contribute its own resources in trade. Moreover, sites that contribute more resources receive more benefit in return, because they can make more trades. Third, the system is robust in the face of failure. Because the trading network is composed of myriad binary trading links, individual links or sites can fail without crashing the whole network. Instead, the "broken" trading links can be replaced with freshly negotiated links between surviving peers.

In this position paper, we argue that these advantages make trading a useful component of the infrastructure of peer-to-peer systems and of community-

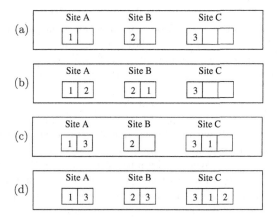

Fig. 1. Data trading example.

based redundancy systems based on a peer-to-peer architecture. Specifically, we describe our current work in developing algorithms and policies for trading, and then discuss how the generality of a trading mechanism makes it widely applicable to a variety of problems.

2 Data Trading

In our current work, we have focused on trading for the application of reliable preservation through replication. The basic framework for trading in this application can be tuned in different ways to achieve the highest reliability. In this section, we illustrate an example of a trading session and then discuss the reliability optimizations we have studied.

Consider the example shown in Figure 1. Figure 1(a) shows sites A and B, each of which have two gigabytes of space, and site C, which has three gigabytes of space. (A gigabyte is represented in the figure as a box.) Site A owns a collection of data labeled "1," site B owns collection "2," and site C owns collection "3." (A collection is an application unit, e.g., a set of technical reports, or a set of census files.) Each collection requires one gigabyte of space. Sites A and B can trade their collections, resulting in the configuration of Figure 1(b). Collections 1 and 2 are now stored more reliably, because if one site goes out of business, goes on strike or burns down, another copy is available. However, now site C cannot trade with either A or B since neither site has free space for collection 3. Thus, collection 3 is not stored reliably.

A different trading order can result in a more desirable scenario. For example, say that from the initial configuration, site A first contacts C and offers a trade. The result is shown in Figure 1(c). Now there is still enough space to make another trade, this time between sites B and C. The resulting situation, in

Figure 1(d), has all three collections reliably stored with two copies. A trading scheme should be effective enough so that sites can make local decisions about which sites to trade with, while still allowing other sites to replicate their collections. At the same time, trading must be flexible enough to deal with the appearance of new sites, new collections, and even new free storage added at an existing site.

This example illustrates that a great deal of care must be put into the local decisions that are made by each site. Although it is often impossible to make optimal decisions, especially without knowledge of future events (such as new sites, new storage space added to an existing site, etc.) we can study useful heuristics that tend to improve the overall reliability of the system. We can encapsulate these heuristics in *trading policies* that guide the local decision making at each peer. In this way, the continuous process of offering and accepting trades can be automated. The system, once configured with appropriate trading policies, autonomously replicates information to ensure high reliability.

2.1   Trading Policies

We have studied several different policies that determine the the behavior of a trading peer. Examples of policies include:

Deed trading. One possibility is for peers to trade collections directly, and this is the approach assumed in the example presented above. This approach has the disadvantage that trades may not be very symmetric; for example, Stanford's collection may be much larger than MIT's, which results in a situation where MIT gives away more storage than it gets in return. A fairer scheme (and, it turns out, more reliable scheme) is one in which blocks of space are traded. For example, Stanford may give MIT 10 GB of space and in return get 10 GB of MIT's space. Each site can then use the space it has acquired as it sees fit. If MIT's collection is smaller than 10 GB, it may be able to use the space it acquired at Stanford to replicate several collections. The bookkeeping mechanism for tracking these trades is called *deeds*: a deed represents the right to use space at another site. Once MIT has acquired a deed for space at Stanford, it can use the deed, save it for the future, split it into smaller pieces, or trade the deed away to another site, as it sees fit.

Advertising policy. A site advertises the amount of storage space it is willing to trade away. In the simplest case, a site advertises all of the space it has free. However, higher reliability over the long term can be achieved by reserving some space for future use. Then, a site only advertises a fraction of its available resources at any one time. This ensures that there is always some space to trade away, which may be needed if the site gets a new collection that it must replicate by proposing new trades.

Remote site selection strategy. When a site wishes to make trades, it must decide which remote sites to offer trades to. One possibility is to choose the remote site that has the lowest probability of failure, estimated based on previous history, reputation, or the quality of components at the site. However, this policy is counterproductive if every peer uses it, because the "high reliability" sites quickly become overloaded. A more effective policy is for peers to choose a small set of "trusted" trading partners, and trade repeatedly with those partners.

Bidding policy. When Stanford asks MIT for a trade, the two sites may exchange equally sized blocks of space. An alternative is for Stanford to offer a trade by saying that it needs a certain amount of space at MIT (say, 10 GB), and asking how much space MIT would want in return. If MIT is eager to trade, because it has many collections to replicate, it may offer a low *bid*, asking for only 5 GB in return. On the other hand, if MIT is reluctant to trade, say because its local space is becoming scarce, then it may offer a high bid, asking for 15 GB in return. In this way, Stanford can contact multiple sites, get bids from each one, and then accept the most attractive offer. In this scenario, each site must decide, based on its local circumstances, what bid to offer for each trade.

These and other policies have been examined in more detail in [11,12,10]. These papers also describe our trading simulator, a system we have built to simulate trading sessions using different policies. Our simulator has allowed us to identify policies which provide the highest reliability in different circumstances.

3   Generalizing Trading in Peer-to-Peer Systems

Although we have studied data trading specifically in the context of trading storage space to replicate collections, we believe it is a general mechanism for several applications. Trading can serve as a part of the infrastructure of a peer-to-peer, community-based redundancy system. In this section we outline some other possible uses for trading.

Trading can be used to exchange resources besides storage space. For example, processing cycles for searching collections can also be traded. Once collections are distributed in a community-based replication system, users will attempt to find collections or individual documents within collections by performing searches. In a highly reliable, highly available system, users should still be able to perform searches even in the presence of site failures. If one site is charged with handling searches for a particular collection, and that site fails or is unreachable due to a network partition, then users will not be able to search the collection even though copies may still be available in the network. On the other hand, the search load may be replicated and distributed in the same way that the physical collections are replicated. If Stanford agrees to process searches over MIT's collections, and in return MIT agrees to process searches over Stanford's collections, then collections can always be searched despite a failure at either site.

In addition to trading processing for processing, it is possible that a trading infrastructure can be used to trade one type of resource for another. For example, the site that "owns" a collection may not have the processing capacity or bandwidth to support all of the searches submitted by users. On the other hand, this site may have an excess of storage space. The site may try to shed some of the query load by contracting with other sites that will take over some of the search processing. The mechanism for contracting may be based on trading, in which the site gives away some of its storage space in return for processing cycles at other sites.

Trading may also be extended for exchanging more abstract resources. One example is trading access to content. A site may have a limited budget, and is not able to directly purchase access to important collections. However, that site may also have collections of its own that are desired by other sites. Then, the site can gain the right to use a valuable collection owned by another site by trading away the right to use its own collections.

In general, any redundancy systems that allocate limited resources can use a trading mechanism as an infrastructure component. Some existing systems allocate redundant resources in a fixed, static way. Although it is possible to reason about good or even optimal policies for certain configurations, it is difficult to do so in a distributed system with autonomous peers. Moreover, if the configuration is highly dynamic then the fixed allocation may no longer be appropriate. In contrast, other existing distributed and peer-to-peer systems allocate resources in response to user demand, or even randomly. Allocating in response to user requests may mean that less popular collections are not preserved at all. Allocating randomly may make inefficient use of community resources. If the goal is to ensure redundancy and high reliability, then trading provides a way to achieve effective allocation while dynamically adapting to changes in user requirements and network configuration.

Finally, in a general trading system, it may be difficult to distinguish trustworthy trading partners from less reliable or malicious sites. In some cases, it may be sufficient to implement a reputation system to identify (and ostracize) peers that do not fulfill their responsibilities. However, reputation systems only operate after a node has misbehaved. It may be necessary to implement a security policies to prevent or at least mitigate malicious attacks. Preliminary work in this area is described in [8].

3.1   Related Work

Our work draws upon concepts developed in related systems. Traditional data management schemes, such as replicated DBMS's [5,18], replicated filesystems [15] and RAID disk arrays [22] utilize replication to protect against failures in the short term. A peer-to-peer trading system provides more autonomy and fairness for individual storage components than traditional solutions. Another difference is that traditional solutions are concerned with load distribution, query time and update performance, as well as reliability [14,23,24]. Here, we are primarily concerned about preservation (given the constraint of preserving site autonomy). In contrast, traditional replicated databases tend to trade some reliability for increased performance [20]. Similarly, replicated filesystem schemes such as Coda [19] or Andrew [16] use caching to improve availability. Our goal is different: long term preservation despite failures, rather than short term preservation in the face of network partitions.

Many existing peer-to-peer such as Freenet [1] or Gnutella [2] use "trading" as a model, although these systems trade content (such documents or audio files), not necessarily resources (such as storage). Also, these systems are focused on finding resources within a dynamic, ever-changing collection, and not on reliability. While popular items may become widely replicated, less popular or frequently accessed items are deleted. Thus, systems like Gnutella provide searching but do not guarantee preservation. A searching and resource discovery mechanism could be built on top of our data trading system; however, our primary focus is surviving failures over the long term.

Other peer-to-peer systems have focused on reliable storage using an economic model similar to trading. FreeHaven [13] uses a trading system very similar to our work. However, anonymity and peer accountability are primary goals of FreeHaven. As a result, trades occur in order to obscure the true owner of a document, and trading partners are chosen based on reputation. Our model aims for a different goal, that of long term reliability; at the same time, we examine (and scientifically evaluate) a wider range of policies for choosing trading partners. MojoNation [4] also used trading, but transactions were made via an intermediate currency called "mojo." As with any currency-based mechanism, the system is vulnerable to fluctuations in the money supply and manipulations of the currency value. Barter, such as in our work, attempts to avoid these problems.

Systems such as the Archival Intermemory [6] and OceanStore [17] are very good at preserving digital objects through replication. Our trading techniques could serve as the storage allocation and replica placement mechanism for these systems, increasing reliability and providing site autonomy.

The problem of optimally allocating data objects given space constraints is well known in computer science. Distributed bin packing problems [21] and the File Allocation Problem [7] are known to be NP-hard. This is one reason we have not sought to find an optimal placement for data collections. Moreover, these problems are even harder when the number of sites and number and sizes of collections are not known in advance.

4   Conclusion

A peer-to-peer infrastructure is useful for a variety of applications. One important application is reliability through redundancy. The large number of individual resources in the community, the geographical and administrative distribution, and the heterogeneity of peers are all advantages to a peer-to-peer architecture for community-based redundancy system. In such a system, it is vital that peers can use a dynamic, flexible and effective mechanism for allocating resources. Data trading provides such a mechanism. Because it preserves autonomy, ensures fairness and is robust in the face of failure, trading is a good mechanism for providing community-wide replication through decisions made locally by sites in their own self interest.

We have focused on trading as a way to allocate storage space for data replication. Our research has examined several heuristic policies that can be used to achieve high reliability in such a trading system despite the limited information available to each local site. We have also argued that the trading framework can be extended for other purposes and applications. The advantages of trading make it a valuable component of the peer-to-peer infrastructure.

References

1. The Freenet Project. http://freenet.sourceforge.net/, 2001.
2. Gnutella. http://gnutella.wego.com, 2001.
3. Lots of copies keeps stuff safe (LOCKSS). http://lockss.stanford.edu/, 2001.
4. MojoNation. http://www.mojonation.net/, 2002.
5. F. B. Bastani and I-Ling Yen. A fault tolerant replicated storage system. In *Proc. ICDE*, May 1987.
6. Yuan Chen, Jan Edler, Andrew V. Goldberg, Allan Gottlieb, Sumeet Sobti, and Peter N. Yianilos. A prototype implementation of archival intermemory. In *Proc. ACM Int'l Conf. on Digital Libraries*, 1999.
7. W. W. Chu. Multiple file allocation in a multiple computer system. *IEEE Transactions on Computing*, C-18(10):885–889, Oct. 1969.
8. B. F. Cooper, M. Bawa, N. Daswani, and H. Garcia-Molina. Protecting the PIPE from malicious peers. http://dbpubs.stanford.edu/pub/2002-3, 2002. Technical report.
9. B. F. Cooper, A. Crespo, and H. Garcia-Molina. Implementing a reliable digital object archive. In *Proc. European Conf. on Digital Libraries (ECDL)*, Sept. 2000. In LNCS (Springer-Verlag) volume 1923.
10. B. F. Cooper and H. Garcia-Molina. Bidding for storage space in a peer-to-peer data preservation system. http://dbpubs.stanford.edu/pub/2001-52, 2001. Technical Report.
11. B. F. Cooper and H. Garcia-Molina. Creating trading networks of digital archives. In *Proc. 1st Joint ACM/IEEE Conference on Digital Libraries (JCDL)*, June 2001.
12. B. F. Cooper and H. Garcia-Molina. Peer-to-peer data trading to preserve information. *ACM Transactions on Information Systems*, to appear.
13. R. Dingledine, M.J. Freedman, and D. Molnar. The FreeHaven Project: Distributed anonymous storage service. In *Proceedings of the Workshop on Design Issues in Anonymity and Unobservability*, July 2000.
14. X. Du and F. Maryanski. Data allocation in a dynamically reconfigurable environment. In *Proc. ICDE*, Feb. 1988.
15. B. Liskov et al. Replication in the Harp file system. In *Proc. 13th SOSP*, Oct. 1991.
16. J. H. Morris et al. Andrew: A distributed personal computing environment. *CACM*, 29(3):184–201, March 1986.
17. J. Kubiatowicz et al. OceanStore: An architecture for global-scale persistent storage. In *Proc. ASPLOS*, Nov. 2000.
18. J. Gray, P. Helland, P. O'Neal, and D. Shasha. The dangers of replication and a solution. In *Proc. SIGMOD*, June 1996.
19. J. J. Kistler and M. Satyanarayanan. Disconnected operation in the Coda file system. *ACM TOCS*, 10(1):3–25, Feb. 1992.
20. E. Lee and C. Thekkath. Petal: Distributed virtual disks. In *Proc. 7th ASPLOS*, Oct. 1996.
21. S. Martello and P. Toth. *Knapsack Problems: Algorithms and Computer Implementations*. J. Wiley and Sons, Chichester, New York, 1990.

22. D. Patterson, G. Gibson, and R. H. Katz. A case for redundant arrays of inexpensive disks (RAID). *SIGMOD Record*, 17(3):109–116, September 1988.
23. H. Sandhu and S. Zhou. Cluster-based file replication in large-scale distributed systems. In *Proc. SIGMETRICS*, June 1992.
24. O. Wolfson, S. Jajodia, and Y. Huang. An adaptive data replication algorithm. *ACM TODS*, 2(2):255–314, June 1997.

Erasure Coding Vs. Replication: A Quantitative Comparison

Hakim Weatherspoon and John D. Kubiatowicz

Computer Science Division
University of California, Berkeley
{hweather, kubitron}@cs.berkeley.edu

Abstract. Peer-to-peer systems are positioned to take advantage of gains in network bandwidth, storage capacity, and computational resources to provide long-term durable storage infrastructures. In this paper, we quantitatively compare building a distributed storage infrastructure that is self-repairing and resilient to faults using either a replicated system or an erasure-resilient system. We show that systems employing erasure codes have mean time to failures many orders of magnitude higher than replicated systems with similar storage and bandwidth requirements. More importantly, erasure-resilient systems use an order of magnitude less bandwidth and storage to provide similar system durability as replicated systems.

1 Introduction

Today's exponential growth in network bandwidth, storage capacity, and computational resources has inspired a whole new class of distributed, peer-to-peer storage infrastructures. Systems such as Farsite[2], Freenet[4], Intermemory[3], OceanStore[8], CFS[5], and PAST[7] seek to capitalize on the rapid growth of resources to provide inexpensive, highly-available storage without centralized servers. The designers of these systems propose to achieve high availability and long-term durability, in the face of individual component failures, through replication and coding techniques.

Although wide-scale replication has the potential to increase availability and durability, it introduces two important challenges to system architects. First, system architects must increase the number of replicas to achieve high durability for large systems. Second, the increase in the number of replicas increases the bandwidth and storage requirements of the system.

This paper makes the following contributions: First, we briefly quantify the availability gained using erasure codes. Second, we show that erasure-resilient codes use an order of magnitude less bandwidth and storage than replication for systems with similar *mean time to failure* (MTTF). Third, we show that employing *erasure-resilient codes* increase the MTTF of the system by many orders of magnitude over simple replication with the same storage overhead and *repair*[1] times. The contributions of this work over [3, 12] are the addition of bandwidth as a comparison.

---

[1] Data is periodically repaired to replace lost redundancy in both replicated and erasure encoded systems.

P. Druschel, F. Kaashoek, and A. Rowstron (Eds.): IPTPS 2002, LNCS 2429, pp. 328–337, 2002.
© Springer-Verlag Berlin Heidelberg 2002

2 Background

Two common methods used to achieve high durability of data are complete replication[2, 7] and parity schemes such as RAID[9]. The former imposes extremely high bandwidth and storage overhead, while the latter does not provide the robustness necessary to survive the high rate of failures expected in the wide area.

An *erasure code* provides redundancy without the overhead of strict replication. Erasure codes divide an object into m fragments and recode them into n *fragments*, where $n > m$. We call $r = \frac{m}{n} < 1$ the *rate* of encoding. A rate r code increases the storage cost by a factor of $\frac{1}{r}$. The key property of erasure codes is that the original object can be reconstructed from *any* m fragments. For example, using an $r = \frac{1}{4}$ encoding on a block divides the block into $m = 16$ fragments and encodes the original m fragments into $n = 64$ fragments; increasing the storage cost by a factor of *four*.

Erasure codes are a superset of replicated and RAID systems. For example, a system that creates four replicas for each block can be described by an ($m = 1$, $n = 4$) erasure code. RAID level 1, 4, and 5 can be described by an ($m = 1$, $n = 2$), ($m = 4$, $n = 5$), and ($m = 4$, $n = 5$) erasure code, respectfully.

Data Integrity: Erasure coding in a malicious environment requires the precise identification of failed or corrupted fragments. Without the ability to identify corrupted fragments, there is potentially a factorial combination of fragments to try to reconstruct the block; that is, $\binom{n}{m}$ combinations. As a result, the system needs to detect when a fragment has been corrupted and discard it. A secure verification hashing scheme can serve the dual purpose of identifying and verifying each fragment. It is necessarily the case that any m *correctly verified* fragments can be used to reconstruct the block. Such a scheme is likely to increase the bandwidth and storage requirements, but can be shown to still be many times less than replication.

3 Assumptions

We assume that replicated and erasure encoded systems consist of a collection of *independently, identically distributed* failing disks – same assumption made by both *A Case for RAID*[9] and disk manufacturers – and that failed disks are immediately replaced by new, blank ones[2]. During dissemination, each replica (or *fragment*) for a given block is placed on a unique, randomly selected disk. Finally, we postulate a global sweep and repair process that scans the system, attempting to restore redundancy by reconstructing each block and redistributing lost replicas (or fragments) over a new set of disks – Repair in *RAID*[9] is *triggered* when a disk fails, which is fundamentally different than sweep and repair. Some type of repair is required; otherwise, data would be lost in a couple years regardless of the redundancy. We denote the time period between sweeps of the same block an *epoch*.

---

[2] We are ignoring other types of failures such as software errors, operational errors, configuration problems, etc., for this simple analysis.

4   Availability

Availability gained using erasure codes is a result of exploiting the statistical stability of a large number of components. The availability of a block can be computed as follows

P_o probability that a block is available
n total number of fragments
m number of fragments needed for reconstruction
N total number of machines in the world
M number of currently unavailable machines

$$P_o = \sum_{i=0}^{n-m} \frac{\binom{M}{i}\binom{N-M}{n-i}}{\binom{N}{n}} \tag{1}$$

where the probability a block is available is equal to the number of ways in which we can arrange unavailable fragments on unreachable servers multiplied by the number of ways in which we can arrange available fragments on reachable servers, divided by the total number of ways in which we can arrange all of the fragments on all of the servers.

With a million machines, ten percent of which are currently down, simply storing two complete replicas provides only two nines (0.99) of availability. A rate $\frac{1}{2}$ erasure coding of a document into 32 fragments gives the document over eight nines of availability (0.999999998), yet consumes the same amount of storage and bandwidth, supporting the assertion that *fragmentation increases availability*.

5   System Comparison

We use the same system size (*total blocks*) and write rate ($\frac{wBlocks}{s}$) to compare systems based on replication to that of erasure codes. In this section we make three comparisons. First, we fix the *mean time to failure* (MTTF) of the system and *repair epoch*. Second, we fix the storage overhead and repair epoch. Finally, we fix the MTTF of the system and the storage overhead.

We compare replicated and erasure encoded systems (denoted by x) in terms of total storage S_x, total bandwidth (leaving the source or entering the destination) BW_x, and the total number of disk seeks required to sustain rate (repair, write, and read) D_x. We do not compare reads when considering storage and bandwidth because the amount of data required to read a block is the same for both systems; that is, m fragments is equivalent to one replica in storage and bandwidth requirements.

5.1   Fix MTTF and Repair Epoch

In this subsection we compare replicated systems to erasure encoded systems that have the same fixed system MTTF and repair epoch.

Assuming that we store and do not delete data, the total system size in terms of the total number of fixed size blocks B that will be reached throughout the systems lifetime can be computed using the total number of users N as follows

$$B = N \cdot \frac{wBlocks}{s} \cdot \text{total seconds}$$

More generally, given a system size (defined by the number of users) we focus on answering the question, what are the resources required to store data in a system long-term. We define the *durability of the system* to be the expected MTTF of losing *any* block is sufficiently larger than the expected lifetime of the system given some number of users. That is

$$MTTF_{system} = \frac{MTTF_{block}}{B} \gg \text{total seconds}$$

We derive how to compute storage, bandwidth, and disk seeks required by solving the following equations.

$$S_x = \text{total bytes stored in system } x$$
$$BW_x = BW_{x_{write}} + BW_{x_{repair}}$$
$$D_x = D_{x_{write}} + D_{x_{repair}} + D_{x_{read}}$$

S_x is the total storage capacity required of the system x, BW_x is a function of the bandwidth required to support both writes and repair of the total storage every repair epoch, and D_x is the number of disk seeks required to support repair, writes, and reads. The repair bandwidth is computed by dividing the total bytes stored by the repair epoch. Next, we compute the storage for both systems

$$S_{repl} = b \cdot R \cdot B$$
$$S_{erase} = \frac{b}{m} \cdot n \cdot B = b \cdot \frac{1}{r} \cdot B$$

where R is the number of replicas, $r = \frac{m}{n}$ is the rate of encoding, and b is the block size. We now compute the bandwidth in terms of storage as follows

$$BW_{repl} = b \cdot R \cdot N \cdot \frac{wBlocks}{s} + \frac{S_{repl}}{e_{repl}}$$
$$BW_{erase} = b \cdot \frac{1}{r} \cdot N \cdot \frac{wBlocks}{s} + \frac{S_{erase}}{e_{erase}}$$

where e_x is the repair epoch of system $x \in \{replica, erasure\}$. We show now that the bandwidth due to only the original data being written and repaired can be expressed as *DataRate*

$$DR_x = N \frac{wBlocks}{s} + \frac{B}{e_x}$$

Further, we compute the number of disk seeks required to support writes, repair, and reads. db_{sz} is the size of a disk block.

$$D_{repl} = \left(R \cdot N \cdot \frac{wBlocks}{s} + R \cdot \frac{B}{e_{repl}} + 1\right) \cdot \frac{b}{db_{sz}}$$
$$D_{erase} = \left(n \cdot N \cdot \frac{wBlocks}{s} + n \cdot \frac{B}{e_{erase}} + m\right) \cdot \frac{b}{m \cdot db_{sz}}$$

The above equation states that the number of disk seeks required is dependent on the number of replicas (or total number of fragments), throughput, system size, repair epoch, the number of replicas (or fragments) needed to reconstruct the block, and the number of replicas (or fragments) that can fit in a disk block.

Finally, a replicated system can be compared to a similar erasure encoded system with the following bandwidth, storage, and disk seek ratios

$$\frac{S_{repl}}{S_{erase}} = R \cdot r \qquad (2)$$

$$\frac{BW_{repl}}{BW_{erase}} = \frac{R \cdot DR_{repl}}{\frac{1}{r} \cdot DR_{erase}} = R \cdot r \qquad (3)$$

$$\frac{D_{repl}}{D_{erase}} = \frac{(R \cdot DR_{repl} + 1) \cdot \frac{b}{db_{sz}}}{(n \cdot DR_{erase} + m) \cdot \frac{b}{m \cdot db_{sz}}} \approx R \cdot r \qquad (4)$$

We make the abstract numbers concrete using the following parameters as appropriate. Bolosky *et. al*[2] measured that an average workstation produces $35 \frac{MB}{hr}$ of data. We associate a workstation with a user. We set $b = $ 8kB blocks, $db_{sz} = $ 8kB disk blocks, $N = 2^{24}$ users, $e_{repl} = e_{erase} = 4$ months, and $MTTF_{system} > 1000$ years. As a consequence of the former parameters we calculate $B = 10^{17}$ total blocks; hence, $MTTF_{block} = 10^{20}$ years. Finally, using the analysis described in [12] and reprinted in Appendix A, we solve for the number of replicas and rate and compute that $R = 22$ and $r = \frac{32}{64} = \frac{1}{2}$ satisfy above constraints, respectively.

Applying these parameters to equations 2, 3, and 4 we produce the following result

$$\frac{BW_{repl}}{BW_{erase}} = 11$$

$$\frac{S_{repl}}{S_{erase}} = 11$$

$$\frac{D_{repl}}{D_{erase}} = 11$$

These results show that a replicated system requires an order of magnitude more bandwidth, storage, and disk seeks as an erasure encoded system of the same size.

5.2   Fix Storage Overhead and Repair Epoch

The same formulas from subsection 5.1 above can be used to verify durability of system calculations presented in [3, 12]. For example, using our simple failure model presented in section 3 and parameters in section 5.1, we set repair time of $e_{repl} = e_{erase} = four$ months, $R = two$ replicas[3], and rate $r = \frac{32}{64}$. Both the replicated and erasure encoded systems have the same apparent storage overhead of a factor of *two*. Using Appendix A, we compute the $MTTF_{block}$ of a block replicated onto two servers as 74 years and the $MTTF_{block}$ of a block using a rate $\frac{1}{2}$ code onto $n = 64$ servers as 10^{20} years! It is this difference that highlights the advantage of erasure coding.

---

[3] In section 5.1 $R = 22$ to attain the same durability

5.3   Fix MTTF and Storage Overhead

As a final comparison, we can fix the MTTF and storage overhead between a replicated and erasure encoded system. This implies that the storage and bandwidth for writes are equivalent for these two systems. In this case erasure encoded systems must be repaired less frequently, and hence, require less repair bandwidth.

For example, we can devise systems that have a $MTTF_{block} = 10^6$ years, a factor four storage overhead, $B = 1000$ blocks, and a $MTTF_{system} = 1000$ years. The replicated system meets the above requirements using $R =$ four replicas and a repair epoch of $e_{repl} =$ one month. The erasure encoded system meets the same requirements using an $r = \frac{16}{64} = \frac{1}{4}$ code and a repair epoch of $e_{erase} = 28$ months. The replicated system uses 28 times more bandwidth than erasure encoded system for repair.

If, instead, the $MTTF_{block} = 10^{20}$ years, $B = 10^{17}$ blocks, $MTTF_{system} = 1000$ (as described in subsection 5.1), and still using a factor of *four* storage overhead, the erasure encoded system meets the requirements using an $r = \frac{16}{64} = \frac{1}{4}$ code and a repair epoch of $e_{erase} = 12$ months, but a replicated system with $R = 4$ replicas would have to repair all blocks almost instantly and continuously.

6   Discussion

The previous section presented the advantages of erasure codes, but there are some caveats as well. Two issues that we would like tob highlight are the need for intelligent buffering of data and the need for caching.

Each client in an erasure-resilient system sends messages to a larger number of *distinct* servers than in a replicated system. Further, the erasure-resilient system sends smaller "logical" blocks to servers than the replicated system. Both of these issues could be considered enough of a liability to outweigh the results of the last section. However, we do not view it this way. First, we assume that the storage servers are utilized by a number of clients; this means that the additional servers are simply spread over a larger client base. Second, we assume intelligent buffering and message aggregation. Although the outgoing fragments are "smaller", we simply aggregate them together into larger messages and larger disk blocks, thereby nullifying the consequences of fragment size. These assumptions are implicit in our exploration via metrics of total bandwidth and number of disk blocks in the previous section.

Another concern about erasure-resilient systems is that the time and server overhead to perform a read has increased, since multiple servers must be contacted to read a single block. The simplest answer to such a concern is that mechanisms for *durability* should be separated from mechanisms for *latency reduction*. Consequently, we assume that erasure-resilient coding will be utilized for durability, while replicas (i.e. caching) will be utilized for latency reduction. The nice thing about this organization is that replicas utilized for caching are *soft-state* and can be constructed and destroyed as necessary to meet the needs of temporal locality. Further, prefetching can be used to reconstruct replicas from fragments in advance of their use. Such a hybrid architecture is illustrated in Figure 1. This is similar to what is provided by OceanStore [8].

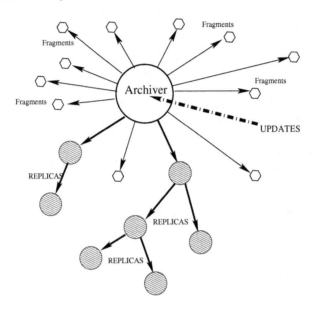

Fig. 1. Hybrid Update Architecture: *Updates are sent to a central "Archiver", which produces archival fragments at the same time that it updates live replicas. Clients can achieve low-latency read access by utilizing replicas directly.*

7   Future Work

We present some open research issues that affect both replicated and erasure encoded systems alike.

Failure Independence: The most troubling assumption of the previous sections are that failures are *independent* and *identically distributed*. This is not true for all sets of storage servers. We list two possible techniques to address independence. First, most routing overlay networks, such as CAN[11], Chord[13], Pastry[7], and Tapestry[14], provide a location and routing infrastructure that permits fragments to be distributed to geographically diverse locations, eliminating a large class of correlations caused by natural disasters, denial of service attacks, and administrative boundaries. Second, sophisticated measurement and modeling techniques could be used to choose a set of nodes that are maximally independent during fragment dissemination.

Efficient Repair: Action must be taken to maintain replicas (or fragments) despite failure; otherwise, all replicas (or fragments) will be lost. The sweep and repair is simplistic because it assumes that all data in the world is reconstructed on some periodic basis. While this is not entirely implausible (every object is independent and could be repaired in parallel), it does consume many resources.

8   Related Work

The idea of a global-scale, distributed, persistent storage infrastructure was first motivated by Ross Anderson in his proposal for the Eternity Service[1]. To our knowledge the tradeoffs between bandwidth, storage, and disk seeks when comparing a replicated system to an erasure coded system have not been discussed in literature. A discussion of the durability gained from building a system from erasure codes first appeared in Intermemory[3]. The authors describe how their technique increases an object's resilience to node failure, but the system does not incorporate a repair mechanism that would also increase objects durability. More recently, there has appeared a large body of work on the subject of wide-scale, distributed storage. FreeHaven[6] is a system for anonymous publishing that uses an information dispersal algorithm, in a manner analogous to erasure codes.

Our discussion in Section 6 motivated the need for a hybrid system. OceanStore[8] is a distributed storage system that uses the notion of *promiscuous caching*, where replicas are soft-state and only for read benefit, while erasure codes are used for durability.

Other systems with similar goals include PAST[7] and Farsite[2]. PAST is a large-scale peer-to-peer storage utility. Farsite seeks to provide an organizational-scale distributed file system comprised of cooperating, but not trusting, machines. Both rely on replication for durability and availability.

9   Conclusion

In this paper we have described the availability and durability gains provided by an erasure-resilient system. We quantitatively compared systems based on replication to systems based on erasure codes. We showed that the *mean time to failure* (MTTF) of an erasure encoded system can be shown to be many orders of magnitude higher than that of a replicated system with the same storage overhead and repair period. A novel result of our analysis showed that erasure-resilient codes use an order of magnitude less bandwidth and storage than replication for systems with similar MTTF. Finally, if care is taken to take advantage of temporal and spatial locality erasure encoded systems can use an order of magnitude less disk seeks than replicated systems.

References

[1] ANDERSON, R. The eternity service. In *Proceedings of Pragocrypt* (1996).

[2] BOLOSKY, W., DOUCEUR, J., ELY, D., AND THEIMER, M. Feasibility of a serverless distributed file system deployed on an existing set of desktop PCs. In *Proc. of Sigmetrics* (June 2000).

[3] CHEN, Y., EDLER, J., GOLDBERG, A., GOTTLIEB, A., SOBTI, S., AND YIANILOS, P. Prototype implementation of archival intermemory. In *Proc. of IEEE ICDE* (Feb. 1996), pp. 485–495.

[4] CLARK, I., SANDBERG, O., WILEY, B., AND HONG, T. Freenet: A distributed anonymous information storage and retrieval system. In *Proc. of the Workshop on Design Issues in Anonymity and Unobservability* (Berkeley, CA, July 2000), pp. 311–320.

[5] DABEK, F., KAASHOEK, M. F., KARGER, D., MORRIS, R., AND STOICA, I. Wide-area cooperative storage with CFS. In *Proc. of ACM SOSP* (October 2001).

[6] DINGLEDINE, R., FREEDMAN, M., AND MOLNAR, D. The freehaven project: Distributed anonymous storage service. In *Proc. of the Workshop on Design Issues in Anonymity and Unobservability* (July 2000).

[7] DRUSCHEL, P., AND ROWSTRON, A. Storage management and caching in PAST, a large-scale, persistent peer-to-peer storage utility. In *Proc. of ACM SOSP* (2001).

[8] KUBIATOWICZ, J., ET AL. Oceanstore: An architecture for global-scale persistent storage. In *Proc. of ASPLOS* (Nov. 2000), ACM.

[9] PATTERSON, D., GIBSON, G., AND KATZ, R. The case for raid: Redundant arrays of inexpensive disks, May 1988.

[10] PATTERSON, D., AND HENNESSY, J. *Computer Architecture: A Quantitative Approach.* Forthcoming Edition.

[11] RATNASAMY, S., FRANCIS, P., HANDLEY, M., KARP, R., AND SCHENKER, S. A scalable content-addressable network. In *Proceedings of SIGCOMM* (August 2001), ACM.

[12] RHEA, S., WELLS, C., EATON, P., GEELS, D., ZHAO, B., WEATHERSPOON, H., AND KUBIATOWICZ, J. Maintenance free global storage in oceanstore. In *Proc. of IEEE Internet Computing* (2001), IEEE.

[13] STOICA, I., MORRIS, R., KARGER, D., KAASHOEK, M. F., AND BALAKRISHNAN, H. Chord: A scalable peer-to-peer lookup service for internet applications. In *Proceedings of SIGCOMM* (August 2001), ACM.

[14] ZHAO, B., JOSEPH, A., AND KUBIATOWICZ, J. Tapestry: An infrastructure for fault-tolerant wide-area location and routing. Tech. Rep. UCB//CSD-01-1141, University of California, Berkeley Computer Science Division, April 2001.

A   Appendix: Durability Derivation

In this appendix we describe the mathematics involved in computing the *mean time to failure* (MTTF) of a *particular* erasure encoded block.

Considering the server failure model and repair process as described in Section 3, we can calculate the MTTF of a block as follows. First, we calculate the probability that a given fragment placed on a randomly selected disk will survive until the next epoch as

$$p(e) = \int_{e}^{\infty} \frac{l p_d(l)}{\mu} \frac{l - e}{l} dl \qquad (5)$$

$$= \frac{1}{\mu} \int_{e}^{\infty} p_d(l)(l - e) dl \qquad (6)$$

where e is the length of an epoch, μ is the average life of a disk, and $p_d(l)$ is the probability distribution of disk lives. This equation is derived similarly to the equation for the residual average lifetime of a randomly selected disk. The term $\frac{l-e}{l}$ reflects the probability that, given a disk of lifetime l, a new fragment will land on the disk early enough in its lifetime to survive until the next epoch. The probability distribution $p_d(l)$ was obtained from disk failure distributions in [10], augmented by the assumption that all disks still in service after five years are discarded along with their data.

Next, given $p(e)$, we can compute the probability that a block can be reconstructed after a given epoch as

$$p_b(e) = \sum_{m=rn}^{n} \binom{n}{m} [p(e)]^m [1 - p(e)]^{n-m} \qquad (7)$$

where n is the number of fragments per block and r is the rate of encoding. This formula computes the probability that at least rn fragments are still available at the end of the epoch.

Finally, the MTTF of a block for a given epoch size can be computed as

$$\mathrm{MTTF}_{block}(e) = e \cdot \sum_{i=0}^{\infty} i[1 - p_b(e)][p_b(e)]^i \qquad (8)$$

$$= e \cdot \frac{p_b(e)}{1 - p_b(e)}. \qquad (9)$$

This last equation computes the average number of epochs a block is expected to survive times the length of an epoch.

Author Index

Lecture Notes in Computer Science

For information about Vols. 1–2419
please contact your bookseller or Springer-Verlag